AF688644

Sparse Phased Array Antennas

Theory and Applications

For a complete listing of titles in the
Artech House Antennas and Propagation Library,
please turn to the back of this book.

Sparse Phased Array Antennas

Theory and Applications

Ashutosh Kedar

ARTECH
HOUSE

BOSTON | LONDON
artechhouse.com

Library of Congress Cataloging-in-Publication Data
A catalog record for this book is available from the U.S. Library of Congress.

British Library Cataloguing-in-Publication Data
A catalog record for this book is available from the British Library.

ISBN 13: 978-1-63081-878-4

Cover design by Charlene Stevens

© 2022 Artech House
685 Canton Street
Norwood, MA 02062

All rights reserved. Printed and bound in the United States of America. No part of this book may be reproduced or utilized in any form or by any means, electronic or mechanical, including photocopying, recording, or by any information storage and retrieval system, without permission in writing from the publisher.

All terms mentioned in this book that are known to be trademarks or service marks have been appropriately capitalized. Artech House cannot attest to the accuracy of this information. Use of a term in this book should not be regarded as affecting the validity of any trademark or service mark.

10 9 8 7 6 5 4 3 2 1

Contents

7

Planar Sparse Array Antennas 205

Preface

Antenna technology is a revolutionary field being explored continuously since Hertz's first experiment in the 1880s, where he demonstrated radio wave propagation involving a loop antenna that he had discovered. After World War II, a 27-volume set of the Massachusetts Institute of Technology (MIT) Radiation Laboratory Series was created that compiled the hard work of various eminent scientists, among which the first-ever book published on antennas, *Microwave Antenna Design* by Samuel Silver, found its place in Volume 12 of the series. Later, there were continuous proposals made regarding antenna designs; however, due to the unavailability of adequate computational resources and manufacturing techniques, many concepts remained at the hypothesis level. In modern times, with rapid advancements in computer engineering and computer-aided design (CAD) tools, these earlier hypotheses began to be realized, which led to some remarkable antenna designs. Numerous books have been written on the subjects related to antennas and antenna array technology to date. Most of these books are unique in their own right and represent the respective authors' perspectives to understand and explain the underlying concepts. Among these concepts, phased array antennas (PhAAs) is an important area serving many critical defense and commercial requirements such as radars, satcom on the move, sonars, weather monitoring, fifth generation (5G) and sixth generation (6G) for mobile communication, fault and crack detection in buildings and underground pipelines, automotive anti-collision mechanisms in automobiles, mine detection, and through-wall imaging. PhAAs offer multifunctionality and multifaceted applications that are extremely useful; they involve complex and costly circuitry and hardware that limit their applications to very specific areas where cost does not matter. I have been working in the area of phased arrays since 2003 after completion of my doctorate and was fortunate enough to work on many state-of-the art antenna systems at Electronics and Radar Development Establishment (LRDE), which offered me a great experience and joy to work with many experts within and outside the country.

The very thought of reducing the cost and complexity of antenna systems forced me to search for other viable solutions. After many discussions with my seniors, students, and experts in the area, I became interested in the topic of shared aperture antennas. While reading numerous texts and articles on the topic, I came across sparse array antennas as a promising solution to realize the shared aperture antenna systems in an efficient and cost-effective manner. I again discussed with my seniors and experts and organized an intensive training program on sparse array antennas by Professor L. P. Ligthart, Delft University, for my organization in 2010 with the support from my lab director. During the course, I became even more convinced of the potential as introduced by the professor and became even more interested in pursuing the topic. I started working in the field and read every possible text that I found in order to understand the technology and the various design principles. I am thankful to my wife, Sujata, who was a constant source of inspiration, and many thought-provoking suggestions with my wife and Prof. Ligthart made me realize that there is no single text available on this topic that showed the high potential of phased array antennas. Although my wife is not in the same field, but is a psychologist, she helped me a lot during my numerous moments of frustration when I got stuck on various problems, while managing my work-life balance at home and at the office. I am also thankful to my children, Suhaani and Ishaan, for sacrificing their time with me to allow me to work on this Herculean project. It took me almost 10 to 12 years of hard work to understand and investigate this topic and have a few publications with illustrated peer-reviewed journals to finally arrive at the baseline for the manuscript of this book. The publications are very important to give the confidence that one is going in the right direction and the work is useful for the community. The aim was to develop a few design techniques that are very simple and intuitive for engineers and students to follow and adopt in their design practices. I am thankful to the engineers at Altair India Pvt. Ltd. and Ansys working on FEKO and HFSS who allowed me to utilize their high-performance computing platforms to validate my designs with electromagnetic modeling and simulation using these software. I am really thankful to Dr. Merlin Fox at Artech House, whom I met at the IEEE Radar Conference 2020 and who motivated me to write this book and make my research accessible to any interested readers.

This book is divided into three parts. Part I presents the motivation behind this book and the importance of this topic. Part II presents basic fundamentals on antenna and antenna arrays theory. The topic is well-presented in numerous earlier texts; henceforth, here I have focused more on the fundamentals rather than mathematics involved in the various formulations. Having gained from my teaching experience with graduate and postgraduate students and young scientists while involved in various lectures, courses, and dissertations, I realized that a book is required to present these concepts in an innovative way that is very simple and interesting for engineers to follow and visualize. Part II is derived from my teaching experience and presents the various concepts in a comprehensive manner, which is easy to understand and follow. Part III presents the actual focus of this book (i.e., sparse antenna arrays mainly focused on wide scanning phased array antennas). It discusses various types of arrays, their design strategies, and many illustrative numerical test cases to understand the design mechanism. I hope that my effort yields good results and comments to further expand on in the future versions of this book.

PART I

Introduction

1

The Beginning

1.1 Introduction

Steve Jobs once said, "Design is not just what it looks like and feels like. Design is how it works."

This book covers the design aspects of a special class of antenna arrays, and sparse array antennas, vastly studied by the various researchers, but not presented in a collective form of a book until now. This book focuses upon the fundamentals and the design aspects of this special class of antenna arrays, well describing the various design and analysis techniques for these antenna types. This book focuses mainly on my indigenous work.

1.2 Past and Present of Sparse Antenna Arrays

Modern communication system serving various fronts of commercial as well as defense applications such as radars, automotive radars, fifth generation (5G) mobile communications, upcoming sixth generation (6G) mobile communications, satellite communication (satcom) on the move, weather monitoring, fault detection in pipes and buildings, and multiple input multiple output (MIMO) require antenna arrays necessarily at their front ends [1–8] serving as the eyes and ears for the system. These systems require longer ranges of communication, narrow angular resolution, and low sidelobe levels (SLL), which are served well by a properly designed antenna array having appropriate gain, half-power beamwidth (HPBW), and SLL characteristics. The antenna arrays require wide scanning capabilities to serve the modern spatial coverage requirements [1, 2] such as in phased array antenna-based radars (also called electronically scanned arrays) and 5G MIMO applications. Usually, the conventional dense or uniform antenna arrays have performance constraints due to the mutual coupling [3] between the array elements and the onset of grating lobes [5] on scanning the main beam of these arrays. These aspects are covered in detail in Chapter 3 of this book. These design restrictions are easily overcome by using sparse array antennas instead of the conventional antenna arrays. Part III of this book covers the sparse array antenna-related aspects in detail with lots of illustrations and design strategies.

Sparse array antennas are also referred to as aperiodic or nonuniform antenna arrays [9] and the later chapters in this book present exact classification of these arrays. The first instance of an occurrence of sparse phased or simply sparse array antennas dates back to 1958 by Lo [10] and later during the 1960s it included lots of initial works was presented in this area by lots of researchers [11–32]. However, due to the lack of computational resources, the design techniques were focused on mainly deterministic techniques yielding sublocal optima solutions most of the time. Later, with the advent of computational resources and the dawn of the stochastic or meta-heuristic techniques, lots of analysis and design of sparse arrays were presented by various researchers based on these techniques yielding globally optimum solutions [33–40]. The researchers have also proposed hybridization of meta-heuristic and deterministic techniques as well [38–41]. These techniques are dealt with elaboratively in the upcoming chapters with numerical test case studies illustrating their usage.

1.3 Advantages and Limitations of Sparse Array

Sparse antenna arrays have their own advantages as well as disadvantages, but wise designers can use these with proper trade-offs to tap their maximum advantages leading towards a successfully efficient design serving their requirements. The advantages are:

- Disruption of periodicity in the placement of antenna elements in the array alleviates the onset of the grating lobes.
- Larger separations between the array elements leads to a reduction in the mutual coupling, thus having a minimal impact on the antenna performance.
- Reduction of SLL in uniformly fed sparse array antennas with reference to conventional uniformly fed antenna arrays helps in achieving high taper efficiency closer in comparison to the amplitude tapered conventional antenna arrays.
- The reduction in the number of elements in the fixed array aperture compared to the conventional antenna arrays leads to reduction in size, weight, aperture, and power-cost (SWAP-C) resources.

The disadvantages are:

- The reduction in the number of elements leads to the reduction in the directive gain of array.
- Irregular array elements' placement complicates the realization of radio frequency (RF) feed circuitry in the back end of the array in case of phased array antennas.
- SLL reduction is proportional to the number of elements; henceforth, for ultralow sidelobe level (SLL) (lesser than –30 to –40 dB), a large number of elements leading to a bigger array aperture are required.

However, with an efficient design of sparse array antennas, these disadvantages can be managed well while leading to an efficient antenna array systems [42]. This book presents lots of such design strategies focusing upon these aspects.

1.4 Application Scenario for Sparse Array Antennas

Sparse antenna arrays have wide spread applications in various fields, namely:

- Mobile communications: MIMO for 5G and 6G;
- Phased array antennas for radar;
- Satellite communication;

- Square kilometer array (SKA);
- Frequency modulated continuous wave (FMCW) radars;
- Shared aperture antenna arrays.

The book will shed light on a few of these applications and readers are encouraged to refer to the vast list of references provided in this book for details of these applications.

1.5 Organization of the Book

The book is organized in to three major parts:

- Part I: Introduction;
- Part II: Antenna and Antenna Arrays;
- Part III: Electromagnetics of Sparse Array Antennas.

Part I introduced the topic and its relevance and applications.

Part II discusses the basic fundamentals of antennas and antenna arrays required by the readers to understand the various concepts presented in the book clearly. A different physics-based approach is adopted in this book to explain these concepts with ease to the readers without dwelling more into the mathematics as the mathematical part is presented in numerous books.

Part III dwells upon the sparse antenna arrays' electromagnetics in details organized into various chapters on linear, planar, and concentric ring sparse array antennas while discussing the involved fundamentals and the efficient design strategies for these antennas. The numerous test cases help to understand the concepts presented in the book.

References

[1] Fourikis, N., *Phased Array-Based Systems and Applications*, New York: John Wiley & Sons, 1997.

[2] Weber, R. H., and R. Weber, *Internet of Things*, Vol. 12, New York: Springer, 2010.

[3] Balanis, C. A., *Antenna Theory: Analysis and Design*, Second Edition, New York: John Wiley & Sons, 1997.

[4] Stutzman, W., "Bibliography for Antennas: A List of Every English-Language Antenna Book Ever Written," *IEEE Antennas and Propagation Magazine*, Vol. 50, No. 4, August 2008.

[5] Elliott, R. S., *Antenna Theory and Design*, New York: Wiley-IEEE Press, 2003.

[6] Oweis, A. I., et al., "A CoPrime Array of Patch Antennas for DOA Estimation in Mobile Handheld Devices," *2018 IEEE International Symposium on Antennas and Propagation & USNC/URSI National Radio Science Meeting*, Boston, MA, 2018, pp. 1305–1306.

[7] Hoydis, J., S. Ten Brink, and M. Debbah, "Massive MIMO: How Many Antennas Do We Need?" *2011 49th Annual Allerton Conf. Communication, Control, and Computing (Allerton)*, Monticello, IL, 2011, pp. 545–550.

[8] Pinchera, D., et al., "Antenna Arrays for Line-of-Sight Massive MIMO: Half Wavelength Is Not Enough," *Electronics*, Vol. 6, No. 3, 2017, p. 57.

[9] Toso, G., and R. Mailloux, (eds.), "Special Issue on Innovative Phased Array Antennas Based on Non-Regular Lattices and Overlapped Subarrays," *IEEE Transactions on Antennas and Propagation*, Vol. 62, No. 4, April 2014, pp. 1546–1795.

[10] Lo, Y., "A Spacing Weighted Antenna Array," *1958 IRE International Convention Record*, 1958, pp. 191–195.

[11] Sandler, S., "Some Equivalences Between Equally and Unequally Spaced Arrays," *IRE Transactions on Antennas and Propagation*, Vol. 8, No. 5, September 1960, pp. 496–500.

[12] Lindsay, J., "A Circular Loop Antenna with Nonuniform Current Distribution," *IRE Transactions on Antennas and Propagation*, Vol. 8, No. 4, July 1960, pp. 439–441.

[13] Skolnik, M., "Resolution of Angular Ambiguities in Radar Array Antennas with Widely-Spaced Elements and Grating Lobes," *IRE Transactions on Antennas and Propagation*, Vol. 10, No. 3, May 1962, pp. 351–352.

[14] Ishimaru, A., "Theory of Unequally-Spaced Arrays," *IRE Transactions on Antennas and Propagation*, Vol. 10, No. 6, November 1962, pp. 691–702.

[15] Maffett, A., "Array Factors with Nonuniform Spacing Parameter," *IRE Transactions on Antennas and Propagation*, Vol. 10, No. 2, March 1962, pp. 131–136.

[16] Lo, Y., "A Spacing Weighted Antenna Array," *1958 IRE International Convention Record*, New York, 1962, pp. 191–195.

[17] Lo, Y., "High Resolution Antenna Arrays with Randomly Spaced Elements," *1963 Antennas and Propagation Society International Symposium*, Boulder, CO, 1963, pp. 212–218.

[18] Maher, T., and D. Cheng, "Random Removal of Radiators from Large Linear Arrays," *IEEE Transactions on Antennas and Propagation*, Vol. 11, No. 2, March 1963, pp. 106–112.

[19] Sherman, J., and M. Skolnik, "Thinning Planar Array Antennas with Ring Arrays," *1958 IRE International Convention Record*, New York, 1963, pp. 77–86.

[20] Lo, Y., "A Mathematical Theory of Antenna Arrays with Randomly Spaced Elements," *IEEE Transactions on Antennas and Propagation*, Vol. 12, No. 3, May 1964, pp. 257–268.

[21] Galejs, J., "Minimization of Sidelobes in Space Tapered Linear Arrays," *IEEE Transactions on Antennas and Propagation*, Vol. 12, No. 4, July 1964, pp. 497–498.

[22] Snover, A., and A. Ferraro, "An Investigation of the Properties of Synthesized Nonuniformly Spaced Antenna Arrays," *IEEE Transactions on Antennas and Propagation*, Vol. 12, No. 1, January 1964, pp. 132–133.

[23] Skolnik, M., G. Nemhauser and J. Sherman, "Dynamic Programming Applied to Unequally Spaced Arrays," *IEEE Transactions on Antennas and Propagation*, Vol. 12, No. 1, January 1964, pp. 35–43.

[24] Galindo, V., "Nonlinear Antenna Arrays," *IEEE Transactions on Antennas and Propagation*, Vol. 12, No. 6, November 1964, pp. 782–783.

[25] Ma, M., "Application of Bernstein Polynomials and Interpolation Theory to Linear Array Synthesis," *IEEE Transactions on Antennas and Propagation*, Vol. 12, No. 6, November 1964, pp. 668–677.

[26] Ma, M., "Another Method of Synthesizing Nonuniformly Spaced Arrays," *IEEE Transactions on Antennas and Propagation*, Vol. 13, No. 5, September 1965, pp. 833–834.

[27] Larson, R., J. Ferris, and V. Powers, "Nonuniformly Spaced Linear Arrays of Unequal-Width Elements," *IEEE Transactions on Antennas and Propagation*, Vol. 13, No. 6, November 1965, pp. 974–976.

[28] Ishimaru, A., and Y. Chen, "Thinning and Broadbanding Antenna Arrays by Unequal Spacings," *IEEE Transactions on Antennas and Propagation*, Vol. 13, No. 1, January 1965, pp. 34–42.

[29] Tang, C., "An Approximate Method of Designing Nonuniformly Spaced Arrays," *IEEE Transactions on Antennas and Propagation*, Vol. 13, No. 1, January 1965, pp. 177–179.

[30] Lo, Y., and S. Lee, "A Study of Space-Tapered Arrays," *IEEE Transactions on Antennas and Propagation*, Vol. 14, No. 1, January 1966, pp. 22–30.

[31] Chow, Y., "On the Error Involved in Poisson's Sum Formulation of Nonuniformly Spaced Antenna Arrays," *IEEE Transactions on Antennas and Propagation*, Vol. 14, No. 1, January 1966, pp. 101–102.

[32] Coman, C. I., I. E. Lager, and L. P. Ligthart, "A Deterministic Solution to the Problem of Interleaving Multiple Sparse Array Antennas," *2005 European Microwave Conference*, Paris, France, 2005, pp. 4–1970.

[33] Dai, D., et al., "An Asymmetric Mapping Method for the Synthesis of Sparse Planar Arrays," *IEEE Antennas and Wireless Propagation Letters*, Vol. 17, No. 1, January 2018, pp. 70–73.

[34] Zhao, X., Q. Yang and Y. Zhang, "A Hybrid Method for the Optimal Synthesis of 3-D Patterns of Sparse Concentric Ring Arrays," *IEEE Transactions on Antennas and Propagation*, Vol. 64, No. 2, February 2016, pp. 515–524.

[35] Siakavara, K., "Hybrid-Fractal Direct Radiating Antenna Arrays with Small Number of Elements for Satellite Communications," *IEEE Transactions on Antennas and Propagation*, Vol. 58, No. 6, June 2010, pp. 2102–2106.

[36] Bekers, D. J., et al., "A Ka-Band Spaceborne Synthetic Aperture Radar Instrument: A Modular Sparse Array Antenna Design," *IEEE Antennas and Propagation Magazine*, Vol. 61, No. 5, October 2019, pp. 97–104.

[37] Chen, K., et al., "Synthesis of Sparse Planar Arrays Using Modified Real Genetic Algorithm," *IEEE Transactions on Antennas and Propagation*, Vol. 55, No. 4, April 2007, pp. 1067–1073.

[38] Jiang, Y., S. Zhang, and Q. Guo, "An Effective Two-Step Approach to the Synthesis of Uniform Amplitude Linear Arrays," *IEEE Antennas and Wireless Propagation Letters*, Vol. 16, 2017, pp. 437–440.

[39] Prisco, G., and M. D'Urso, "Maximally Sparse Arrays Via Sequential Convex Optimizations," *IEEE Antennas and Wireless Propagation Letters*, Vol. 11, 2012, pp. 192–195.

[40] Yan, C., et al., "Synthesis of Planar Sparse Arrays with Minimum Spacing Constraint," *IEEE Antennas and Wireless Propagation Letters*, Vol. 17, No. 6, June 2018, pp. 1095–1098.

[41] Zhao, X. W., Q. S. Yang, and Y. H. Zhang, "Synthesis of Sparse Concentric Ring Antenna Arrays by Convex Optimization," *IET International Radar Conference 2015*, Hangzhou, 2015, pp. 1–5.

[42] Vigano, M. C., et al., "Sparse Array Antenna for Ku-Band Mobile Terminals Using 1 Bit Phase Controls," *IEEE Transactions on Antennas and Propagation*, Vol. 62, No. 4, April 2014, pp. 1723–1730.

PART II

Antenna and Antenna Arrays

2

Antenna Fundamentals

2.1 Introduction

The IEEE Standard Definitions for Terms related to an antenna (IEEE Std. 145–1983) [1] defined it as a "means for radiating or receiving radio waves." Irrespective of the type of the antenna, the basic underlying principle of its radiation is that the "time changing current radiates and the accelerated charge radiates." In other words, the antenna can be defined as a transitional or an impedance-matching structure between the free-space and an RF energy guiding device, as shown in Figure 2.1. Antennas find their applications in numerous applications such as radar, mobile communication, broadcast receivers, missile guidance, deep space networks, and near-field communications for the Internet of Things (IoT) [2–6].

Figure 2.1 shows a flared waveguide structure connected to an RF signal source and the inherent wave propagation inside the structure. The structure is analyzed using a finite element method (FEM)-based solver, Ansys HFSS [7]. It is clearly shown in the figure how the wave escapes from the open and flared end of the waveguide structure into the free space. The simulations also show the formation of big loops of the radiated emissions illustrating the *em* propagation in free space or any medium depending upon its material properties. A plethora of books are existing describing the antenna theory and the various types of antennas with few pioneering ones as listed in [8–12]. This chapter briefly revisits the

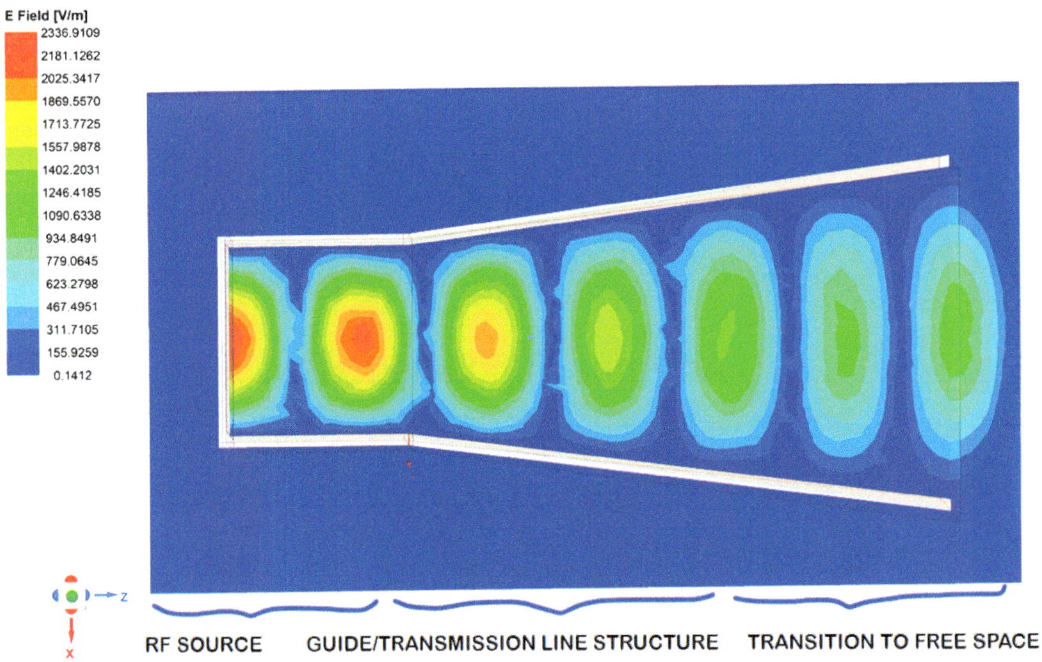

Figure 2.1
The radiation energy flow mechanism in a flared open-ended waveguide structure.

basics to refresh the readers' minds about various topics of the basic antenna theory before proceeding towards the discussion on the sparse antenna arrays (SAAs) in the following chapters in Part III, the main theme of this book.

2.2 Broad Classification of Antennas

Antennas may be broadly classified into three categories: isotropic, omnidirectional, and directional antennas (see Figure 2.2). The isotropic antenna is a hypothetical concept describing radiation from an infinitely small point source uniformly in all directions as shown in Figure 2.2(a). It provides a figure of merit to assess the gain performance of any antenna and the unit used for gain of an antenna is dBi where 'i' denotes isotropic. The closest practically feasible realization of an isotropic antenna is an omnidirectional antenna giving almost constant gain in one of the principal planes (E- or H-planes).

Figure 2.2(b) shows a directional pattern generated from a monocone antenna with a ground plate. Figure 2.2(c) shows an omnidirectional antenna's pattern generated from a helix antenna, which is highly desirable for various applications such as mobile communications, radars, and satellite communications, giving highest gain along an angle in an angular domain.

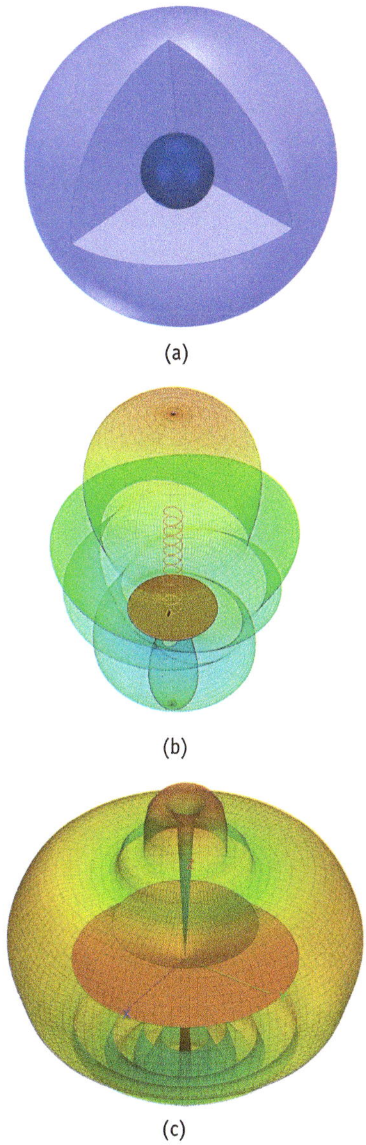

Figure 2.2
Antenna types: (a) isotropic, (b) directional, and (c) omnidirectional.

2.3 Antenna Characterization

In this section, the characteristic parameters of antennas are discussed that determine the quality of the antenna design and its type as well. An antenna element is characterized by the various parameters as given here:

- *Antenna radiation pattern:* It is the angular spread of *em* radiation around the antenna. A typical antenna radiation pattern is shown in Figure 2.3 of an antenna lying in the *xy*-plane centered at the origin, *O*. $P(\theta, \phi)$, is an arbitrary point of observation in space. It shows a plot of radiation field intensity, $F(\theta, \phi)$, as a function of angular coordinates θ and ϕ with its components F_θ and F_ϕ. The intensity is maximum in the main lobe and other lobes are classified as side lobes and back lobes as shown in Figure 2.3 and the points of zero emission are shown as nulls. The field pattern can be characterized by the cut-planes ($\phi = \phi_o$) through the main-lobe axis (herein it is the

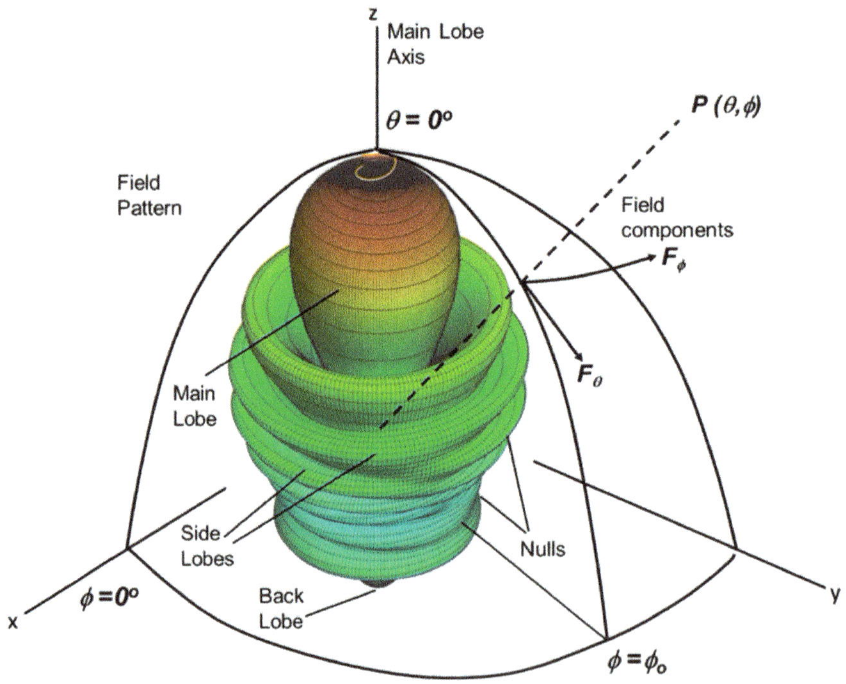

Figure 2.3
A typical 3-D radiation pattern of an antenna.

z-axis). Figure 2.4(a) shows the radiation pattern polar plot in the principal planes (ϕ = 0° and 90°), at right angles to each other in the *xz* and *yz* planes and are also called azimuthal and elevation planes of the antenna (also shown in Figure 2.4(c)). Figure 2.4(b) shows the various alternatives of analyzing the antenna radiation pattern plot (i.e., as a field pattern or as a power pattern). The –3-dB or half-power points with respect to the peak of the main beam defines the half-power beamwidth (HPBW) of the antenna.

- *Directivity:* It is the ratio of the power density in the direction of the antenna's beam peak to the average power density, denoted by *D*.

$$D(\theta,\varphi) = \frac{4p\left\|F(\theta,\varphi)_{\max}\right\|^2}{P_0} \tag{2.1}$$

where P_o is the total radiating power available from the antenna.

- *Gain:* The gain is the reduced measure of directivity after considering the inherent losses in the antenna due to the impedance mismatch, dielectric material losses, radiation, and metallic portions (copper loss) losses as given in (2.2).

$$G(\theta,\varphi) = \left(1 - \left\|\Gamma_a\right\|^2\right)\varepsilon_{\text{loss}} D(\theta,\varphi) \tag{2.2}$$

Γ_a is the reflection coefficient at the input of the antenna and $\varepsilon_{\text{loss}} = \varepsilon_{\text{rad}}\varepsilon_d\varepsilon_{\text{Cu}}$ is the total antenna efficiency, a product of losses due to radiation (*rad*), dielectric (*d*), and copper loss (*Cu*) [6, 9, 10].

- *Impedance:* It is the input impedance at the input of the antenna terminal.
- *Polarization and axial ratio:* The geometric figure traced out by the instantaneous electric field vector (**E**) with respect to the time while transmitting describes its sense of *polarization* (see Figure 2.5). The polarization types are linear and elliptical (circular is the special case of elliptical polarization). Axial ratio (AR) is the mathematical parameter to measure the polarization purity. AR = 1 for circularly polarized (CP), AR = 0 for

horizontally polarized (HP) antenna, and AR = ∞ for vertically polarized (VP) antennas.

- *Bandwidth:* Antenna bandwidth may be categorized as radiation (RBW) and impedance (IBW) bandwidth. The range of frequencies over which desired radiation and impedance characteristics are achieved defines the formal and the later categories, respectively. For instance, an antenna achieves a desired gain > 3 dB over a bandwidth (BW) of 1 GHz and has VSWR < 2:1 over a BW of 500 MHz centered at f_o = 10 GHz. Such an antenna has RBW = 1 GHz (or BW/f_o ~ 10%) and IBW = 500 MHz (or BW/f_o ~ 5%). The scenario presented can be opposite and IBW may be greater than RBW as well. A particular bandwidth criterion can be chosen based on the tradeoff with respect to the application intended for the particular antenna.

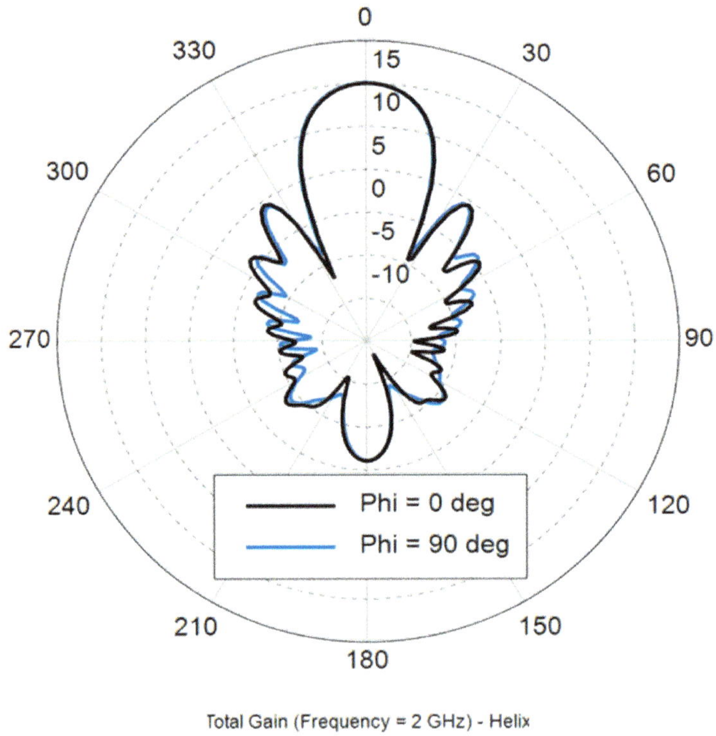

(a)

Figure 2.4
Radiation patterns of an antenna in two principal planes: (a) polar plot in principal planes and (b) radiation pattern plot types. (c) Plot in terms of $\theta\varphi$ space coordinates.

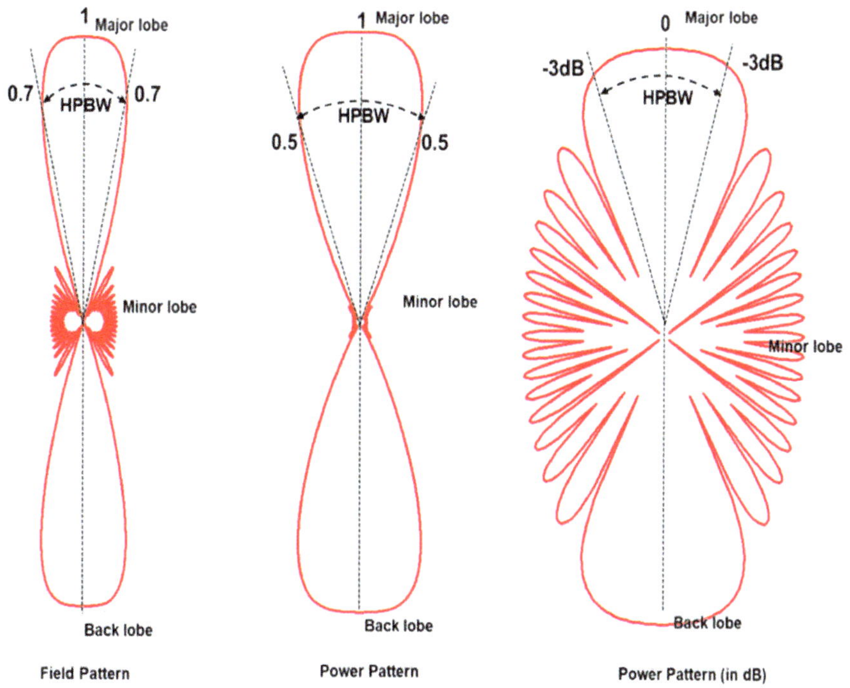

(b)

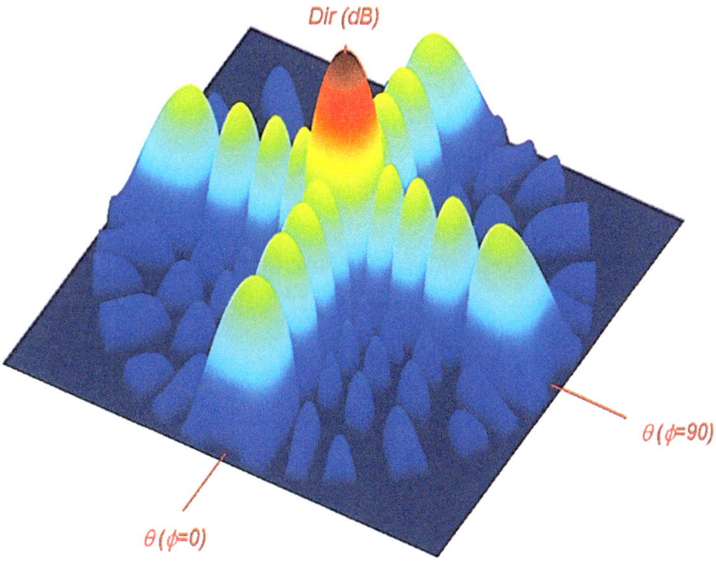

(c)

Figure 2.4
(Continued)

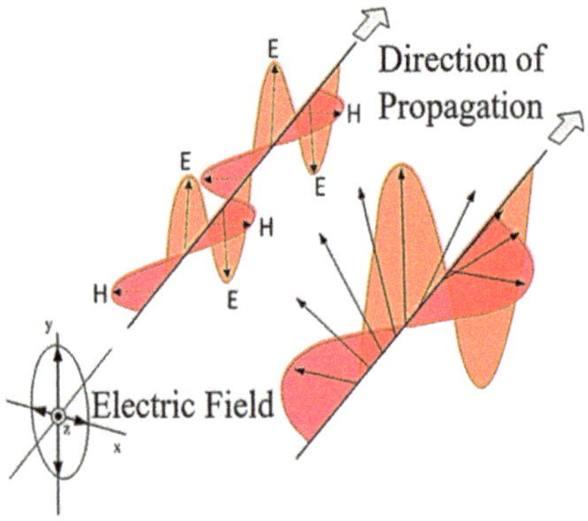

Figure 2.5
Antenna polarization.

2.4 Types of Antennas

The antennas exist in the various forms depending upon the intended applications and are categorized herein based on their principle of operation and the physical form factor (see Figure 2.6):

- Wire antennas such as dipole, monopole, TV aerials;
- Traveling-wave antennas such as helices, Vivaldi antenna, spirals, series-fed slotted waveguide antenna array;
- Reflector antennas such as a Gregorian reflector and a parabolic reflector;
- Microstrip antennas;
- Log-periodic antennas;
- Aperture antennas such as a horn antenna;
- Self-scaling antennas such as a spiral, tapered slot antenna or a fractal antenna.

The individual antenna element's gain may vary from 0 dBi (monopole) to 10–12 dBi (tapered slot antenna and helical) and may be chosen for certain applications depending upon their various characteristics such as the power-handling capability, gain, polarization, bandwidth, size, weight, and volume. For instance, the astronomical telescopes require higher gain and power-handling capabilities along with an installation in a large open area, exposed to a severe and different topological environmental conditions. This

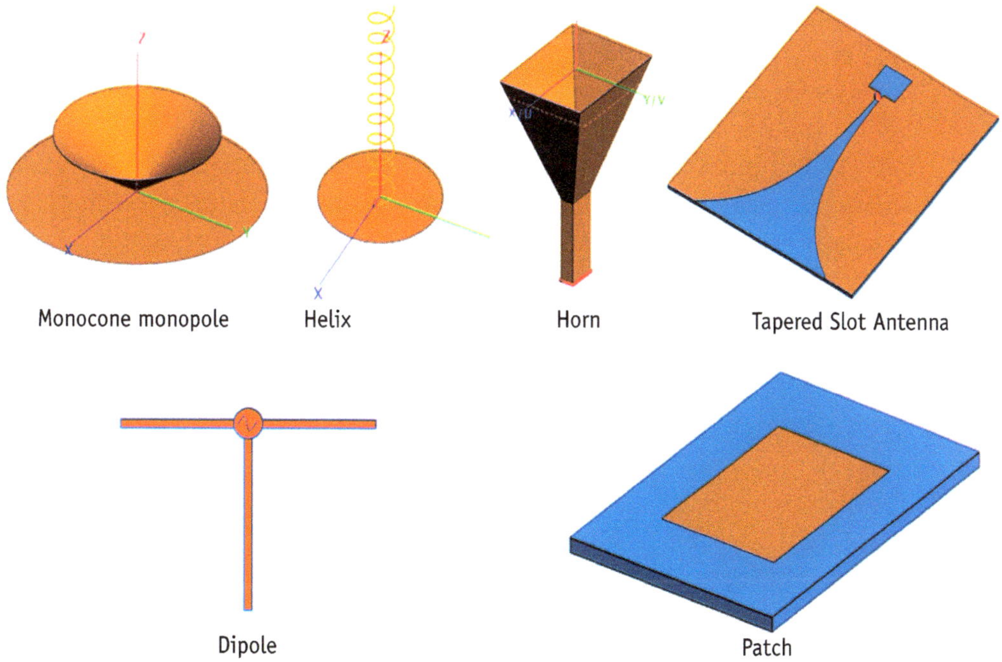

Figure 2.6
Various types of antenna examples.

requirement can be easily satisfied by reflector antenna arrays. For limited real estate platforms as in the case of an airborne platforms such as high-altitude platforms (HAPS) [13] and fighter aircraft, the microstrip antennas are very useful due to their light weight, low profile, and conformability [11, 12]. Traveling-wave antennas and log-periodic antennas are most suitable for ultrawideband and high-power handling applications [6, 10–12].

These antennas may be resonant or nonresonant antennas depending upon their IBW characteristics. Resonant antennas have usually narrow IBW and have a broad main beam with low to a moderate gain. Dipoles and microstrip patches belong to this category. For various applications, it requires that the antenna should possess a wide or ultrawide bandwidth, which is achievable from nonresonant structures such as a tapered slot antenna and a log-periodic dipole. These antennas also fall under the category of self-scaling structures, which lead to frequency independence because the antenna's physical dimensions cater towards different electrical lengths corresponding to the frequency of operation.

These antennas have high gain and narrow main beam (narrow HPBW).

A group of identical or nonidentical antennas arranged in a particular regular or nonregular (aperiodic) arrangement in space is called an antenna array. An antenna array provides a very high gain and an extremely narrow beam as often desired for application such as radars and satcom. A group of 100 elements may give a gain typically of the order of $10\log_{10}(100) \sim 20$ dBi. The topic on the antenna arrays is entertained in detail in the next chapter.

2.5 Antenna Design Techniques

There does not exist any general or universal antenna design technique that may be ubiquitously applied to design any type of antenna. In the initial days, the antenna designs were optimized usually by trial and error along with the experimentations based on the designer's experience. However, with the advent of computers and the various computational electromagnetic techniques (CEM), there are various em tools available for the accurate and efficient design of the antennas. The theory for the computer-aided design (CAD) of antennas has been derived mainly based on Maxwell's equations using either vector equations or diffraction theory or both combined. Lots of theoretical and experimental research in the computational electromagnetics field has led to the postulation of:

- Antenna reciprocity theorem;
- Transmission-reflection concept;
- Field comparison;
- Theory of the antenna system;
- Antenna measurement techniques;
- Great diversity in antenna applications.

Before starting any antenna design process, the particular input specifications required are:

- Electrical properties of an antenna (frequency of operation, radiation characteristics);
- Mechanical properties of an antenna (2-D/3-D structure, metallic, printed);

- Antenna support and mounting structures (types of platform, antenna platform interaction).

Antennas can be classified based on their electrical length typically as:

1. Antennas with cross-section: $< \lambda$.
2. Horn antennas with dielectrics: $< 10\ \lambda$.
3. Traveling-wave antennas: $< 100\ \lambda$.
4. Reflector antennas: $< 1{,}000\ \lambda$.
5. Antenna arrays: $>> \lambda$.

There are various antenna modeling techniques available and used for solving integral equations such as:

1. Geometrical Optics (GO);
2. Physical Optics (PO);
3. Local plane wave reflection method;
4. Spherical wave expansion method (SWE);
5. Numerical methods such as method of moments (MOM), finite element method (FEM), and finite difference time domain (FDTD).

The numerical approaches for solving the wave equations by means of integral equations mainly are mainly:

1. Vector potential method;
2. Green's theorem and Lorentz' reciprocity theorem.

An illustrative and detailed description of these techniques is available in the books [10, 14], to which the readers may be referred for more details.

2.6 Field Regions

The radiation space containing electromagnetic fields surrounding an antenna is usually categorized into three different regions, namely, reactive near-field, radiating near-field (Fresnel), and far-field (Fraunhofer) regions as shown in Figure 2.7. These regions are

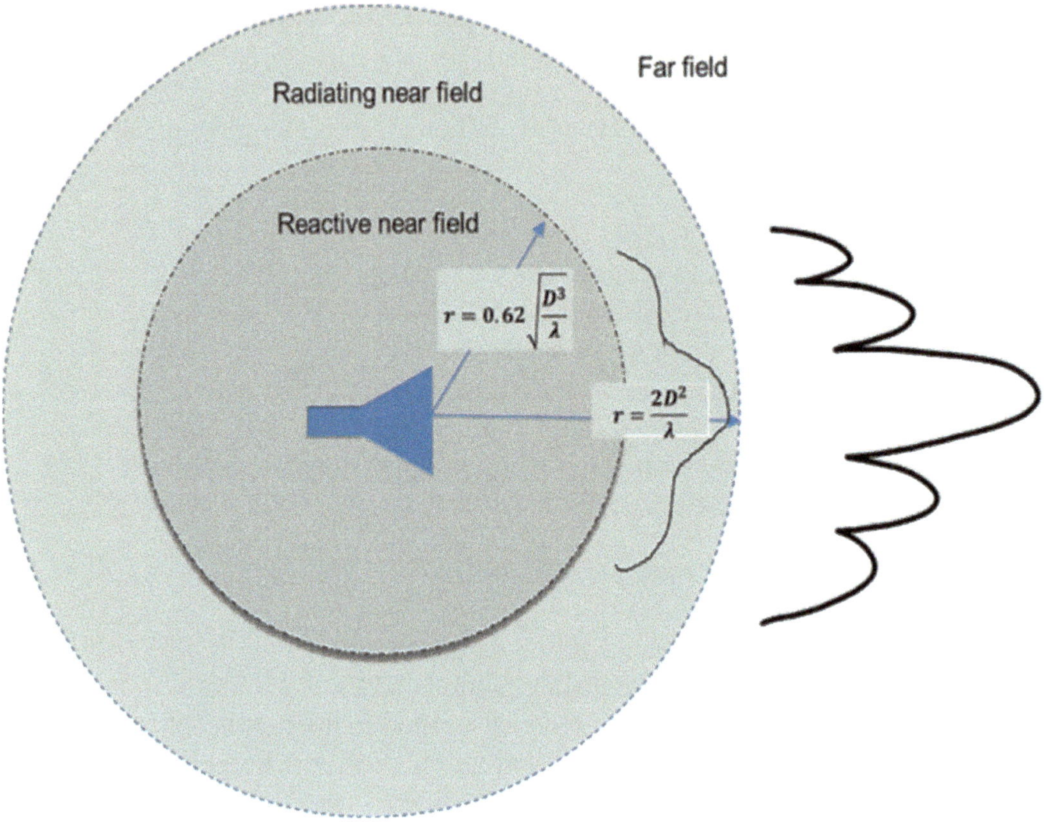

Figure 2.7
Field regions of an antenna.

so designated to identify the field radiation pattern in each of these regions. These regions have no clear distinct changes while their boundaries are crossed; however, still there are distinct differences among them. Although the boundaries separating these regions are not unique, there are various criteria that have been established to commonly identify these regions.

- *Reactive near-field region:* This is defined as a "portion of the near-field region immediately surrounding the antenna with a dominant reactive field" [9, 15]. It is assumed to be prevalent at a distance $R < 0.62\sqrt{D^3}/\lambda$ from the antenna's surface, where λ is the wavelength and D is the largest dimension of the antenna. In the case of small antennas, the outer boundary exists $\lambda/2\pi$ distance away from the antenna surface.

- *Radiating near-field (Fresnel) region:* This is defined as "the field region between the reactive near-field and the far-field regions wherein the radiation field predominates and the angular field distribution depends upon the distance from the antenna" [9, 15]. It is also referred to as the Fresnel region. If the antenna's maximum overall dimension is very small with respect to the wavelength, this field region may not exist. The inner boundary is at the distance, $R < 0.62\sqrt{D^3}/\lambda$ and the outer boundary is at a distance $R < 2D^2/\lambda$, where $\boldsymbol{D}$ is the largest dimension of the antenna based on a maximum phase error of $\pi/8$ [9, 10].
- *Far-field (Fraunhofer) region:* This is defined as "the region of the field of an antenna with the angular field distribution essentially independent of the distance from the antenna" [9, 15]. It is commonly assumed to exist at a distance, $R > 2D^2/\lambda$. The far-field patterns of certain antennas, such as multibeam reflector antennas, are sensitive to variations in the phase over their apertures and this criterion may be inadequate. In practice, if the antenna has a maximum overall dimension, $D >> \pi/|\gamma|$, the far-field region may be assumed to exist at a distance equal to $|\gamma| D^2/\pi$ from the antenna (γ is the propagation constant in the medium). The far-field region is sometimes referred as the Fraunhofer region. Here, the field components are essentially transverse and the angular distribution is independent of the radial distance at which the measurements are made. The inner boundary is taken to be the radial distance $R > 2D^2/\lambda$ and the outer one is taken to be at infinity.

The amplitude pattern of an antenna varies as the observation distance from the reactive near-field to the far-field region. It changes in shape because of variations of the fields, both in the magnitude and the phase. The reactive near-field region has a widely spread pattern that is nearly uniform with the slight variations. As the observation point is moved to the radiating near-field region, the pattern smooths and forms lobes in the pattern. In the far-field region, the pattern is well formed and consists of few minor lobes and one, or more, major lobes.

2.7 Antenna Measurements

The need for the antenna measurements arises because the theoretical investigation of antennas is often limited by their (structural and electrical) complexity. Henceforth, the experimental results are used to validate the theoretically predicted behavior of these (complex) radiators. The antenna pattern measurements are required because in order to completely characterize the radiation field of an antenna, by measuring its relative amplitude, phase, polarization, and the power gain on the surface of the hypothetical sphere, the center of which is located at the antenna under test (AUT). The sphere should extend up to the far-field zone (see Figure 2.8). A basic assumption is made that the AUT is treated as a passive, linear, and reciprocal device. Hence, its properties can be measured in either transmitting or receiving mode. A key limitation of the antenna measurement technique is the operating environment in which AUT is installed, which may substantially modify the intrinsic pattern of an antenna. Therefore, measurements in situ may be required.

The instrumentation of a typical antenna-range measuring system consists of mainly [15]:

- Source antenna;
- Transmit (Tx) and receive (Rx) systems;
- Antenna positioning systems;

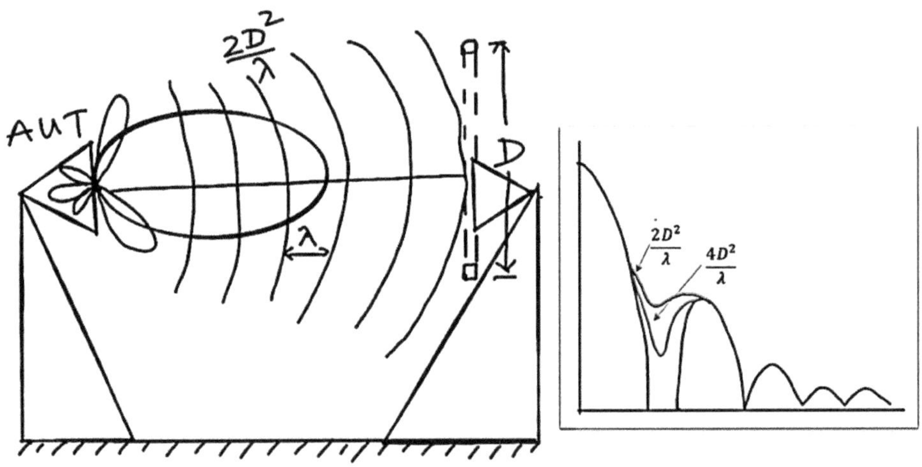

Figure 2.8
Far-field measurement setup.

- Recording systems;
- Data-processing units.

The antenna measurement ranges are designed in order to enable the measurement of antenna patterns independently from their operational environment. There are two main types of antenna ranges typically:

- Outdoor measurement range (open test range):
  - Guarantee a plane-wave incidence on the antennas under test as the measurements are done at distance;
  - Suffer from the weather conditions and nearby structures.
- Indoor test range (near-field anechoic chambers):
  - Guarantee an all-weather operation;
  - For low frequencies may require complicated systems for the far-field operation;
  - Elevated test ranges (ETR).

A typical elevated open test range setup is shown in Figure 2.9.

ETR is primarily used for:

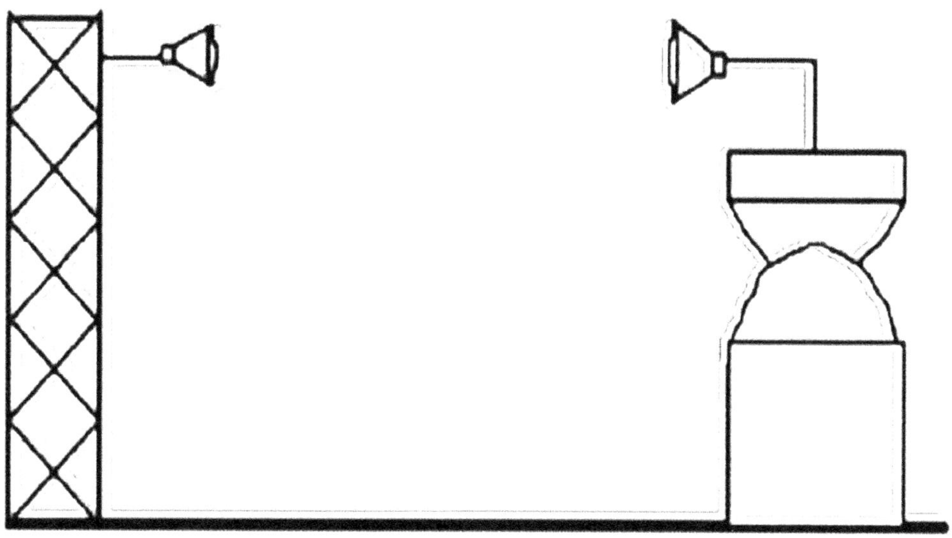

Figure 2.9
Elevated test range.

- Physically large antennas;
- Antennas mounted on towers or tall buildings;
- Mostly operate over rather smooth terrains.

Disadvantages of ETR are:

- Range width has to be sufficiently wide to accommodate the source antenna beam so to reduce spurious contributions.
- Reflections from the ground are difficult to be removed (time gating techniques can be used).

The ground-reflection test ranges (GRTR) is another type of open test range in an elevated test range setup as shown in Figure 2.10.

GRTR is primarily used for:

- Used for physically large antennas;
- Antennas placed over a flat reflecting surface;
- The specular reflection is accounted for; it can be associated to an image antenna.

The disadvantages of GRTR are:

- The height of the source antenna depends on the frequency.
- The correct characterization of the polarization suffers.

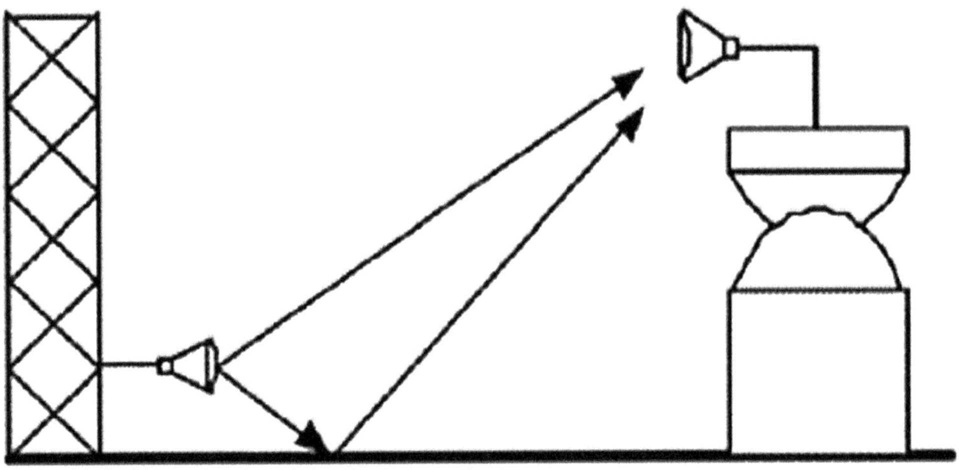

Figure 2.10
Ground reflection test range.

In electromagnetic anechoic chambers (EACs), electromagnetic anechoic chambers are shielded (screened) enclosures (see Figure 2.11) in which internal reflections of electromagnetic waves are reduced to a minimum. Herein:

- Free-space propagation conditions are mimicked.
- External electromagnetic interfaces are minimized.

In EACs:

- The chamber's walls are covered with electromagnetic absorbing material (carbon-loaded polyurethane foam) (Figure 2.11(a)).
- They are mostly used at microwave frequencies (availability of the absorbers) and are not suited for a low frequency of operations (e.g., high frequency/very high frequency (HF/VHF)).

2.8 Summary of the Chapter

This chapter briefly refreshed the basics of antenna theory without dwelling on the mathematics involved herein as the mathematical formulation is already available in the literature in abundance

(a)

(b)

Figure 2.11
(a) Electromagnetic absorbing material. (Source: Internet.) (b) Photograph of an antenna array in an electromagnetic anechoic chamber.

[6–10]. This book intends to just enlist the various basic fundamentals required to understand the main theme of this book, that is, the sparse antenna arrays (SAAs). The various topics are related to SAAs, while covering all major aspects of the involved theory and the mathematical formulation necessary for the synthesis and analysis of these antennas are given in the succeeding chapters.

References

[1] "IEEE Standard Definitions of Terms for Antennas," *IEEE Std 145-1983*, 1983, pp. 1–31.

[2] Weber, R. H., and R. Weber, *Internet of Things*, Vol. 12, New York: Springer, 2010.

[3] Sarkar, T. K., and S. Palma, "A History of the Evolution of Radar," *2014 44th European Microwave Conference*, Rome, Italy, October 6–9, 2014, pp. 734–737.

[4] Munson, R. E., and J. Haddad, "Microstrip Reflectarray Antenna for Satellite Communication and RCS Enhancement," U.S. Patent 4684952, August 1987.

[5] Sarkar, T., et al., *History of Wireless*, New York: John Wiley & Sons, 2006.

[6] Kedar, A., "Tutorial and Research Trends in Antenna Technology," *Microwave Journal*, October 28, 2020.

[7] Ansys HFSS User Manual, www.Ansys.com/hfss.

[8] Visser, H. J., *Antenna Theory and Applications*, New York: John Wiley & Sons, 2012.

[9] Balanis, C. A., *Antenna Theory: Analysis and Design*, Fourth Edition, New York: John Wiley & Sons, 2016.

[10] Stutzman, W. L., and G. A. Thiele, *Antenna Theory and Design*, Third Edition, New York: John Wiley & Sons, 2012.

[11] Milligan, T. A., *Antenna Theory and Design*, Third Edition, New York: John Wiley & Sons, 2005.

[12] Munk, B. A., *Finite Antenna Arrays and FSS*, New York: Wiley-IEEE Press, 2003.

[13] Capstick, M. H., and D. Grace, "High Altitude Platform mm-Wave Aperture Antenna Steering Solutions," *Wireless Personal Communications*, Vol. 32, 2005, pp. 215–236.

[14] Garg, R., *Analytical and Computational Methods in Electromagnetics*, Norwood, MA: Artech House, 2008.

[15] Hansen, J. E., et al., *Spherical Near-Field Antenna Measurements*, London, U.K.: The Institution of Engineering and Technology, 1988.

3

Antenna Arrays

3.1 Introduction

Antennas are indispensable in any type of a wireless system, whether it is a radar, mobile communication system (MCS), satcom, massive multiple input multiple output (M-MIMO), and near-field communication such as radio-frequency identification device (RFID). A single antenna usually has a directivity/gain, G_{iso}, varying from 0 dBi to 8–10 dBi depending upon its type as discussed in Chapter 2. However, the wireless systems such as radars, MIMO, MCS, and satcom require larger gain to meet the expected ranges (in the order of hundreds or thousands of kilometers) as per the standard radar range equation [1, 2], while considering the atmospheric and the system losses. Thus, single antenna being rendered useless for the said objectives is multiplied in the number to form a group of antennas elements, called an antenna array, to provide larger gain values. Typically, the gain achieved is proportional to the total number of elements, say, N_{tot}, assuming that there are no inherent losses in the antenna array environment. There are lots of good monographs and texts written on antenna arrays and their numerous applications [2–14]. An exhaustive list of books written on the topic is presented in [7] and is quite useful. This chapter will discuss the topic succinctly and will elaborate more upon the important topics required to understand the sparse phased array antennas (SPhAAs) or simply sparse array antennas.

3.2 Array Classification

Figure 3.1 shows the typical geometry of various antenna arrays, out of which linear, planar, and circular are the common generic array geometries. The other examples of the array geometry are shown in Figure 3.1, such as a rectangular grid with circular aperture, multifaceted conformal geometry, hexagonal grid, and subarray configurations, which are a few of the variants out of infinite possibilities depending upon the designers' creative abilities. The antenna array's design depends mainly upon the requirements of the array designer, which may be broadly classified as:

- Frequency of operation and operational bandwidth;
- Antenna array gain, G_{arr};
- Array beam resolution (half-power beamwidth (HPBW));
- Scan volume in the case of phased array antennas;
- Mutual coupling and grating lobes consideration;
- Platform for the installation of the antenna array.

These considerations will be discussed in this chapter in detail with the help of a few examples.

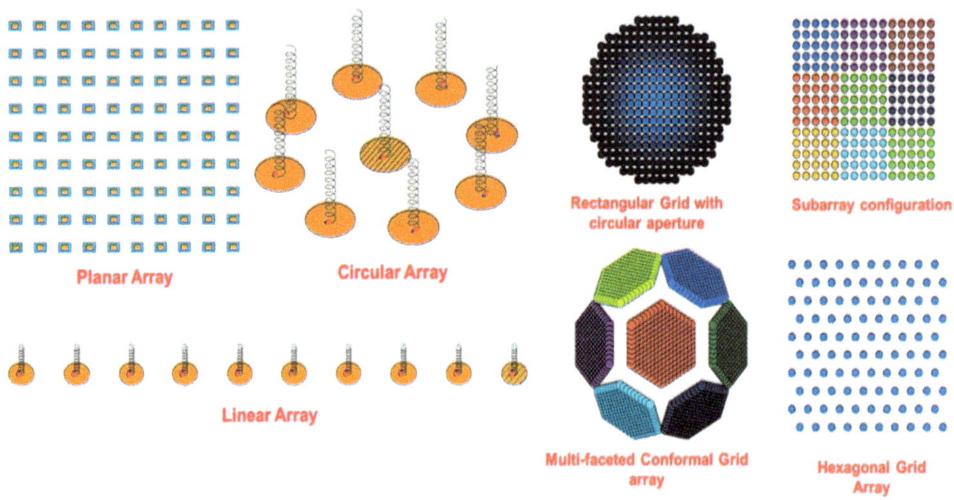

Figure 3.1
Various types of antenna arrays.

3.3 Antenna Array Electromagnetics

An antenna designer needs to understand the basic array electromagnetics to carry out an efficient design of the antenna array. It is required to understand the various parameters that are needed to characterize the suitability of the array design in conformance to the prescribed array specifications based on the various criteria mentioned in the previous section. It is always simple and intuitive to understand these facts about antenna array by starting from the analysis and design of a linear antenna array (LAA) as shown in Figure 3.2.

3.3.1 Linear Antenna Array (LAA)

Figure 3.2 shows an LAA with its elements aligned in a regular grid arrangement along the *z*-axis in a *yz* plane. It consists of N_{tot} (= $2N_n$ or $2N_n + 1$, if the number of elements is even or odd, respectively) identical antenna elements symmetrically placed around the origin, *O*. The variable, *P*, denotes a general observation point for assessing the radiation intensity at a distance, *r*, and angle, θ. The interelement spacing between elements is assumed equal (uniform grid) and denoted as d_n ($n = 0, 1, 2, \ldots, N_n - 1$). The array is assumed to be consisting of isotropic antenna elements (with $G_{\text{iso}} = 1$ or 0 dBi). The radiation pattern, *I*, of an LAA may be defined as:

$$I(\theta) = EP(\theta) \times F(\theta) \tag{3.1}$$

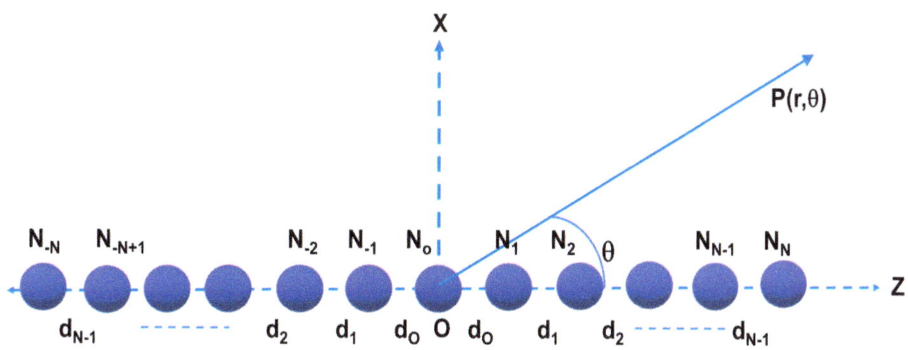

Figure 3.2
LAA.

where *EP* is the element pattern and *F* is the array factor. Since the array is having only one dimension, only θ dependence is considered. Thus, (3.1) implies that the total radiation pattern of an array is dependent on the single antenna element's behavior as well as the array geometry in which the multiple antenna elements are arranged. Thus, the choice of an antenna element is also critical and it will be discussed later how the choice improvises the antenna array performance. In the case of an isotropic element, $EP = \cos\theta$ with $D_o = 1$ (0 dBi).

The array factor (*AF*) for an LAA is given as:

$$F(\theta) = A_0 + \sum_{\substack{n=-N_n \\ n\neq 0}}^{N_n} A_n e^{jkx_n\Delta u} \tag{3.2}$$

In (3.2), A_n denotes the amplitude excitation applied to n^{th} element. Herein, $A_0 = 1$ or 0, if N_n is odd or even. An LAA can be categorized as uniformly fed (UFA) or nonuniformly fed or tapered array (NUFA) based on the excitations fed to its elements. An LAA is UFA, if $A_1 = A_2 = \ldots = A_n$ (equal excitations given to all elements) or NUFA if $A_1 \neq A_2 \neq \ldots \neq A_n$, respectively. There are various options for the amplitude tapers available and the readers may refer for the details of these in a great work written by Elliott [8]. Further, in (3.2),

$$\Delta u = \sin\theta - \sin\theta_s \tag{3.3}$$

$$x_n = x_{n-1} + d_n \tag{3.4}$$

In (3.3), θ_s is the look angle or the scan angle at which the array is gazing, which implies the angle at which the AF has a maximum value (or intensity). The maximum value, $F_{\text{max}} = N_{\text{tot}}$ (G_{arr} ~10 log (N_{tot}) in dB) is achieved while considering no losses are present in the array system. Hence, it is apparent that the maximum value can be increased manifold times in concurrence with the antenna element gain (the maximum value of EP) as $G_{\text{arr}} + EP_{\text{max}}$. Thus, a proper choice of an antenna element (EP_{max}) can increase the total array gain, thereby providing a flexibility to reduce the total number of elements.

For example, an array with G_{arr} ~30 dB gain will become 33 dB in case of an element with EP_{max} = 3 dBi. To achieve a G_{arr} ~30 dB, it requires typically 1,000 elements, which will reduce to a requirement of 507 elements approximately by choosing EP_{max} = 3 dB and G_{arr} = 27 dB (equal to 507 elements). Thus, a proper trade-off between EP_{max}, G_{arr}, and half-power beamwidth (HPBW) leads to an efficient design of an array with minimal number of elements. The reduction in the number of elements reduces the overall resource allocation in an active phased array system where each element is connected to an active circuitry behind [12]. Moreover, the effective power aperture product ($\sim P_{out}G_{arr}$) is another figure of merit to maneuver the trade-off for the antenna array design as $P_{out} \propto N_n$. Henceforth, assuming that P_{out} meets the design requirement, EP_{max} and G_{arr} may be chosen while accounting for the inherent system RF losses in an antenna array system [12, 15].

Figure 3.3 shows a typical plot illustrating AF of a 10-element LAA with interelement spacing of a half-wavelength (0.5λ) with uniform and nonuniform amplitude excitations. The nonuniform excitations are applied as per Taylor's distribution [8]. It is observed that the peak side lobe level (PSLL) goes down from –13.2 dB to –30 dB and then finally to –40 dB in proportion to the applied Taylor's distribution. The lowering of PSLL is accompanied by beam broadening as per the law of conservation of energy, which states

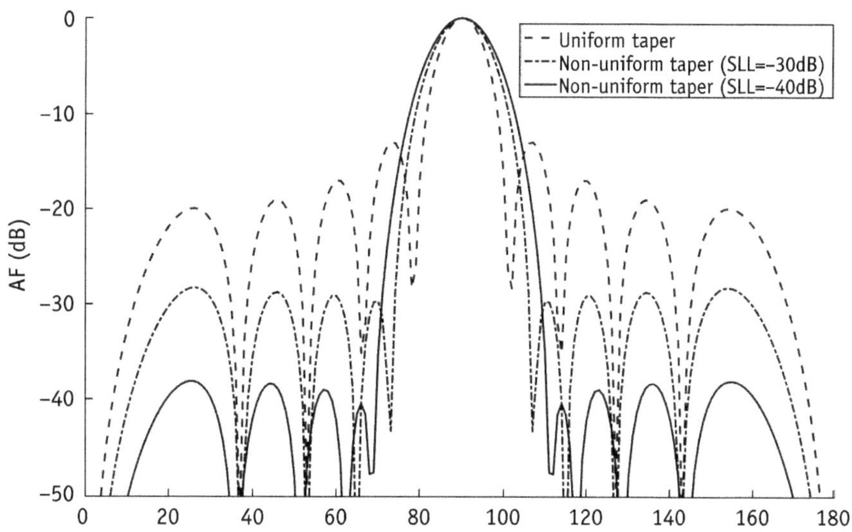

Figure 3.3
Array factor plot of an LAA.

that the total energy stored in an antenna array is same and is equal to the sum of the energy in main lobe and the side lobe regions of the array's radiation pattern. The overall gain of an array is proportional to HPBW and may be typically defined as [3]:

$$G_{\text{arr}} = \frac{27{,}400}{HPBW} \tag{3.5}$$

The expression (3.5) is only valid for UFAs and a more generalized expression of directivity is discussed in detail in Chapter 4. Thus, with the increase in HPBW, G_{arr} will decrease proportionately. It is seen that the antenna array design is a trade-off between several array parameters and a judicious design needs to be carried out keeping all considerations under scanner in order to meet the final goal of the design procedure.

3.3.2 Planar Antenna Array (PAA)

An LAA is quite useful in applications like direction of arrival (DoA) systems, identification friend or foe (IFF), and instrument landing system (ILS), where antenna arrays need directionality along one direction (see 3-D radiation pattern of linear array in Figure 3.4(a))

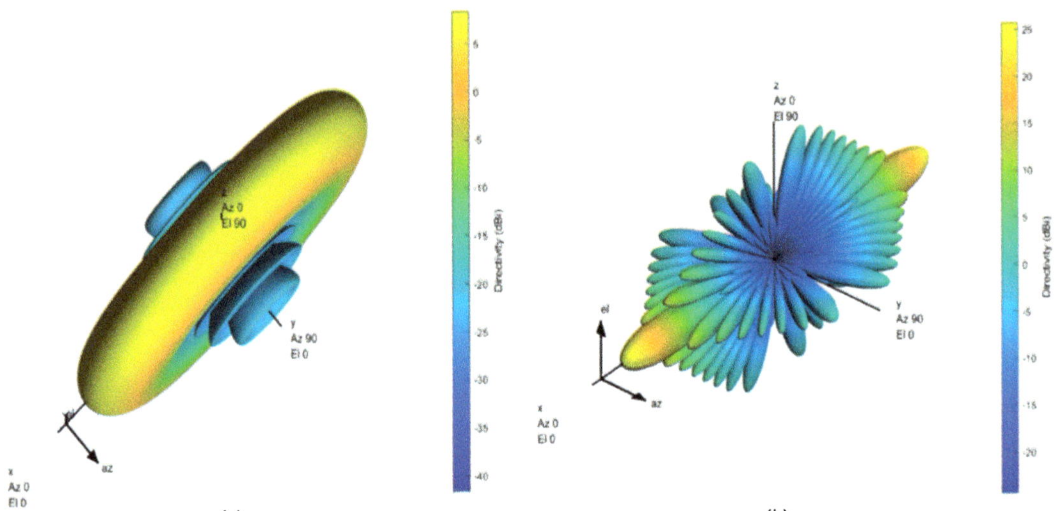

(a) (b)

Figure 3.4
Antenna array patterns: (a) LAA and (b) planar array antenna.

[9–11]. However, it does not suffice when the array beam needs to be a pencil beam-shaped (see 3-D radiation pattern of PAA in Figure 3.4(b)). Henceforth, it is required to arrange antenna elements along both the dimensions (i.e., forming a planar array antenna (PAA)); see Figure 3.5.

Figure 3.5 shows a PAA illustration, lying in *xy*-plane with origin at *O*, radiating in broadside direction along *z*-axis. In Figure 3.5, *P* represents an observation point at a distance, *r*, from the origin at an angle, (θ, φ). The interelement separations along *x*- and *y*-axes are d_x and d_y. The expression for AF of a planar antenna array is given as:

$$F(\theta,\varphi) = \iint_S I(x,y)e^{\left[-ik(x\sin\theta\cos\varphi + y\sin\theta\sin\varphi)\right]}dxdy \tag{3.6}$$

$F(\theta, \varphi)$ is the expression for AF illustrating its angular spread over the complete hemisphere $(\theta, \varphi) = 0°$ to $360°$, where $I(x, y)$ is the complex excitation fed to elements, given as:

$$I(x,y) = I_0(x,y)e^{-\left[-ik(x\sin\theta_0\cos\varphi_0 + y\sin\theta_0\sin\varphi_0)\right]} \tag{3.7}$$

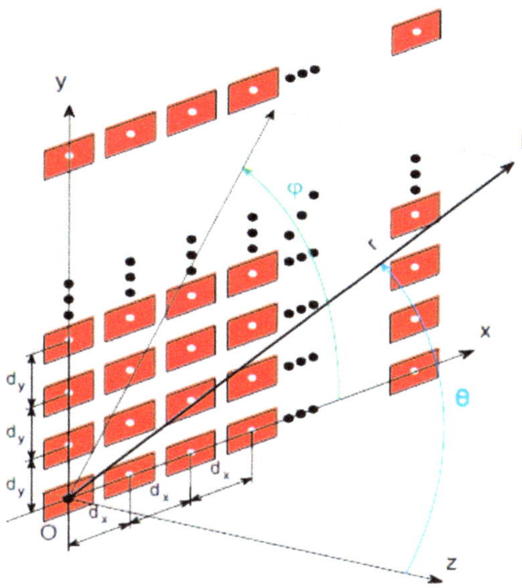

Figure 3.5
A planar uniformly spaced antenna array. (© 2014 IEEE. Reprinted with permission from [34].)

In (3.7), θ_0 and φ_0 represent the look or scan angle along which the main beam is formed, and I_0 is the amplitude of the excitations being fed to an array at (x, y) location. Figure 3.6 shows a beautiful illustration of the current distribution sampled across the aperture of 1,000-element PhAA by discretely located antenna elements. The left side of the figure shows that a maximum current concentration is in the center and it slowly goes down in magnitude towards array edges showing a tapered amplitude distribution being fed to the array. The right side of the figure shows the radiation pattern of the array in two principal planes at boresight showing decrease in SLL due to tapering applied to the elements.

This section concludes the basic array electromagnetics and pattern characteristics necessary to understand the upcoming topics more relevant to sparse phased array antennas. The next section discusses few critical design issues of PhAA while designing phased array antennas that will later highlight the importance of sparse PhAA (SPhAA) systems.

3.4 Phased Array Antennas (PhAA)

A basic question that mystifies the beginners during antenna array design process is how a phased array antenna (PhAA) is different

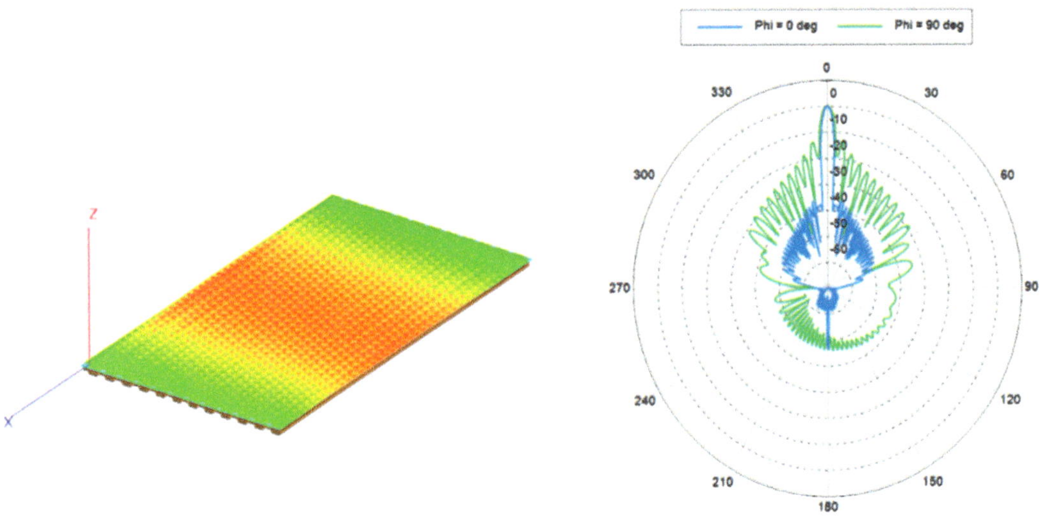

Figure 3.6
A planar uniformly spaced and nonuniformly fed PhAA.

from the conventional PhAAs. To put it simply, there is no difference in the design of antenna arrays for PhAAs. However, there are few important design considerations that need to be taken care of while designing an antenna array for phased beam-steering applications. These will be discussed in the following sections with more focus on the physics behind rather than on the mathematical treatment, which is readily available in the literature [4, 12–14].

3.4.1 Principle of Beam Steering

An array is designed to radiate either in broadside direction (θ_0, φ_0) = (0°, 0°) or in the end-fire direction (θ_0, φ_0) = (0°, 90°) [2–4]. The basic aim for an antenna/antenna array is to cover a dedicated section of space by illuminating it with the radio frequency (RF) energy (energy transmitted by an array and received back by same (monostatic) or another (bistatic) array in the form of echoes or reflections from the illuminated area), greatly explained in theory of radars' operation [1]. Since the antenna beam is highly directional, it cannot illuminate the entire area in a single look and, henceforth, needs to scan or traverse the entire spatial volume by roaming across the region; henceforth, this process is called array scanning. The antenna array can scan across the entire area either by mechanical movement of the antenna pedestal (structure holding the antenna intact) or by just electronically scanning the antenna main beam without any physical movement of the antenna pedestal. The former is called a mechanically scanned antenna array (MSA) and the latter is called an electronically scanned antenna array (ESA). ESAs have the underlying mechanism called as phased beam steering, making them also popularly known as PhAAs [4, 12].

Figure 3.7 illustrates the principle of beam formation and beam steering. It consists of multiple antenna elements (indexed from 1 to K) arranged along an axis at interelement separations, $d_n(n = 1, 2, 3, \ldots, K)$. Each of the elements radiates a primary signal wavefront fed to it from an RF source at its input. In analogy to water waves [12], these wavefronts propagate in the free space (or any medium of propagation) and interact with each other forming constructive or destructive interference. This results into the formation of maximas and minimas, leading to the formation of main beam (maximum), sidelobes (partial construction), and nulls (total destruction) in the radiation pattern. Thus, in order for the constructive beam formation to take place, resulting into an enforced superimposed wave font, at an observation point shown as a dotted blue shaded line (see Figure 3.7), it requires primary wavefronts from each of the sources to reach the observation point at the same instant, t. This implies that there should be no phase lag between these individual wavefronts. It implies $\Delta\Phi = 0°$, where $\Delta\Phi$ is the progressive phase shift between the neighboring elements. The left side of the figure shows the beamforming at the boresight and the right side of the figure shows the case when the beam is focused at a tilted angle, θ (i.e., scan angle). In the tilted case, the distance traveled by the individual wavefront from each of the array elements to reach the observation point (a dotted line) is different. Henceforth, a phase delay, $(n - 1)$ times, needs to be applied to n^{th} element to make its wavefront to reach the observation point at instant t. Thus, a zero delay is applied to the farthest element, (i.e., 0th element). The varying phase delays as applied to the elements form a constructive planar (almost) wavefront at an angle, θ, thus providing the necessary beam scanning electronically. The progressive phase shift in the case of an LAA is given as:

$$\Delta\Phi = \frac{2\pi d_n \sin\theta}{\lambda} \tag{3.8}$$

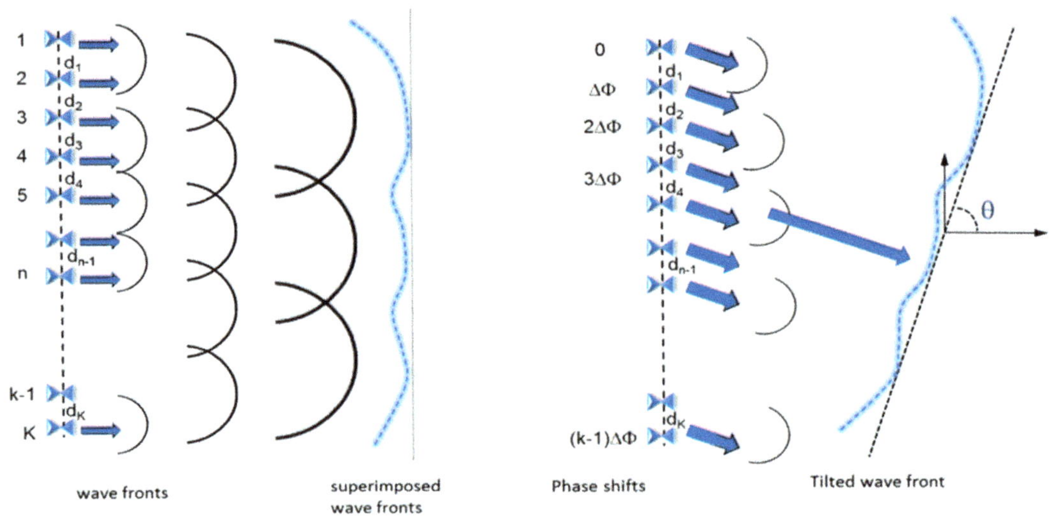

Figure 3.7
Beam-steering mechanism.

The array beamforming mechanism can also be well understood by utilizing the vector calculus as shown in Figure 3.8. Referring to Figure 3.8, it is seen when all phase wavefronts are aligned (forming a maxima) that their vector components are aligned perfectly resulting into a resultant vector with a stronger magnitude, creating a main beam in a particular direction. If there is a slight misalignment in the directions of vectors, it results into a resultant vector with a smaller magnitude, forming sidelobes in the pattern. Lastly, if the vectors are totally in opposite directions, they cancel each other resulting into a null formation.

Thus, the magnitude and direction of the respective vectors (of form $\vec{r_k} = r\angle\theta$, where r is the magnitude and θ is the phasor part of the vector) can be tuned to get requisite beam steering electronically in the particular direction. AF of an LAA can be considered as a superposition of these vectors as $F(\theta) = \sum_{k=1}^{K} r_k \angle\theta_k$, a vector equivalent of (3.7). However, (3.7) is used for all calculations and array designs' assessment.

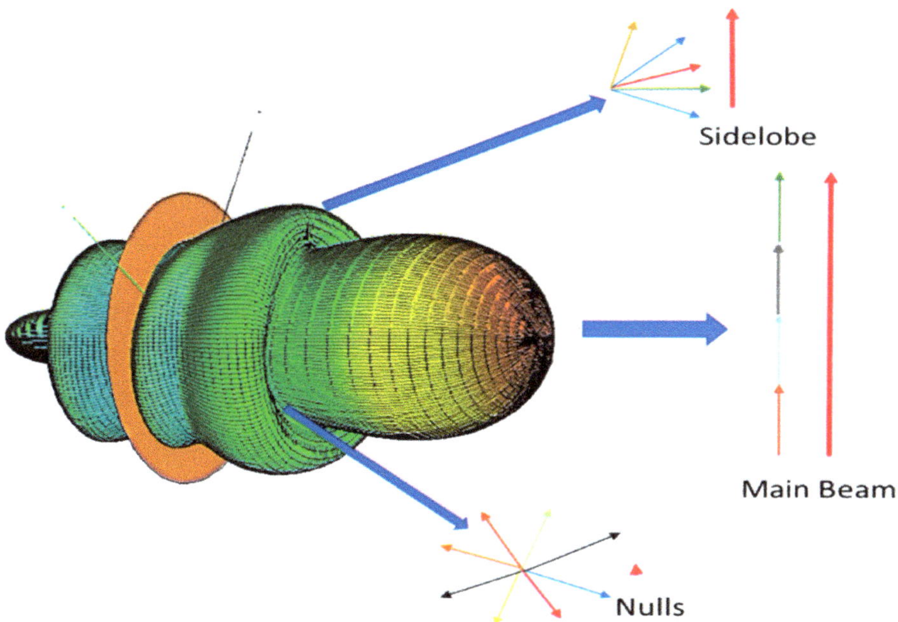

Figure 3.8
Array beamforming.

3.5 Antenna Array Design Constraints

In this section, few critical design constraints are discussed that need to be comprehended before designing the phased array antennas. These also serve as the figures of merit in the following chapters to identify the advantages of the sparse phased array antennas over the regular phased array antennas discussed herein. These constraints are:

- Grating lobes (GL) formation;
- Mutual coupling (MC) phenomenon;
- Scan blindness and limited field of view (LFoV).

3.5.1 Grating Lobes Mechanism

The grating lobe (GL) mechanism or aliasing represents the formation of more than one beam with equal intensity in the spatial far-field region [2, 4]. This phenomenon is not acceptable as the formation of two maxima will lead to the problem of the detection of the same echo or target (in case of radars) from two locations leading to misled target information. Further, it also leads to the reduction in the antenna array gain by half (50% or –3 dB) following the principle of conservation of energy. The condition for avoiding these GLs is given in (3.9), specified by the maximum array interelement separation that can be allowed while designing an antenna array. Ideally, to keep the GLs out of the visible region of an array (–90° to +90°), $d_{max} \leq \lambda/2$. In order to have LFoV, say, less than ±90°, d_{max} can be relaxed, for instance for ±60° FoV or scan volume, it is chosen as $d_{max} \leq 0.536\lambda$.

$$d_{max} \leq \frac{\lambda}{\left|\sin\theta + \sin\theta_o\right|} \tag{3.9}$$

Figure 3.9 shows the GL formation for different cases of interelement spacing in uv space as simulated using MATLAB. Thus, the maximum scan volume of antenna array is $-90° \leq \theta_o \leq +90°$, also called the visible region. Any secondary beams with maximum intensity forming outside this region does not contribute and falls into an invisible region. The green circular region shows the visible space around the main lobe bounded by a unit circle along with other secondary beams forming outside this region (shown as pink) in invisible space.

The top of the first figure shows the case when the interelement separation is less than 0.5λ. In this case, the secondary regions are well separated from the visible region, which is completely free from any influx of GLs while scanning the main beam. The middle on the top of the second figure shows the case for half-wavelength separation. It is a limiting case (see (3.9)), where the boundaries of other lobes just graze the boundary of the visible zone. The third case on the top on the right side showing the case where interelement separation exceeds the half-wavelength limit. It is observed that the secondary lobes start to influx inside the visible zone even without scanning, thus limiting the scan volume (i.e., LFOV) of the main beam. In the bottom of the figure, two cases are shown representing the GLs circle when the main beam is scanned to extreme cases of (45°, 45°) in the case of separations equal to a half-wavelength and more than a half-wavelength, respectively. In the latter

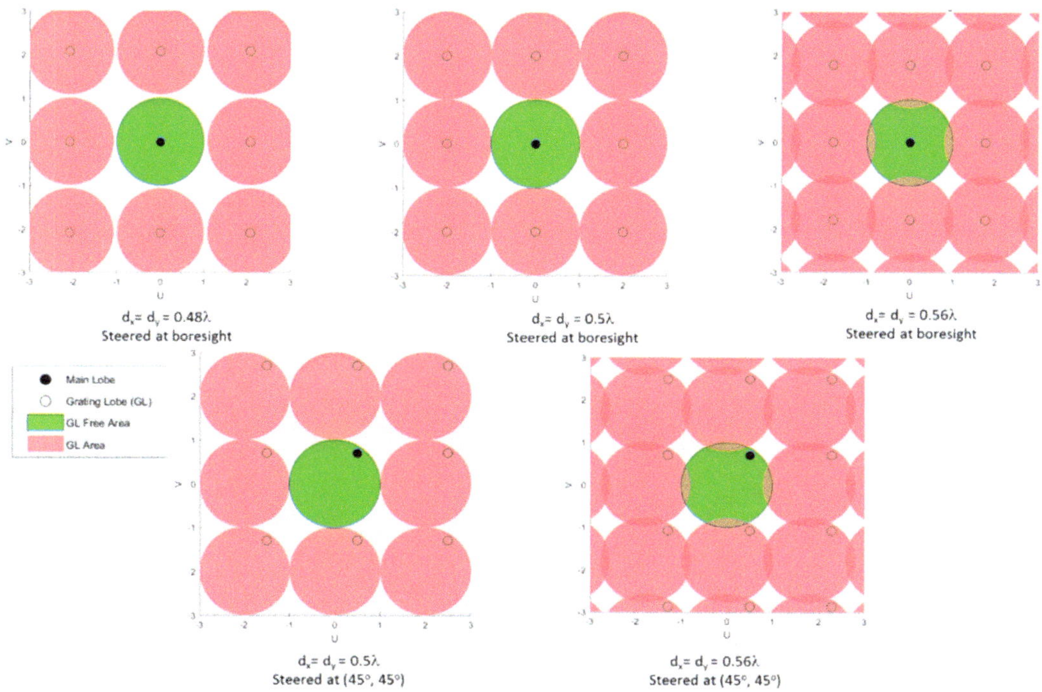

Figure 3.9
Grating lobes formation in the radiation pattern of an antenna array.

case, a limited scan is visible as on scanning the grating lobes enter inside the visible region. This affects the efficiency of the array and also creates a problem of aliasing or target misinterpretation in the case of radars [1]. Figure 3.10 shows the GLs' formation in the radiation pattern for an antenna array with the interelement spacing of one wavelength. Thus, in phased array applications, it is a mandatory requirement to design an antenna array while avoiding any onset of GLs in the desired scan volume region.

3.5.2 Mutual Coupling Phenomenon

Mutual coupling (MC) is defined as an interaction between the two closely spaced antenna elements placed in the near or close vicinity of each other. Two of the most important characteristics of an antenna are its input impedance and its radiation pattern, both being frequency-dependent quantities. The near-field characteristics of an antenna element in the vicinity of another element are much different from the one observed when it is in an unbounded free space (i.e., an isolated environment). The change in boundary conditions leads to the occurrence of new (equivalent) currents

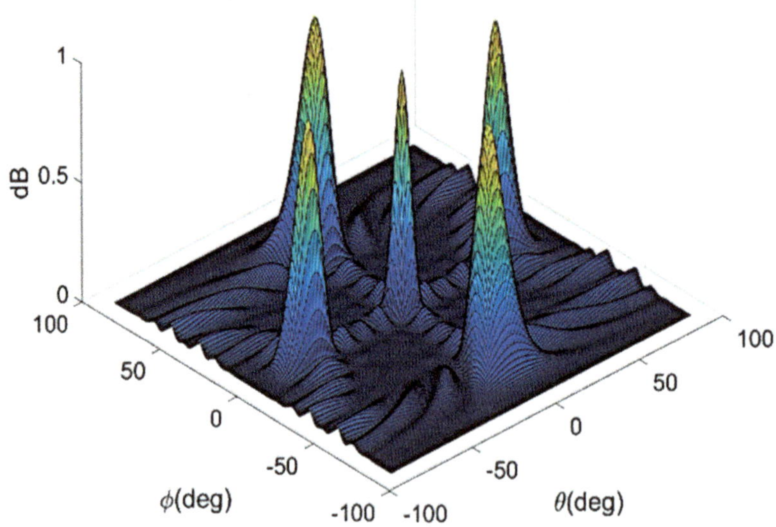

Figure 3.10
GLs' formation in the radiation pattern of an antenna array.

on the neighboring objects. Moreover, the currents on the antenna itself, including its port current, are also modified. This affects both the antenna's radiation pattern and its input impedance behavior. When the neighboring element is another antenna, which is passively terminated, currents may flow through its termination, leading to an energy dissipation through it. This will impose serious constraints on the realized efficiency of an antenna element located in an antenna array environment [13, 14]. The uniformly spaced arrays, that is, consisting of the elements placed at an equal interval, may be called a dense array (to distinguish them from sparse arrays) or regular (conventional) antenna arrays. An extensive treatment of mutual coupling phenomenon in regular arrays is presented by Craeye [16].

Figure 3.11 shows a schematic illustration of mutual coupling phenomenon. It is shown that when an element is radiating in an array environment it gets influenced by the neighboring elements' radiation. The energy may also travel back towards sources connected to the antenna and may cause severe impedance mismatch. This impedance mismatch results into a phenomenon known as

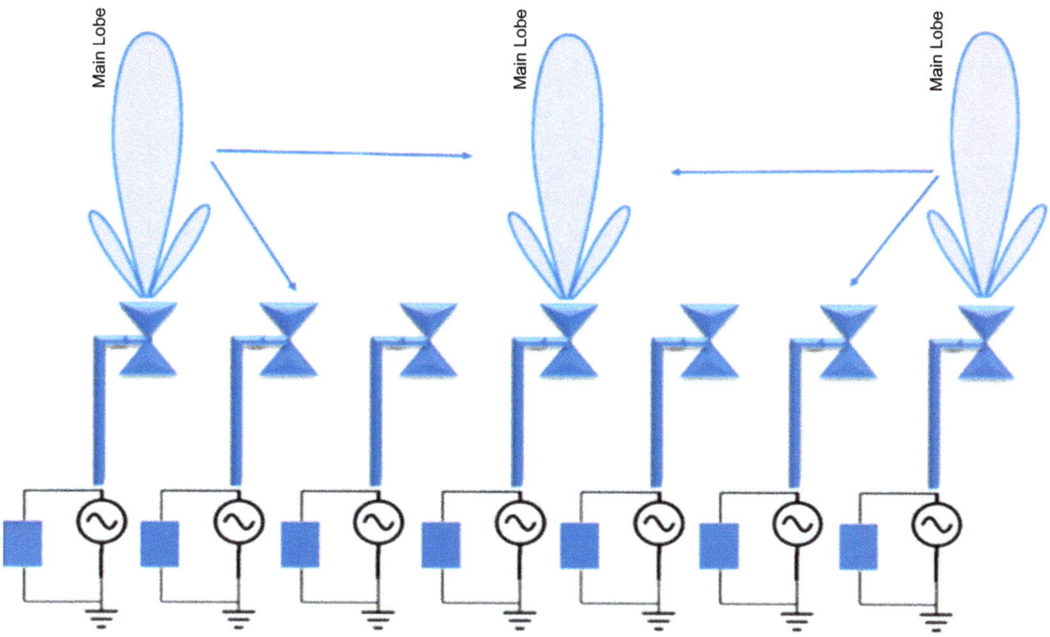

Figure 3.11
Mutual coupling phenomenon in an LAA.

scan blindness [13, 14, 16]. In some extreme cases, it leads to a total reflection (reflection coefficient, $|\Gamma| = 1$), which may result into blind spots (see Figure 3.12) (i.e., a sharp dip in the element radiation pattern).

The mutual coupling effects in planar antenna arrays can be approximated by various techniques like:

- *Ignoring its effect:* This is conventionally followed while conceptualizing the basic design of an antenna element and an antenna array. However, this does not lead to a manufacturing ready design as it may lead to unexpected results' deterioration like onset of blind zones while scanning.
- *Simulation of the entire antenna array:* The simulation of entire antenna array structure using electromagnetic solutions based on the techniques like finite element method (FEM), finite difference time domain method (FDTD), method of moments (MoM), and finite integration technique (FIT) [17]. These techniques take into account the effect of the various practical constraints like mutual coupling [18], scattering from finite edges

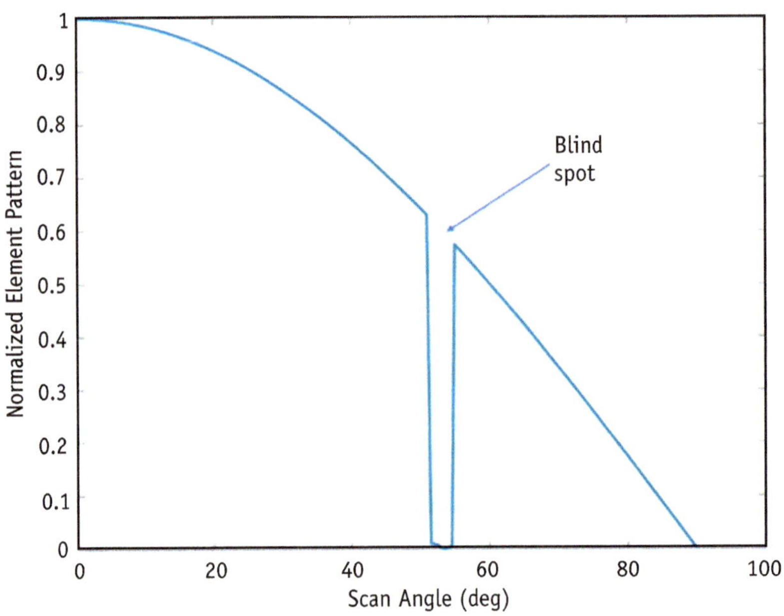

Figure 3.12
Scan blindness phenomenon.

of the array boundaries [19–21], and array boundary shapes [22]. However, the computational burden increases exponentially in proportion to the number of antenna elements in an array and it becomes sometimes practically impossible to solve these within practical limits of computational resources available to us. Thus, this method is applicable to a certain size of an antenna array, especially with smaller sizes electrically [23].

- *Infinite antenna array simulation:* This simulation considers a single antenna element bounded by linked or periodic boundary conditions (PBC) [13], which assumes that an element is embedded in an array with an infinite extent, having a periodic arrangement of elements in it but ignoring the finite array edges. Thus, this technique only accounts for the effect of the mutual coupling and is called as an infinite array analysis [13, 14]. Here, the effect of mutual coupling is considered via Floquet modes [24], which allows accounting for the properties of the elementary radiator inside the array environment, called as an embedded element, wherein the fundamental mode represents the main beam and the higher-order (evanescent) modes represent the horizontal coupling between the neighboring elements. Thus, the aggregate effect accounts for the radiation plus coupling effect on the performance of an antenna element in an array. The limitations of the infinite array analysis are:
    - It is not applicable to edge elements.
    - It is strictly time-harmonic leading to difficulties in (ultra) wideband cases.

It is clear that the Floquet modes approach is not directly applicable to nonuniform or sparse antenna arrays as modeling of the whole array is not feasible. This is due to the basic assumption made in the analysis via Floquet mode technique, that is, the embedded element consists of an infinite number of elements on all of its sides [24–27]. An alternate solution has been proposed in the literature [25] to approximate the mutual coupling behavior between elements of a nonuniform antenna array, not restricted by the size. The appropriate solution to the problem is an approximation of the coupling admittances by the interpolation strategy. After approximating the mutual admittances between the array elements via some polynomial approximations, these values can be used for incorporating the mutual coupling effects in the array analysis.

Figure 3.13 illustrates the effect of the mutual coupling on the performance of an infinite dipole antenna array. It shows the variation of the input resistance (in ohms), input reactance (in ohms), and reflection coefficient (active reflection coefficient's (ARC) magnitude) with respect to the scan angle in three different phi planes (0°, 45°, and 90°). A clear increase in the deterioration is shown at a particular scan angle indicating impedance mismatch or blind spot zone. Henceforth, it is important to consider the effect of the mutual coupling on the array element's performance before going for the final realization of the designed antenna array. The relation between active element pattern (AEP), G_{act}, and the active reflection coefficient (ARC), Γ_{act}, illustrating their inherent dependence on the operating wavelength (λ) and the scan angle (θ_0,φ_0), is shown in (3.10). Figure 3.14 shows the AEP with respect to the scan angle, compared to the case when no mutual coupling (isolated element

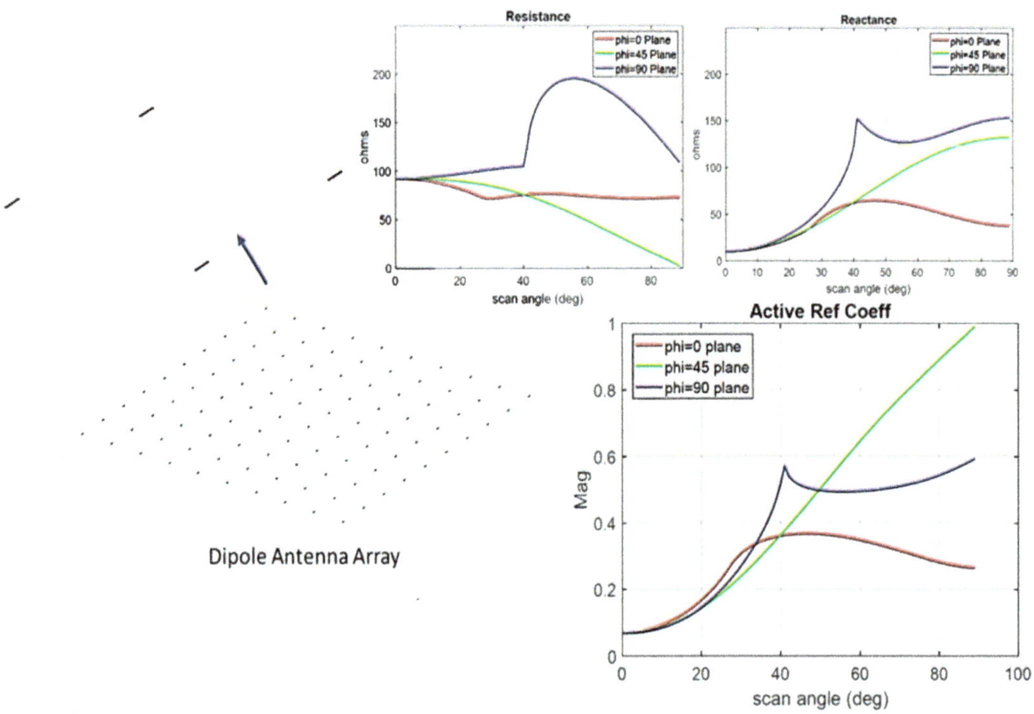

Figure 3.13
Performance of an infinite dipole antenna array.

pattern) is accounted for. As a result of the coupling a blind spot is seen near a 45° scan angle.

$$G_{\mathrm{act}}\left(\theta_o,\varphi_o\right) = \frac{4\pi d_x d_y}{\lambda^2} N\left[1-\left|\Gamma_{\mathrm{act}}\left(\theta_o,\varphi_o\right)\right|^2\right]\cos\theta_o \tag{3.10}$$

Presently, lots of commercial electromagnetic solvers such as HFSS and FEKO have implemented built-in modules to serve the purpose and are very handy and time-saving during the design processes and have been used in the book.

3.5.3 Antenna Array Design Techniques

There are various antenna design techniques that have been presented up to now for the synthesis of antenna arrays, which are applicable for uniform (equally spaced) and nonuniform (nonequal spacing) antenna arrays [3, 4, 8]. Figure 3.15 summarizes these techniques under various categories as applicable to both the types of antenna arrays. These are organized under various categories as

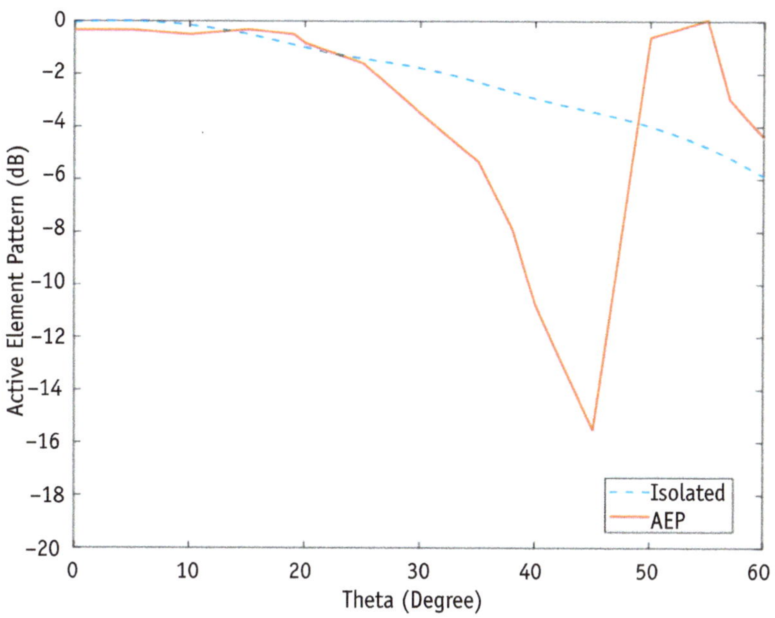

Figure 3.14
Active element pattern.

space tapering, mathematical programming, stochastic optimization, combinatorial mathematics, and fractal engineering. Antenna array analysis and synthesis are important topics that have served as the centroid for many reference books and textbooks [7, 8, 28, 29]. Here we will focus more on the specific techniques that are used in this book for the design of sparse phased array antennas.

Antenna array design is based on certain desired characteristics of a system with which it is intended to be integrated such as a radar, communication system for mobile telephony, and satellite tracking and steering. These desired characteristics are array gain, HPBW, and sidelobe level (SLL) envelope (peak SLL, PSLL, and average SLL, ASLL) over the frequency band and the scan volume of interest. Thus, array beamforming with these characteristics depends on the array aperture design with tailored aperture current distribution, sampled by the antenna elements placed on the aperture. The radiation pattern, $F(\theta, \varphi)$, and the aperture current distribution, $I(x, y)$, of an antenna array are the Fourier transform pairs [30]. Thus, the array synthesis synthesizes a desirable aperture current distribution to get the specific radiation pattern characteristics.

These techniques may be classified into iterative and noniterative (direct) synthesis techniques. The stochastic techniques (genetic algorithm (GA), particle swarm optimizer (PSO), invasive weed optimization (IWO)), mathematical programming (quadratic, dynamic), and iterative fast Fourier transform (IFFT) [30] form the class of iterative techniques requiring multiple iterations to reach the solution. The noniterative techniques include analytical techniques such as Taylor's distribution, Chebyshev's distribution, Bayliss [8], and deterministic techniques such as statistical density tapering (SDT) [30, 31] and combined difference set (CDS) [31–33]. The synthesized continuous aperture current distribution can be sampled through discretely placed antenna elements [8].

A few case studies presented herein illustrate the design procedures in the case of uniform antenna arrays using the few of the above techniques. A 100-element LAA is chosen consisting of isotropic elements with $\lambda/2$ spacing and having an aperture length, 50λ. The array is synthesized for a PSLL < –30 dB using GA. Although this problem can be solved with many of the other techniques as listed in Figure 3.15, but GA is chosen by my preference only for the illustration purpose. Figure 3.16 show the plots of cost function versus iteration, amplitude, and phase coefficients of each

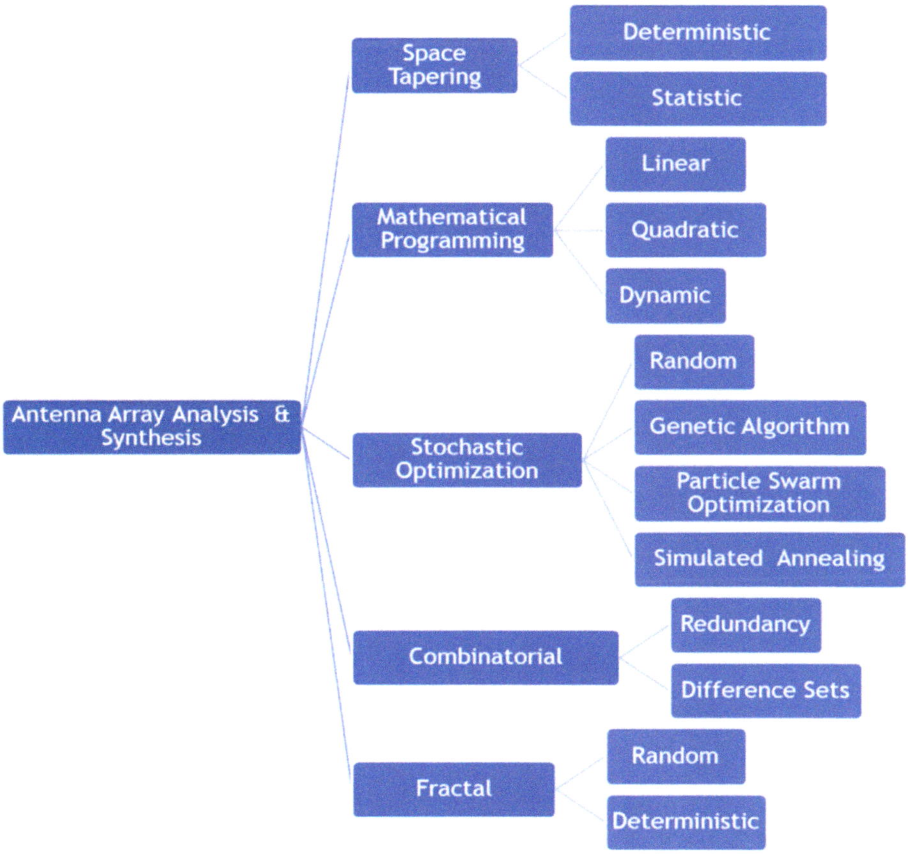

Figure 3.15
Synthesis techniques for the antenna arrays.

element and array factor (where u = sin) plots synthesized using GA. The algorithm converges in less than 400 iterations and amplitude curve indicates high current concentration in the center and gradual decrease in the concentration towards edges of the array.

In the bottom plot of Figure 3.16, the red trace shows the assigned mask (targeted pattern) as the cost function for the GA code written in MATLAB leading to the synthesis of PSLL < −30 dB. Figure 3.17 shows the scanned radiation pattern, a product of AF, and the element pattern, $EP = \cos^{1.5}\theta$. The effect of EP on AF is clearly seen showing that the overall radiation pattern broadens with respect to the scan angle accompanied with a loss in gain in proportion to EP. It is also observed that there is an increase in far SLL region as the beam scans. Figure 3.18 shows the effect of lowering of PSLL

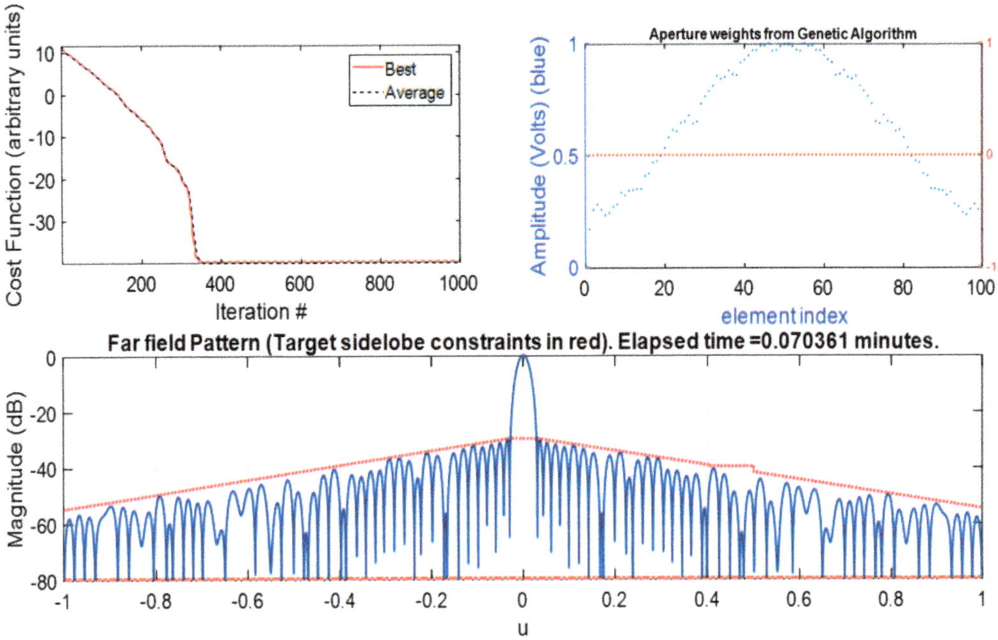

Figure 3.16
A 100-element LAA synthesis using GA.

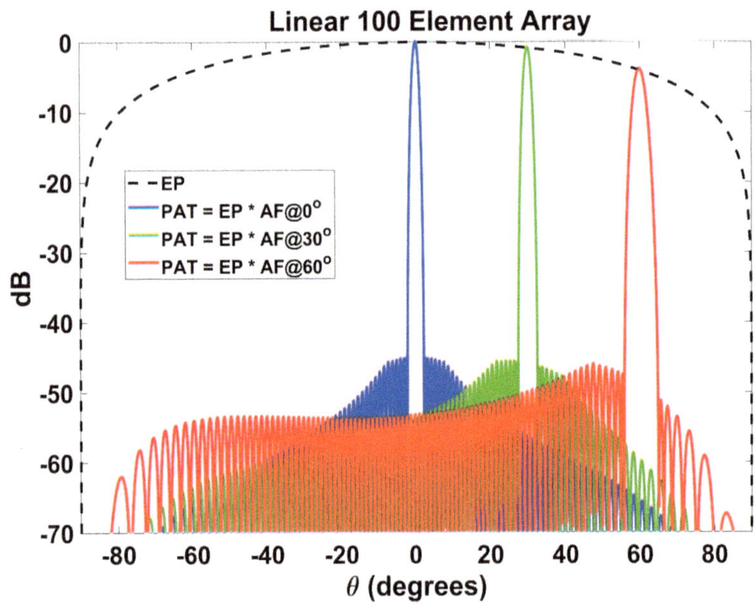

Figure 3.17
A 100-element LAA scanned pattern.

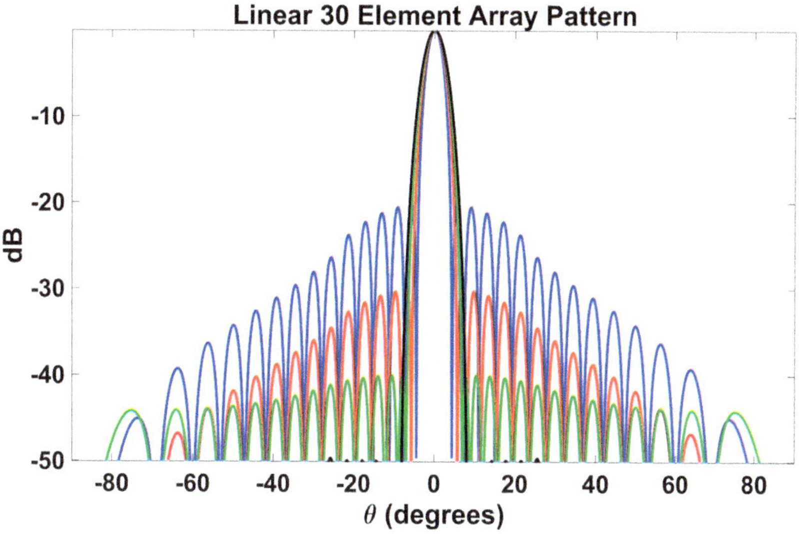

Figure 3.18
HPBW versus SLL in a 30-element LAA.

on the HPBW in the case of a 30-element LAA. It is observed that in concurrence with the principle of conservation of energy the total energy remains constant in an array. As the SLL decreases, the energy in the main lobe increases leading to beam broadening. Thus, SLL lowering is a trade-off with HPBW and thereby the gain and efficiency of an array.

PhAA can be synthesized by designing two LAAs along the two principal planes ($\varphi_o = 0°$ and 90°) and utilizing the array multiplication rule given in (3.11).

$$AF_{\text{planar}} = AF_{\varphi=0^\circ} \times AF_{\varphi=90^\circ} \tag{3.11}$$

A PhAA with 900 (30 × 30) elements is designed for PSLL < –30 dB using two LAAs designed for PSLL < –30 dB utilizing (3.11). Figure 3.19 shows a 3-D array factor plot in uv space ($u = \sin\theta\cos\varphi + \sin\theta_o\cos\varphi_o$; $v = \sin\theta\sin\varphi + \sin\theta_o\sin\varphi_o$). The top figure shows the AF plot at boresight and the bottom figure shows the AF scanned to (45°, 45°) angle. It should be mentioned here that the ASLL of an array accounts for the SLL envelope along intercardinal planes ($\varphi \neq 0°, 90°$). Thus, there should be considerations while designing to lower the SLL along the complete hemisphere.

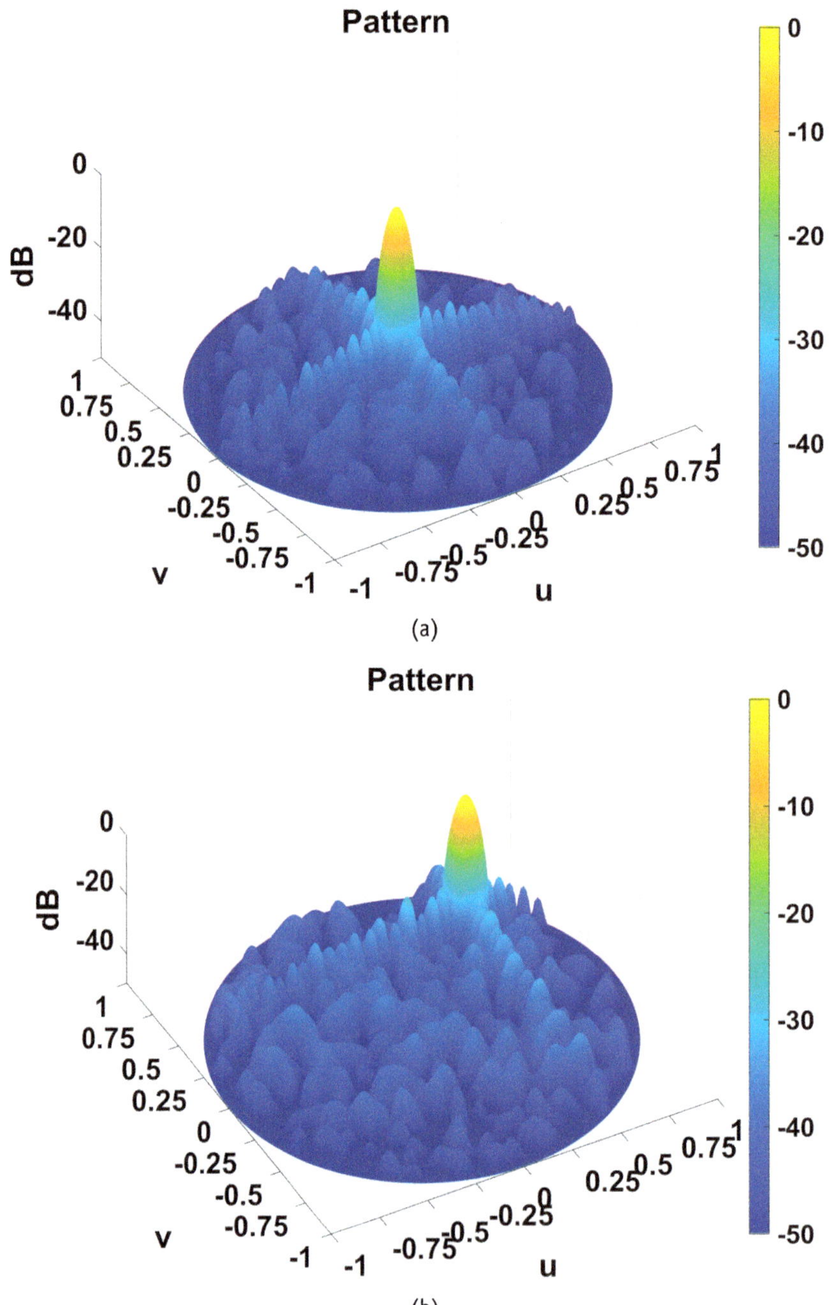

(a)

(b)

Figure 3.19
A 900-element planar antenna array factor plots.

3.5.4 Practical Constraints in Regular or Dense Phased Array Antennas

PhAAs consist of active electronic circuitry at the back end of antenna arrays for beamforming applications [12, 15]. The circuitry provides necessary complex excitations (amplitude and phase weights) to the individual or group of antenna elements to synthesize the desirable aperture current distribution, maximizing the energy in a particular look or scan angle's direction. PhAAs, also called electronically scanning antennas (ESAs), may be classified as passive ESA (PESA) or active ESA (AESA), depending upon the level in ESA at which beamforming circuit is employed. Figure 3.20 shows the ESA classification.

PESA employs a phase-shifting element at the element level in an antenna array, which provides necessary phase shift (utilizing (3.8)) to steer the array main beam electronically in the desired direction. It further consists of array feed network, central receiver (CRx), low noise amplifier (LNA), central transmitter (CTx), and a circulator or diplexer isolating receive and transmit chains. CTx gives the necessary amplified RF power (in the order of kilowatts or megawatts depending upon the requirements of range [1, 12, 15]), which distributes through the fixed feed network to provide necessary fixed amplitude distribution to achieve low SLL. CRx and LNA provide low noise reception and protector isolates the receiver from any high-power leakage from CTx. The disadvantages of PESA are as follows:

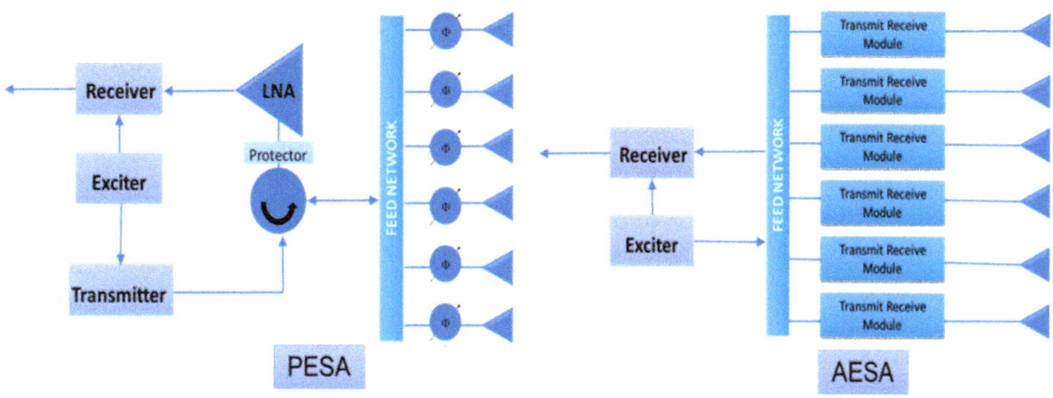

Figure 3.20
Types of PhAAs.

- Single point failure of entire PESA in case of the failure of CTx;
- Maintenance and reliability of high power CTx;
- Cooling system with high reliability to manage heat dissipation from CTx;
- Noise figure of any cascaded system is given by (3.12).

$$NF_{\text{total}} = NF_1 + \frac{NF_2}{G_1 - 1} + \cdots \ldots\ldots \tag{3.12}$$

It shows that the overall noise figure, NF_{total}, is being governed by the noise figure and gain of the first stage of the cascaded system. Thus, in a PESA, the front-end losses due to the feed network, phase shifter, antenna element, and diplexer limit the lowering of the noise figure beyond a certain level. As an example, if NF_{LNA} = 1.0 dB and total losses due to front-end network amounts to 2.2 dB, then the noise figure sets at the most optimum value of 3.2 dB, while ignoring the other losses such as cable and interconnect losses, which may further increase the losses by another 1 dB, making NF_{total} = 4.2 dB.

Another disadvantage is no dynamic control on SLL due to fixed amplitude distribution across the array aperture.

AESA counters the limitations faced by PESA category by distributing the CTx and LNA at the element level, are together known as transmit receive modules (TRMs) [15]. The advantages of AESA are:

- CTX is distributed at the element level; thus, the individual power requirement goes down by an amount proportional to the total number of elements. This reduced thermal load, maintenance, and thermal requirements.
- In the case of distributed CTx, even failure of the few elements (<10%–20%) does not lead to a total failure and the overall system is still operational with compromised performance. This phenomenon is called a graceful degradation.
- The front-end losses are reduced as the LNA is distributed at the element level and the overall noise figure improves greatly. Considering the previous example, NF_{total} = 2 dB is achieved considering only LNA and the cable and interconnects loss.
- Dynamic control of amplitude and phase shift is possible by the virtue of digitally controlled phase shifter and attenuators

employed in TRMs [15]. This allows multiple Rx beams formation in the receiver, facilitating multibeam and multifunctional phased array antenna systems [1].

The disadvantages of AESA are only the higher cost and developmental resources involved. However, with the advent of recent technological development in SiGe, BiCMOS, and liquid crystal polymer (LCP), these problems are also getting solved [15].

The design of PhAAs discussed in the previous section completely ignores the effect of the mutual coupling on the array performance. The mutual coupling tends to change the SLL envelope characteristics deleteriously by filling the nulls and increasing the SLL at random angles. Also, they tend to introduce the blind spots as shown earlier in Section 3.5.2. The effect of the mutual coupling is possible to be minimized by:

- *Increasing the interelement separation:* However, this limits the scanning range of an array before the grating lobes start entering the visible region of an array.
- *Introducing irregularity in the array elements' placement on the aperture:* This mitigates the mutual coupling by increasing the average interelement separation between the array elements and also it mitigates the grating lobes by destroying the periodicity in the array.

Hence, the irregular arrays form the main topic of this book with more attention on the sparse phased array antennas. Thus, the concepts presented in this chapter serves as the basic platform to understand the sparse antenna array concepts presented in the following chapters of Part III of this book.

References

[1] Skolnik, M., *Radar Systems*, Third Edition, New York: McGraw-Hill, 2002.

[2] Hansen, R. C., *Phased Array Antennas*, New York: John Wiley & Sons, 1998.

[3] Balanis, C., *Antenna Theory: Analysis and Design*, New York: Wiley-Interscience, 2005.

[4] Mailloux, R., *Phased Array Antenna Handbook*, Norwood, MA: Artech House, 2005.

[5] Kraus, J. D., *Antennas for All Applications*, Third Edition, New York: McGraw-Hill, 2002.

[6] Volakis, J. L., *Antenna Engineering Handbook*, New York: McGraw-Hill, 2007.

[7] Stutzman, W., "Bibliography for Antennas: A List of Every English-Language Antenna Book Ever Written," *IEEE Antennas and Propagation Magazine*, Vol. 50, No. 4, 2008.

[8] Elliott, R. S., *Antenna Theory and Design*, New York: Wiley-IEEE Press, 2003.

[9] Watts, C. B., "Applications of Slotted Cable Antennas in the Instrument Landing System," *IEEE Aerospace and Electronic Systems Magazine*, Vol. 5, No. 5, 1990, pp. 16–20.

[10] Weber, R. H., and R. Weber, *Internet of Things*, New York: Springer, 2010.

[11] Osman, L., I. Sfar, and A. Gharsallah, "An Overview of Direction-of-Arrival Estimation Using an Antenna Array with Four Elements," *American Journal of Applied Sciences*, Vol. 9, No. 12, 2012, pp. 1979–1984.

[12] Brookener, E., *Practical Phased Array Antenna Systems*, Norwood, MA: Artech House, 1991.

[13] Amitay, N., V. Galindo, and C. P. Wu, *Theory and Analysis of Phased Array Antennas*, New York: Wiley-Interscience, 1972.

[14] Hansen, R. C., *Microwave Scanning Antennas*, New York: Academic Press, 1964.

[15] Sturdivant, R., and M. Harris, *Transmit Receive Modules for Radar and Communication Systems*, Norwood, MA: Artech House, 2015.

[16] Craeye, C., and D. González-Ovejero, "A Review on Array Mutual Coupling Analysis," *Radio Science*, Vol. 46, 2011, pp. 1–25.

[17] Sadiku, M. N. O., *Numerical Techniques in Electromagnetics with MATLAB*, Boca Raton, FL: CRC Press, 2009.

[18] Bailey, M., and C. Bostian, "Mutual Coupling in a Finite Planar Array of Circular Apertures," *IEEE Transactions on Antennas and Propagation*, Vol. 22, No. 2, 1974, pp. 178–184.

[19] Ellgardt, A., and M. Norgren, "A Study of Edge Effects in Triangular Grid Tapered-Slot Arrays Using Coupling Coefficients," *Radio Science*, Vol. 45, 2010, pp. 1–13.

[20] Kedar, A., and K. Beenamole, "Wide Beam Tapered Slot Antenna for Wide Angle Scanning Phased Array Antenna," *Progress in Electromagnetics Research*, Vol. 51, 2011, pp. 235–251.

[21] Kedar, A., "Dielectric Free Wide Scan UWB Low Cross-Pol Metallic Vivaldi Antenna for Active Phased Array Radars," *IETE Journal of Research*, 2019, pp. 1–9.

[22] Bird, T. S., "Analysis of Mutual Coupling in Finite Arrays of Different Sized Rectangular Waveguides," *IEEE Transactions on Antennas and Propagation*, Vol. 38, No. 2, 1990, pp. 166–172.

[23] Holter, H., and H. Steyskal, "On the Size Requirement for Finite Phased-Array Models," *IEEE Transactions on Antennas and Propagation*, Vol. 50, No. 6, 2002, pp. 836–840.

[24] Bhattacharyya, A. K., *Phased Array Antennas: Floquet Analysis, Synthesis, BFNs, and Active Array Systems*, New York: John Wiley & Sons, 2006.

[25] Mehta, M. J., and I. E. Lager, "Two-Dimensional Interpolation for the Numerical Estimation of the Mutual Couplings in Large Antenna Arrays," IRCTR, Delft University of Technology, 2006.

[26] Makarov, S. N., A. Puzella, and V. Iyer, "Scan Impedance for an Infinite Dipole Array: Accurate Theoretical Model Compared to Numerical Software," *IEEE Antennas and Propagation Magazine*, Vol. 50, No. 6, 2008, pp. 132–149.

[27] Makarov, S., *Antenna and EM Modeling with Antenna*, New York: Wiley-Interscience, 2002.

[28] Cheng, D. K., "Optimization Techniques for Antenna Arrays," *Proceedings of the IEEE*, Vol. 59, No. 12, 1971, pp. 1664–1674.

[29] Haupt, R. L., *Antenna Arrays: A Computational Approach*, New York: Wiley-IEEE Press, 2010.

[30] Keizer, W. P. N., "Fast Low-Sidelobe Synthesis for Large Planar Array Antennas Utilizing Successive Fast Fourier Transforms of the Array Factor," *IEEE Transactions on Antennas and Propagation*, Vol. 55, No. 3, 2007, pp. 715–722.

[31] Ball, G. J., "Statistical Density and Amplitude Tapering," *1989 Sixth International Conference on Antennas and Propagation*, Vol. 1, No. 301, Coventry, U.K., 1989, pp. 249–253.

[32] Skolnik, M., J. Sherman, and F. Ogg, "Statistically Designed Density-Tapered Arrays," *IEEE Transactions on Antennas and Propagation*, Vol. 12, No. 4, 1964, pp. 408–417.

[33] Leeper, D. G., "Isophoric Arrays—Massively Thinned Phased Arrays with Well Controlled Sidelobes," *IEEE Transactions on Antennas and Propagation*, Vol. 47, No. 12, 1999, pp. 1825–1835.

[34] Kumar, P., A. Kedar, and A. K. Singh, "Subarray Scheme for Wide Scan Active Phased Array Antennas," *IEEE IMaRC*, 2014, pp. 364{en dash}367.

PART III

Electromagnetics of Sparse Array Antennas

4

Generalized Expression for Directivity

4.1 Introduction

An antenna is globally characterized by a figure of merit (i.e., directivity, D), which classifies its category (i.e., omnidirectional or directional antennas). The various systems such as radars, communication towers for mobiles, direction of arrival (DoA), and satellite communications require highly directive antennas ($D > 15$–20 dBi, typically) with a low sidelobe level (SLL) envelope. This is only realizable by forming an array of the antenna elements. Theoretically, D for an equally fed (isophoric) antenna, neglecting the mutual coupling, equals the total number of elements, N_{tot}, grouped in an array. The SLL envelope is proportional to $1/N_{tot}$. Thus, D is the qualifying parameter that qualifies the suitability of an antenna array for a particular application of interest. The computation of D for a planar antenna array is a time-consuming and challenging task and is currently a topic of continuing research worldwide. In general, D is usually expressed in the form of a double summation approximation or a double integration of the array radiation pattern [1, 2].

An antenna array having a boundary with an arbitrary shape, S, comprising N_{tot} elements, distributed randomly across the array aperture in the xy-plane is shown in Figure 4.1. Each of the elements is known by its spatial coordinates (x_n, y_n) or radial coordinates (θ, φ), respectively. $P(R, \theta, \varphi)$ represents an observation point of the far field assuming a planar wavefront. The arrangement of antennas on

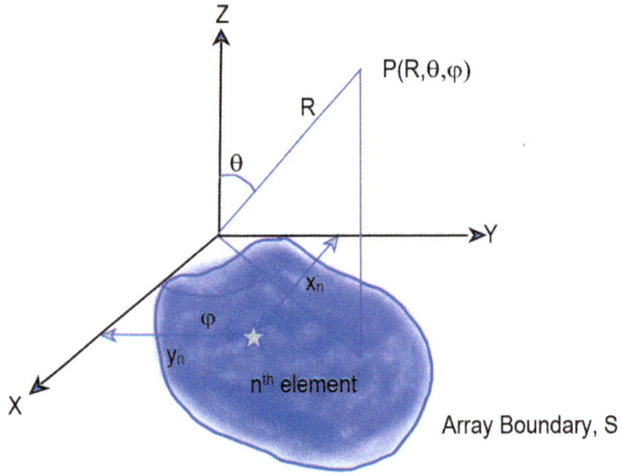

Figure 4.1
Array geometry and the coordinate system.

the array aperture is chosen such that the synthesized continuous aperture current distribution can be sampled optimally in a discrete manner [1]. This can be done either by placing the unequally fed elements at equal intervals (following the Nyquist criterion) separated by a half-wavelength or smaller or by placing equally fed (isophoric) elements at nonuniform separations. The former category is called the dense antenna arrays (DAAs) and the latter one is called the aperiodic or nonuniform antenna arrays. Aperiodic antenna arrays in general may also be referred to as sparse antenna arrays (SAAs) and will be referred as such going forward. The previous chapter discussed DAAs in detail. SAAs will be the focus of attention from this chapter on.

The computation of D for the antenna arrays becomes highly intensive and sometimes nonconvergent in the case of very large N_{tot} (more than hundreds of elements). Thus, the inadvertent time requirement of the process renders it useless for characterizing large PhAAs. An approximate or exact analytical expression for its computation is proposed. Tai [3], Bach [4], and Langston [5] proposed an approximate expression for the directivity of uniformly or equally spaced linear antenna arrays, specifically valid for isotropic and dipole antenna elements only. These expressions did not include the effect of scan angle, arbitrary array geometry, arbitrary antenna element type, complex antenna excitations, and mutual coupling. Hansen [6] proposed an expression for directivity while including the effect of mutual resistance term, antenna array excitations, and SLL. Few researchers also made an attempt to compute the directivity of antenna arrays with a specific element type and excitations [7–13]. Forman [14] was the first one to propose an accurate analytical expression of directivity. The proposed expression considered the effect of arbitrary array geometry, complex antenna excitations, interelement separation, and the scan angle. However, it lacked the consideration for mutual coupling and is restricted to the antenna elements with only cosθ patterns. Ligthart [15] proposed a generalized accurate expression for directive gain considering the effect of arbitrary shape, complex excitations, mutual coupling, scan angle, and arbitrary antenna element type. However, this expression is computationally intensive, especially in the case of very large arrays. Das et al. [16] also presented a similar expression restricted to only the isotropic antenna element. The need for accurate expressions for directivity and the importance of the incorporation of the embedded element pattern (EEP) in it is well explained in [17–20]. Moreover, all of these expressions were utilized to derive the directive estimation of DAAs only.

This chapter presents a generalized expression for directivity of planar antenna arrays. It considers the effect of arbitrary array geometry, arbitrary complex excitations, mutual coupling, scan angle, and embedded element pattern (EEP). In general, computation using such exact analytical expressions is time-consuming in the case of large arrays (number of elements, $N > 100$). Hence, an approximate expression for D assuming aperture with a periodic array grid of infinite extent (i.e., independent of the number of elements) is also presented in this chapter. Furthermore, assumption made in this approximate expression is that the array comprises only isotropic antenna elements. These assumptions do not limit the effectiveness of the proposed expression in providing an accurate estimate of D. It is used to understand the characteristics of:

- Planar aperture antennas (equidistant antenna arrays);
- (Non-) equidistant antenna arrays;
- Antenna arrays with random element allocation.

4.2 Generalized Exact Analytical Expression for Directivity

Figure 4.1 illustrates the array coordinate system *OXYZ* consisting of a planar array in the *OXY* plane. The array aperture is bound with an array boundary, *S*, and it consists of *N* radiating antenna elements with the location of nth element located at (x_n, y_n). These are coherently phased to make the array look in a direction given by (θ_o, φ_o). *R* is the distance of the observation point, *P*, in the far field from the origin, *O*. It is assumed here that the radiation occurs only in the upper half-space. Hence, *D* is defined as [1, 21]:

$$D(\theta,\varphi) = \frac{4\pi\left|F(\theta,\varphi)_{\max}\right|^2}{P_0} \tag{4.1}$$

In (4.1), P_0 is the total radiating power available from the array aperture boundary, *S*, expressed as:

$$P_0 = \int_0^{2\pi}\int_0^{\pi}\left|F(\theta,\varphi)\right|^2 \sin\theta\, d\theta\, d\varphi \tag{4.2}$$

In (4.2), $F(\theta, \varphi)$ is the far-field pattern of the antenna, defined as:

$$F(\theta,\varphi) = \iint_S I(x,y)e^{\left[-ik(x\sin\theta\cos\varphi+y\sin\theta\sin\varphi)\right]}\,dx\,dy \tag{4.3}$$

In (4.3), $I(x, y)$ represents the current distribution over *S*. In (4.1),

$$\left|F(\theta,\varphi)^2\right| = F(\theta,\varphi)\cdot F(\theta,\varphi)^* \tag{4.4}$$

Herein, * denotes a complex conjugate operator, where $F(\theta, \varphi)$ is defined in (4.3) and $F^*(\theta, \varphi)$ is defined as:

$$F^*(\theta,\varphi) = \iint_S I(\xi,\eta)e^{\left[-ik(\xi\sin\theta\cos\varphi+\eta\sin\theta\sin\varphi)\right]}\,d\xi\,d\eta \tag{4.5}$$

In (4.5), (ξ, η) are the coordinates of an arbitrary point on the aperture plane, which plays the same role as (x, y) coordinates. Therefore, $|F(\theta, \varphi)|^2$ becomes:

$$
\begin{aligned}
&|F(\theta,\varphi)|^2 = F(\theta,\varphi) \bullet F^*(\theta,\varphi) \\
&= \iint I(x,y) I^*(x-\xi,\ y-\eta) e^{[ik\{(x-\xi)\sin\theta\cos\varphi + (y-\eta)\sin\theta\sin\varphi\}]}\, dx\, dy\, d\xi\, d\eta
\end{aligned}
\tag{4.6}
$$

An inspection of (4.6) reveals that the exponent represents the phase dependence of an isotropic element in the direction, (θ, φ), at a distance, $d = \sqrt{[(x-\xi)^2 + (y-\eta)^2]}$ which can also be written as $d = \sqrt{(\xi^2 + \eta^2)}$. Thus, the exponent is the only factor having (θ, φ) dependence. By taking another isotropic element located at $\varphi = \varphi + \pi$, we get a two-element array with interelement separation, d. The mutual impedance between two elements is estimated as:

$$
R_0 = \frac{\sin k\sqrt{(x-\xi)^2 + (y-\eta)^2}}{k\sqrt{(x-\xi)^2 + (y-\eta)^2}} \tag{4.7}
$$

Using (4.6) and (4.7), (4.2) gives an expression for P_0 written as:

$$
\begin{aligned}
P_0 &= 4\pi \iint\iint I(x,y) I^*(\xi,\eta) \frac{\sin k\sqrt{(x-\xi)^2 + (y-\eta)^2}}{k\sqrt{(x-\xi)^2 + (y-\eta)^2}} dx\,dy\,d\eta\,d\xi \\
&= 4\pi \iint \chi(\xi,\eta) \frac{\sin k\sqrt{\xi^2 + \eta^2}}{k\sqrt{\xi^2 + \eta^2}}
\end{aligned}
\tag{4.8}
$$

In (4.8), $\chi(\xi, \eta)$ is the convolution of the current distribution on the array aperture and is given as:

$$
\chi(\xi,\eta) = \iint I(x,y) I^*(x-\xi, y-\eta)\, dx\, dy \tag{4.9}
$$

Assuming the current distribution, $I(x, y)$ as given below:

$$
I(x,y) = I_0(x,y) e^{-[-ik(x\sin\theta_0\cos\varphi_0 + y\sin\theta_0\sin\varphi_0)]} \tag{4.10}
$$

The exponential part represents the phase of the current distribution when a beam is scanned at an angle, (θ_0, φ_0). Thus, the expression considers the effect of the amplitude and phase of the arbitrary current distribution applied to the antenna array's constituent antenna elements. Substituting (4.10) into (4.1) and using (4.8) and (4.9), the expression for directivity is derived and given here:

$$D(\theta,\varphi)=\frac{\left|\iint_{s} I_0(\eta,\xi)\,d\xi\,d\eta\right|^2}{\iint_{s}\chi_0(\xi,\eta)\cos\left[k\sin\theta_0\left(\xi\cos\varphi_0+\eta\sin\varphi_0\right)\right]\frac{\sin k\sqrt{\xi^2+\eta^2}}{k\sqrt{\xi^2+\eta^2}}\,d\xi\,d\eta} \tag{4.11}$$

In (4.11), $\chi_0(\xi, \eta)$ is the autocorrelation function of the current distribution at the aperture, defined as:

$$\chi_0=\chi_0(\xi,\eta)=\iint\limits_{s} I_0(x,y)I_0^*(x-\xi,y-\eta)\,dx\,dy \tag{4.12}$$

The denominator in (4.11) accounts for the mutual coupling between two finitely separated elements (isotropic) (through sinc (·)) and the phase distribution along the array aperture (through cos(·) terms). Considering a planar aperture consisting of point sources, the array's current distribution may be represented as:

$$I(x,y)=\sum_{n=1}^{N}\delta(x-x_n)\delta(y-y_n)I_n \tag{4.13}$$

In (4.13), δ represents Kronecker's delta function [22, 23], and I_n is defined as:

$$I_n=I_{0n}\exp\left[-ik\sin\theta_0\left(x_n\cos\varphi_0+y_n\sin\varphi_0\right)\right] \tag{4.14}$$

Thus, (4.11) is rewritten as (see detailed derivation in Appendix 4A)

$$D(\theta,\varphi)=\frac{\left(\sum_{m=1}^{N} I_{0m}\right)^2}{\sum_{m,n=1}^{N} I_{0m}I_{0n}^*\cos\left[kr_{nm}\sin\theta_0\cos\left(\varphi_0-\alpha_{nm}\right)\right]\frac{\sin kr_{nm}}{kr_{nm}}} \tag{4.15}$$

In (4.15), α_{nm} is defined by

$$\cos\alpha_{nm} = \frac{x_n - x_m}{\sqrt{\left(x_n - x_m\right)^2 + \left(y_n - y_m\right)^2}} \tag{4.16}$$

$$\sin\alpha_{nm} = \frac{y_n - y_m}{\sqrt{\left(x_n - x_m\right)^2 + \left(y_n - y_m\right)^2}} \tag{4.17}$$

and the distance $\mathbf{r}_{\mathbf{nm}}$ between the elements with indices m and n is given as:

$$r_{nm} = \sqrt{\left(x_n - x_m\right)^2 + \left(y_n - y_m\right)^2} \tag{4.18}$$

Hence, (4.15) represents an exact analytical expression for D of a finite-sized arbitrarily shaped planar array. It is a generalized expression for directivity, which accounts for:

- Arbitrary array geometry (element locations) and array lattice;
- Mutual coupling;
- Arbitrary complex current distribution on the aperture;
- Antenna look or scan angle or beam-pointing angle.

The proposed expression serves as an efficient method for characterization of dense as well as sparse antenna arrays. The antenna arrays can be classified into different categories based on the choice of interelement separation, d, as:

1. Dense antenna arrays, $d/\lambda < 0.5$: These are antenna arrays consisting of regularly spaced antenna elements with interelement separation lesser than a half-wavelength. This spacing ensures no grating lobe in the visible region of array, as explained in Chapter 3. However, these arrays suffer from the effect of mutual coupling on the array's scan performance. This may lead to the onset of blind spots or scan blind zones in the arrays visible region while scanning [17, 21, 24, 25].
2. Moderately sparse antenna arrays, $0.5 < d/\lambda < 1$: These are antenna arrays consisting of regularly spaced antenna elements with interelement spacing more than a half-wavelength and approaching a wavelength. These arrays possess limited

scanning capability [26–30] due to the onset of grating lobes in a wide scan zone beyond some scan angle. However, a mutual coupling level is reduced.

3. Highly sparse antenna arrays, $d/\lambda > 1$: These are antenna arrays consisting of regularly spaced antenna elements with interelement spacing more than one wavelength. The mutual coupling level is almost negligible in these arrays leading to a wide scan zone free from scan blindness. However, the effect of grating lobes is seen prominently.
4. Random or aperiodic antenna arrays: These types of arrays are based on the concept that the regularity in the array grid leads to the formation of grating lobes (discussed in Chapter 2) and, henceforth, the destruction of periodicity in the arrays completely removes the possibility of the rise of grating lobes. Also, the average interelement separation in these arrays is achieved more than a half-wavelength, thereby reducing the levels of the mutual coupling. Thus, the scan performance is also enhanced.

In this chapter, the generalized expression of directivity is used to demonstrate the behavior of the first three categories of antenna arrays. The last category is discussed in more detail in the subsequent chapters.

In the case of antenna arrays with a large value of d ($\sim d/\lambda > 1$), the mutual coupling contribution between the elements is negligible and the expression for D can be written as:

$$D(\theta,\varphi) = \frac{\left(\sum_{n=1}^{N} I_{on}\right)^2}{\sum_{m,n=1}^{N} I_{om} I_{on}} = N \tag{4.19}$$

Thus, D becomes equal to N in absence of the mutual coupling in the categories (2), (3), and (4). However, it is not so in the case of DAA, that is, category (1), due to the dominance of mutual coupling, $D < N$. This is an important characteristic of SAA making them desirable for the development of phased array antennas with wide scan volume and wide bandwidth.

The effect of embedded element pattern (EEP) [24, 25] can be included in (4.15) as:

$$G_a(\theta,\varphi) = \left(1 - \left|\Gamma_a\right|^2\right)\varepsilon_{\text{loss}} D(\theta,\varphi) \tag{4.20}$$

In (4.20), G_a is the gain including effect of EEP, Γ_a is active reflection coefficient [20], and $\varepsilon_{\text{loss}}$ is the dissipative loss due to the antenna and its associated transmission feed network [31]. A factor of 2 needs to be multiplied to (4.15) in case radiation is assumed only in the upper hemisphere. In this case, the maximum D limit achievable by optimally designed SAA is $D_{sp} = 2N$. Expression (4.15) is time-consuming for large arrays (~$N > 100$). In order to facilitate a quick design aid to antenna array designers to estimate D and their scan performance, an approximate analytical expression is proposed in Section 4.3.

Thus, the proposed directivity expressions form the basis of the directive characteristics of the antenna arrays presented in this book.

4.2.1 Case Studies

4.2.1.1 LAA

This section discusses the characteristics of LAAs utilizing the exact expression, (4.15), derived previously. The various cases discussed highlight the prominent distinctions between dense and sparse LAAs and bring out the distinct advantages of sparse counterparts. An LAA is chosen with 30 and 100 elements placed along the $\varphi_c = 0°$ plane assuming that the array comprises only isotropic antenna elements. The different sizes are chosen to show the effect of the number of elements and the effectiveness of the derived expression in predicting the directivity estimates irrespective of the size of an array. A normalized value (normalized with respect to the total number of elements) of directivity, g, is considered in the plots shown in Figure 4.2. This will help to observe similar characteristics of LAAs irrespective of the number of elements.

In general, for the interelement separation, d, of order of $n\lambda/2$ (n = 1, 2, 3, ...), g is always unity for all scan angles, implying that the directivity is equal to the total number of elements in the case of LAAs. However, for the cases where d is different from $n\lambda/2$, there is a variation observed as in the case of $d = 0.75\lambda$ and 1.2λ. The pattern observed for g is identical irrespective of the size of array. It is further to be observed that as there is an increase in d, the variation

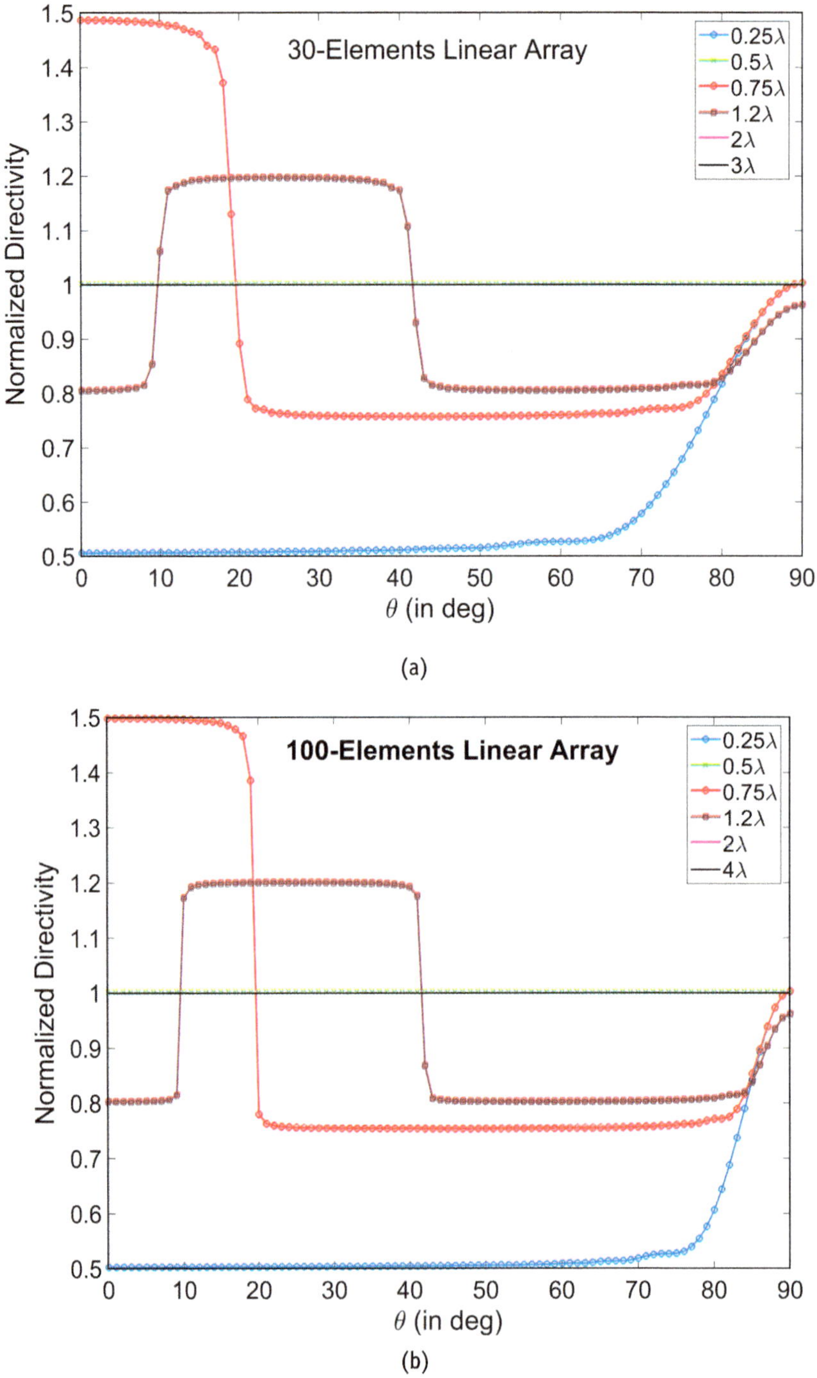

Figure 4.2
Normalized directivity of the LAA versus the scan angle. (© 2019 IEEE. Reprinted with permission from [21].)

in g decreases. In the cases where $d > 2\lambda$, there is no variation at all in g. Thus, for larger interelement separations, a wide scan performance of sparse LAAs is evident.

Figure 4.3 shows the effect of array size on the magnitude of directivity for $d = 0.5, 1.2\lambda$ for various LAA sizes, $N = 10, 20, 30$, and 100 elements. It is observed that for $d = 0.5\lambda$, directivity is equal to N and for a noninteger value such as 1.2λ, it peaks more than N and it oscillates around the value of N. Table 4.1 shows the validation of the proposed expression with the published literature [16] showing a remarkable similarity of the results.

4.2.1.2 Planar Antenna Array

The study is extended to the planar antenna arrays (PAAs) with square and rectangular boundaries to observe the variation of directivity with respect to the scan angle and the size of an array. The normalized directivity, g, is independent of the array size and is considered to analyze the effect of the interelement separation on the scan performance of the antenna arrays. Considering a 100-element

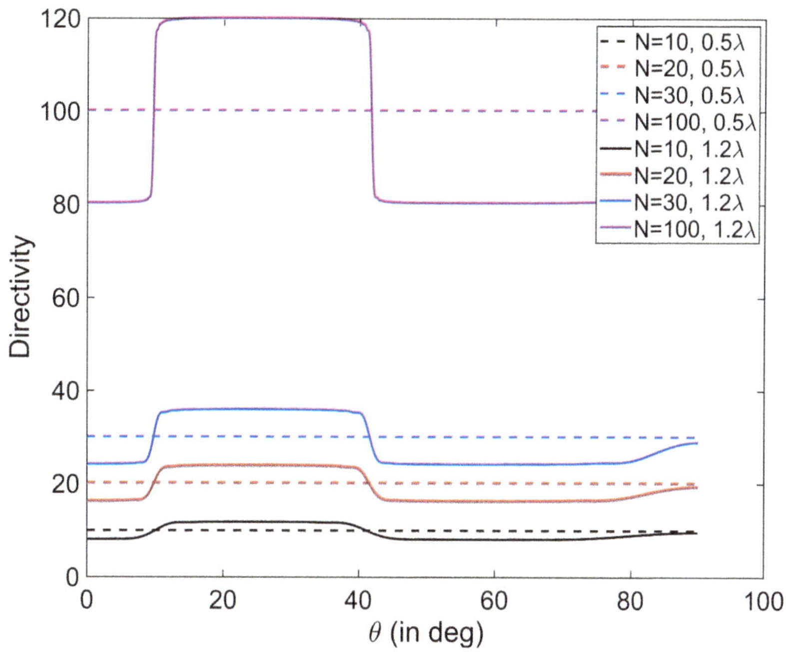

Figure 4.3
Directivity of LAA versus scan angle for various array sizes.

Table 4.1
Analysis of *N*-Element Linear Array (Directivity at Boresight)

Case	*N*	Location (λ)	Amplitudes	*D* [16] (dB)	*D* (dB)
1	14	0.26 0.82 1.36 2.05 2.87 3.72 4.46	0.90 0.78 0.96 0.97 1.00 0.91 0.75	11.6	11.9
2	18	0.28 0.82 1.42 2.04 2.69 3.47 4.32 5.16 5.90	0.91 0.88 0.86 1.00 0.95 0.85 0.84 0.74 0.69	12.7	13.69
3	22	0.24 0.74 1.24 1.75 2.25 2.81 3.36 4.04 4.86 5.70 6.44	0.78 0.95 1.00 0.89 0.83 0.95 0.92 0.93 0.81 0.76 0.73	13.58	14.04
4	26	0.28 0.81 1.37 1.92 2.49 3.08 3.66 4.33 4.98 5.66 6.43 7.25 7.99	0.97 1.00 0.91 0.90 0.90 0.85 0.93 0.90 0.78 0.70 0.73 0.75 0.68	14.30	14.9

PAA ($M_x = 10$ and $M_y = 10$) comprising uniformly spaced isotropic elements separated by distances d_x and d_y along the x- and y-axes. Figure 4.4 shows the plots of g (magnitude) versus the scan angle for different values of d_x and d_y. The demonstrated behavior is not the same as LAA due to the planar current distribution on the array aperture. There is a more pronounced effect of 2-D mutual coupling on the array performance due to planar distribution of radiators on the array aperture. The array with $d/\lambda = 0.25$ has very tight mutual coupling resulting into negative directivity (10 log(0.4) at 0° ~ −3.97 dB) showing deterioration in array efficiency.

The phenomenon of directivity being equal to the number of elements for interelement spacing equal to the integer multiples of a half-wavelength is not observed in PAAs as is evident from the Figure 4.4. This shows the stronger effect of the mutual coupling due to an element surrounded by several elements in an array. The directivity is shown to vary with respect to the scan angle for any value of interelement separation. In the case of dense PAAs (i.e., $d \leq 0.5\lambda$), the variation of g versus a scan angle (= θ) follows a trend proportional to cosθ, that is, g falls by 50% of its maximum value at around $\theta = 60°$, with the appearance of a grating lobe as $\theta_0 \rightarrow 90°$. As we move into the domain of a medium SAA ($0.5 < d/\lambda < 1$), there is an increase in g due to an increase in the array aperture. The variation in directivity no longer follows the cosθ pattern (see curve for $d/\lambda = 0.7$) and is approximately flat in nature with respect to the scan angle. However, due to periodicity in the array grid, there is an

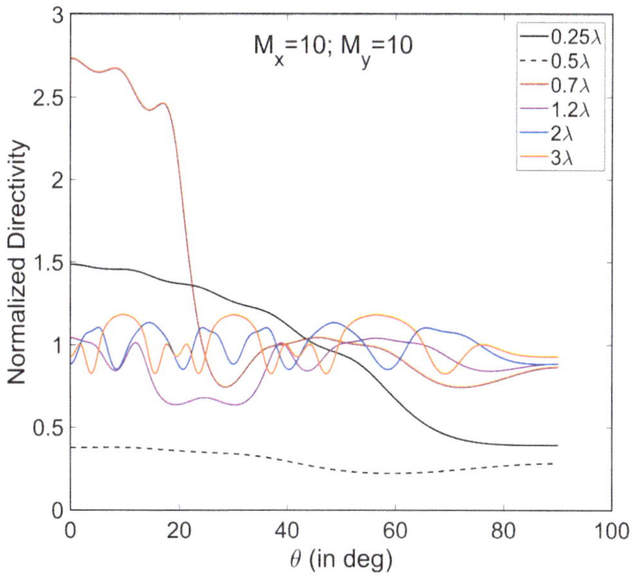

Figure 4.4
Directivity of PAA versus a scan angle.

early onset of grating lobe near scan angle, $\theta_0 \sim 30°$. Another grating lobe is seen between 80° and 90°. The presence of the grating lobes is shown by large dips in the directivity curve. This problem can be tackled by introducing aperiodicity in the array, as shown in later chapters. However, the main point to be emphasized is the visibility in the variation of the directivity curve becoming flatter. It is observed from Figure 4.4 that, for $d > 0.5\lambda$, the variation in g with regard to the scan angle decreases significantly. As separation increased beyond $d/\lambda = 1$ (i.e., highly sparse array region), it is seen that the variation in g reduces. A variation by a factor of 3.8 is observed in the case of $d = 0.7\lambda$ and this variation factor goes down to 1.5 in the case of $d = 3\lambda$ due to a significant decrease in the mutual coupling levels at these separations. Thus, a wide scan performance with the least variation in gain beyond a scan angle of 80° is observed. This highlights the advantages of the sparse antenna arrays. This is very promising for applications such as phased array antenna radar systems, MIMO radars, and low probability of intercept (LPI) radars. This also results into achieving wide bandwidths due to decrease in mutual coupling.

Figure 4.5 shows g in decibels versus the scan angle for different φ-cuts to show the effect of sparsity on array's performance along

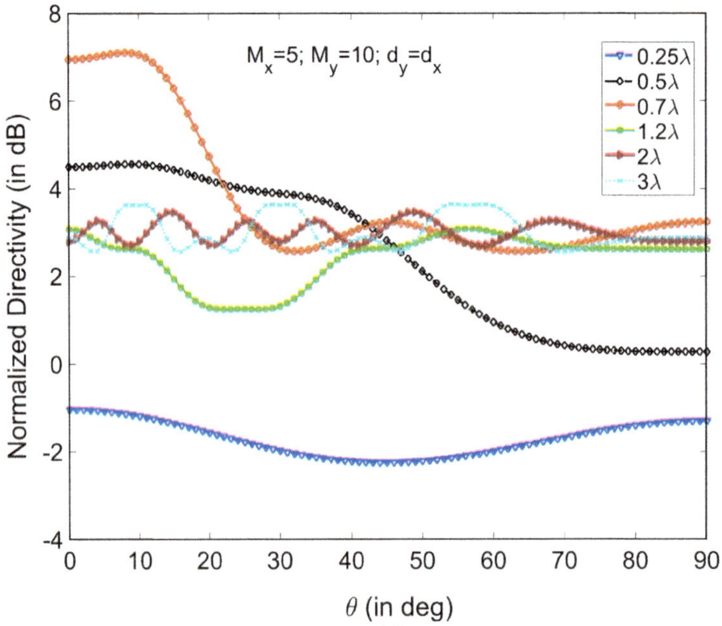

(a)

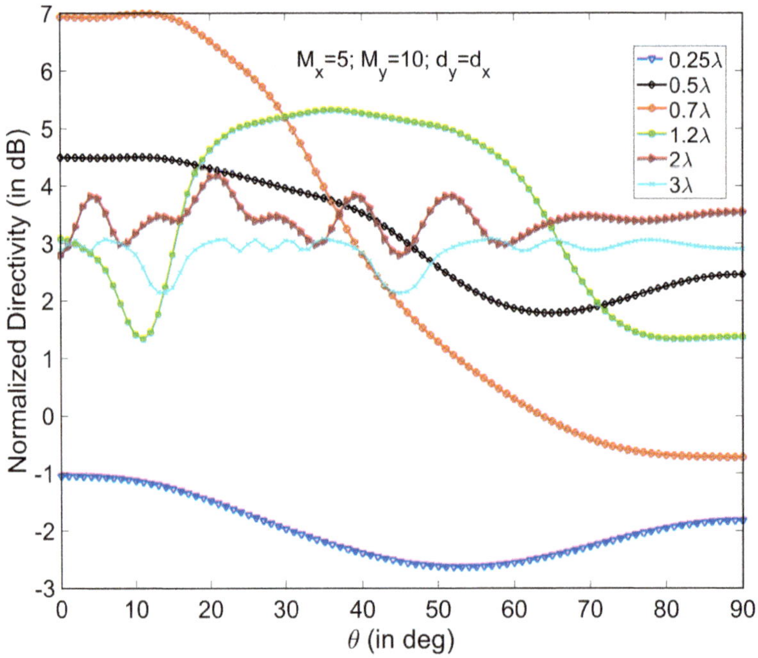

(b)

Figure 4.5

Variation of g with respect to the scan angle for a 50-element array along different φ-cuts. (a) φ= 0°. (b) φ = 45°. (c) φ = 90°.

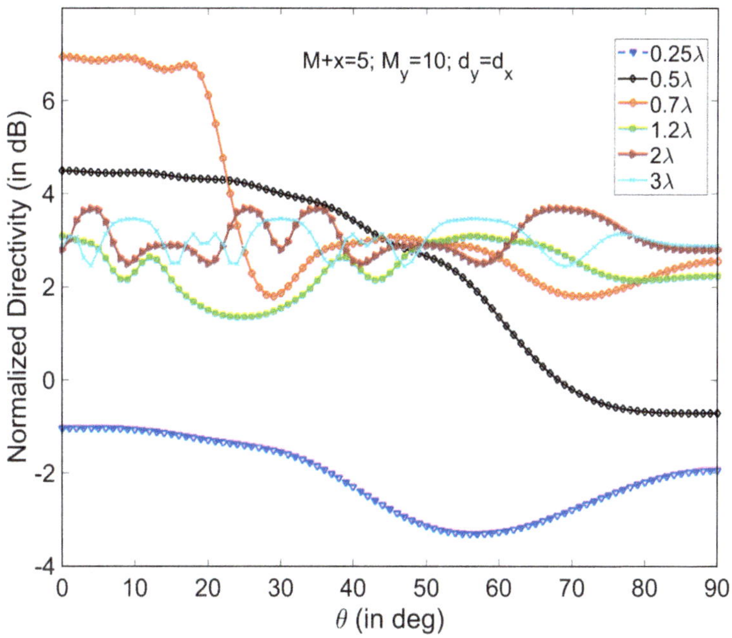

(c)

Figure 4.5
(Continued)

the intercardinal planes. The plots in Figure 4.5 (a–c) shows the variation in three φ-planes (i.e., 0°, 45°, and 90°). In general, similar trends are observed and g varies with the scan angle. The variation in g reduces with the increase in the interelement separation. The minimal variation with respect to the scan angle is observed in all of the φ-planes.

In Figure 4.6, the variation of g with respect to the interelement separation for different array sizes at boresight is plotted. A strong variation in the directivity estimates is observed for $d < 1$, but as d becomes larger than 1, the value of directivity stabilizes and becomes constant.

In medium SAAs (shaded region), the directivity is more compared to dense arrays along with wider scan zone. Hence, it provides a region in which array designers may focus for realizing wideband wide scan antenna array systems with higher-power aperture products. The area to be avoided for design is $d/\lambda \sim 0.25$ or less as tighter mutual coupling renders the realization of such arrays difficult. In medium SAAs, the variation in directivity is

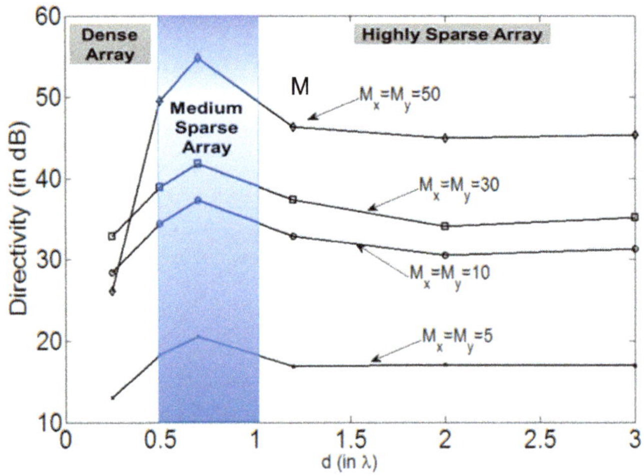

Figure 4.6
Variation of normalized directivity with respect to the interelement separations for different array sizes.

rapid with respect to the interelement separation. Thus, the performance of such arrays will vary drastically with different phi-cuts. A PAA may be treated as amalgamation of different linear arrays along different phi-cut planes. Hence, the mean or average interelement spacing along the plane will vary, which may lead to a variation in the performance of the array with respect to the azimuth angle (φ-cuts). However, in highly sparse PAAs, the variation in directivity is minimal (almost nil) with respect to the interelement separation. Hence, I recommend a highly sparse PAA for the realization of antenna arrays with constant performance along different phi planes while providing wide scan and wideband performance.

4.3 Approximate Expression for Directivity

An approximate expression [21] for directivity of antenna arrays is presented to provide a quick design aid for the array designer. It provides a computationally fast way to prepare a preliminary design of an antenna array based on its directive estimation and scan performance assessment. It considers all possible interference modes or grating lobes that may exist in the visible region due to the chosen interelement separation in an antenna array. The proposed expression assumes an isophoric (equally fed elements) and periodic infinite planar antenna array. However, it does not consider the finite edge effects in an array. It definitely does consider the effect of mutual coupling, scan angle, and interelement spacing. It also helps in predicting the scan performance of the arrays by accurately locating the angular locations for the appearance of blind spots in its scan zone.

The total number of elements in a planar array are assumed as $N_{tot} = M_x M_y$, where M_x and M_y are the total number of elements along the principal axes (Figure 4.1). Going forward, normalized directivity, D_{norm}, is considered for analysis and drawing various inferences from the study presented herein and is expressed as (see the detailed derivation in Appendix 4B):

$$D_{norm} = \frac{D(\theta,\varphi)}{M_x M_y} = \frac{2\pi d_x d_y}{\lambda^2 \sum_{p,q} \left[1 - \left(\frac{\lambda}{d_x} p - \sin\theta_0\right)^2 - \left(\frac{\lambda}{d_y} q\right)^2\right]^{-1/2}} \tag{4.21}$$

The indices p and q correspond to pq^{th} grating lobe entering into the real (visible region) space by the virtue of the choice of d_x and d_y. A particular pq^{th} mode contributes towards the summation in denominator of (4.21), if it satisfies the condition given here:

$$1 - \left(\frac{\lambda}{d_x} p - \sin\theta_0\right)^2 - \left(\frac{\lambda}{d_y} q\right)^2 \geq 0 \tag{4.22}$$

This condition ensures that the grating lobe is existing in the real space and is not existing in the imaginary space, as it does not contribute to antenna array's radiation pattern. Thus, the choice for the total number of pq modes is critical while computing D_{norm}. This helps to include the effect of all possible different grating lobe modes that may exist in the visible region. The corresponding expression for the radiated power is:

$$P_0 = \sum_{p_1,p_2}^{M_x} \sum_{q_1,q_2}^{M_y} \cos\left[kd_x \sin\theta_0 (p_1 - p_2)\right] \frac{\sin\left(kr_{p_1,p_2,q_1,q_2}\right)}{kr_{p_1,p_2,q_1,q_2}} \tag{4.23}$$

In (4.23), r_{p_1,p_2,q_1,q_1} is given as:

$$r_{p_1,p_2,q_1,q_2} = \sqrt{d_x^2(p_1 - p_2)^2 + d_y^2(q_1 - q_2)^2} \quad (4.24)$$

After summation over two indices (4.23) becomes (see Appendix 4B):

$$\begin{aligned} P_0 &= M_x M_y \sum_{p=-(M_x-1)}^{M_x-1} \sum_{q=-(M_y-1)}^{M_y-1} \left(1 - \frac{|p|}{M_x}\right)\left(1 - \frac{|q|}{M_y}\right) \\ &\bullet \frac{\sin\left(k\sqrt{d_x^2 p^2 + d_y^2 q^2}\right)}{k\sqrt{d_x^2 p^2 + d_y^2 q^2}} \bullet \left(\cos\left(kd_x m \sin\theta\right)\right) \end{aligned} \quad (4.25)$$

In (4.25), assuming a large array ($M_x \to \infty$, $M_y \to \infty$) and considering the oscillating nature of the function $\cos(kd_x m\sin\theta)$, it is assumed that $(1 - |p|/M_x)(1 - |q|/M_y) \approx 1$. The use of Poisson's formula and the properties of modified Bessel function [22], a simplified expression for P_0 is obtained as:

$$P_0 = \frac{M_x M_y \pi}{kd_x} \sum_{m=-\infty}^{\infty} \sum_{l} I_0\left(kd_y m\sqrt{1 - \left(\cos\theta_0 - \frac{l\lambda}{d_x}\right)}\right) \quad (4.26)$$

restricted to those values of l for which $1 - (\cos\theta_0 - l\lambda/d_x) \geq 0$.

In (4.26), I_0 denotes modified Bessel's function of the first kind of order zero [23]. The expression of P_0 may be rewritten as:

$$P_0 = \frac{M_x M_y \lambda}{2\pi d_x d_y} \sum_{p,q} \left[1 - \left(\frac{\lambda}{d_x} p - \sin\theta_0\right)^2 - \left(\frac{\lambda}{d_y} q\right)^2\right]^{-1/2} \quad (4.27)$$

for those values of p and q, which satisfies the condition (4.22). Considering the dense arrays (d_x/λ and $d_y/\lambda < 0.5$), (4.21) becomes:

$$D_{\text{norm}} = \frac{2\pi d_x d_y}{\lambda} \cos\theta_0 \quad (4.28)$$

which is valid at any angle except for those satisfying the condition

$$\cos\theta_0 < \frac{3}{4}\sqrt{2\frac{\lambda}{L}} \tag{4.29}$$

(L denotes length of array). Hence, a conclusion is drawn that in case the distance between the elements in the equidistant array is small, the directivity varies according to the cosine law, except in the sector where the main beam is near the aperture plane following (4.27). In (4.27), the expression within brackets is modified for 2-D scanning as

$$1-\left(\frac{\lambda}{d_x}p-\sin\theta_0\cos\varphi_o\right)^2-\left(\frac{\lambda}{d_y}q-\sin\theta_0\sin\varphi_o\right)^2 \tag{4.30}$$

Thus, the effect of 2-D scanning on the onset of grating lobes is accounted using this expression.

Table 4.2 lists the various possible pq modes' combinations for different values of d. It also lists the angles of onset for grating lobes corresponding to these modes. A similar study can be carried out for any other arbitrarily shaped antenna array using (4.21). It is seen that the behavior observed for variation in g is similar to that observed using exact expression earlier. The only difference from the previous section is that deeper dips (blind spots) are observed, due to the infinite nature of the array extent assumed herein. The accuracy of the results depends upon the choice of indices (p, q) and, henceforth, needs to be chosen properly; see Figure 4.7.

Figure 4.8(a) depicts graphically the lobes predicted in the Table 4.2. It may be observed that the directivity varies according to cos for $d = 0.5\lambda$, but in the case of larger interelement spacing, $d > 1$, the directivity is almost constant with respect to the variaion in the scan angle except for sharp dips because of the grating lobes due to periodicity in the array grid.

In Figure 4.8(b) for $\varphi_c = 45°$, for $d = 1.2$, a flat plateau kind of structure in the angular range of 20° to 70° is observed, implying a constant directivity pattern with respect to the scan angle. The grating lobes are observed in the angular range 10° to 13° and at 73.8° and 90°. However, in the case of $d = 2$, $d_{avg} > 1\lambda$ leads to the onset of

Table 4.2
Various Possible *pq* Combinations and Corresponding GL Locations for Different *d* for a Square Lattice ($d_x = d_y = d$)

d (in λ)	(p, q)	θ_o (deg)
0.5	(0,0) and (1,0)	90
1.0	(−1,0), (0,1), and (1,0)	0
	(1,1), (0,0), and (2,0)	90
1.2	(−1,0)	9.6
	(1,1)	16.3
	(0,1)	33.5
	(2,0)	41.8
	(0,0)	90
1.5	(−1,1)	4.5
	(−1,0) and (2,0)	19.5
	(0,1)	36
	(0,0)	48
	(3,0)	90
2.0	(−2,0), (0,2), and (2,0)	0
	(1,2), (−1,0), (1,2), and (3,0)	30
	(0,1)	60
	(2,2)	80
	(0,0), (2,2), and (4,0)	90
2.5	(−2,1)	6.7
	(−1,2)	11.5
	(3,1)	16.5
	(−1,1)	31
	(−1,0)	37
	(0,2)	37
	(3,2)	37
	(4,0)	37
	(4,1)	43
	(0,1)	66
	(1,2), (4,2), (0,0), and (5,0)	90
3.0	(−3,0), (0,3), and (3,0)	0
	(3,1)	3.3
	(−2,2)	4.5
	(−2,1)	−15.9
	(1,3), (4,0), and (−2,0)	19.5
	(4,1)	23
	(−1,2)	24.3
	(4,2)	36.1
	(5,0), (2,3), and (−1,0)	41.8
	(5,1)	46.5
	(0,2)	48.15
	(5,2)	67.2
	(0,1)	70.5
	(3,3) and (6,0)	90

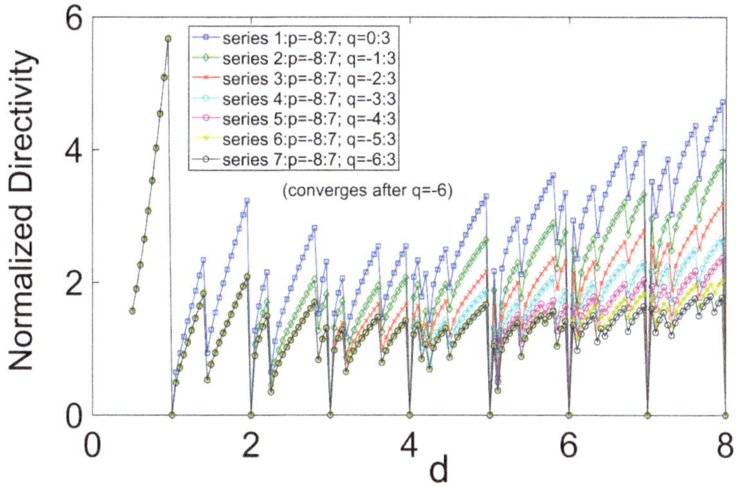

Figure 4.7
Normalized directivity versus interelement separation for the various choice of (p, q).

more of a number of grating lobes as seen in Figure 4.8(b). In Figure 4.8(c), for the case of $\varphi_c = 90°$, being a square array, the directivity behavior observed is the same as that observed in the case of $\varphi_c = 0°$ cut. In general, the directivity remains almost constant for $d > 1$ except for sharp dips due to the onset of grating lobes, which can be removed by altering the periodicty in the array grid, as will be discussed in the successive chapters. Hence, it may be stated that sparse antenna arrays can be used for designing ultrawideband scanning phased array antennas, which can scan up to the horizon.

In general, in the case of periodic sparse arrays, due to the number of grating lobes, the energy in the sidelobes gets redistributed and confined in regions other than the dips. This leads to the higher values of directivity as compared to the denser counterpart at angles other than boresight. However, (4.21) is obliged by the sparse array as well.

Having studied the properties of equally fed, equidistant, dense, and sparse antenna arrays with a square-shaped grid ($d_x = d_y$), the next point of consideration is the arrays with rectangular grids. Figure 4.9 illustrates the variation of g, with respect to the scan angle for different combinations of the interelement spacing in the array, at different phi-cuts. All of the results are compared with the behavior of a dense array with a square grid ($d_x = 0.5\lambda$, $d_x = d_y$).

Figure 4.9(a) shows the variation of g with respect to the scan angle for $d_x = 0.5\lambda$ and $d_x = 1.2\lambda$ in $\varphi_c = 0°$, 45°, and 90° cut-planes. The traces are compared with the behavior of the dense array $d_x = d_y = 0.5\lambda$. The behavior of g follows a cos pattern except for a few dips due to a large interelement separation along the y-axis. Figure 4.9(b) shows the variation of g with respect to the scan angle for $d_x = 0.5\lambda$

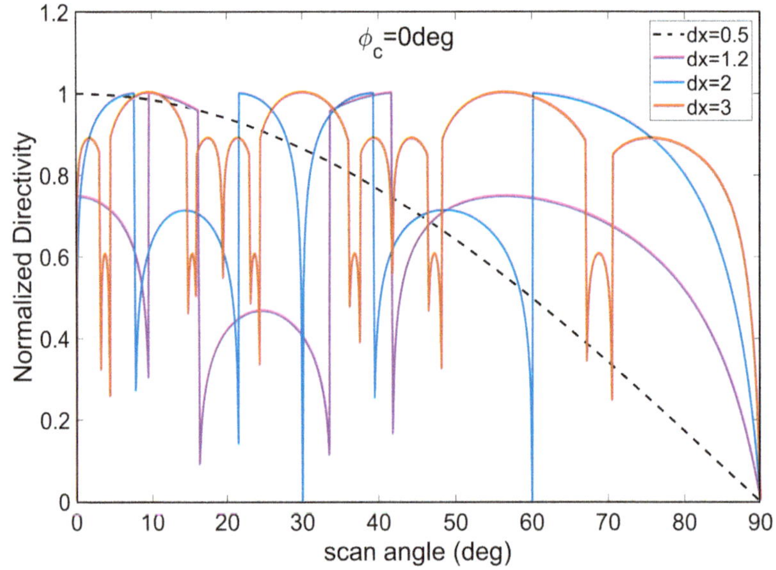

(a)

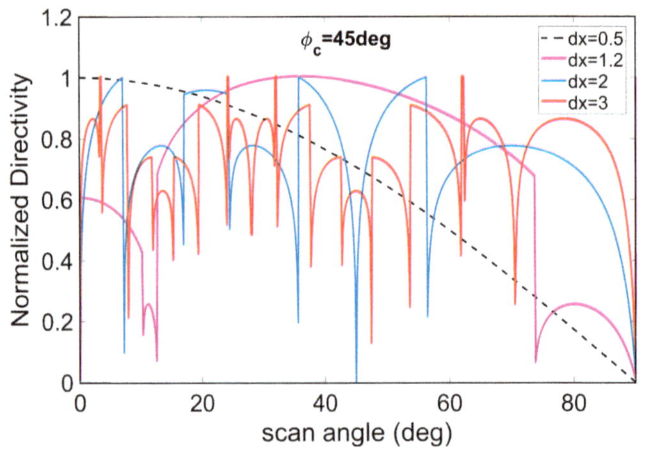

(b)

Figure 4.8

Normalized directivity versus the scan angle for different phi-cuts for a square array.

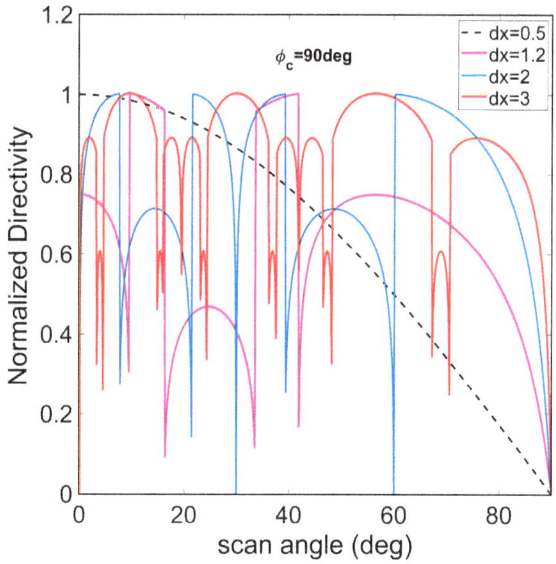

(c)

Figure 4.8
(Continued)

and $d_x = 2\lambda$ in $\varphi_c = 0°$, 45°, and 90° cut-planes. Figure 4.9(c) shows the variation of g with respect to the scan angle for $d_x = 1.2\lambda$ and $d_x = 2\lambda$ in $\varphi_c = 0°$, 45°, and 90° cut-planes. Further, due to different interelement separations along the x- and y-axes, different behavior is observed in the $\varphi_c = 0°$ and 90° cuts. The utility of the approximate expression is demonstrated in this section. This expression is quite useful and is the least time-consuming in the preliminary design process of large antenna arrays while choosing the apt interelement separations. It helps in predicting the probable blind spots over the complete hemispherical coverage a priori and allowing us to devise a technique to avoid these carefully.

4.4 Concentric Ring Antenna Arrays (CRAs)

CRAs (see Figure 4.10) are highly desirable for applications requiring main beam symmetry across all the phi-cuts. It is a known fact that a lower SLL can be obtained with the isophoric CRAs as compared to their rectangular counterpart. Hence, CRAs are considered

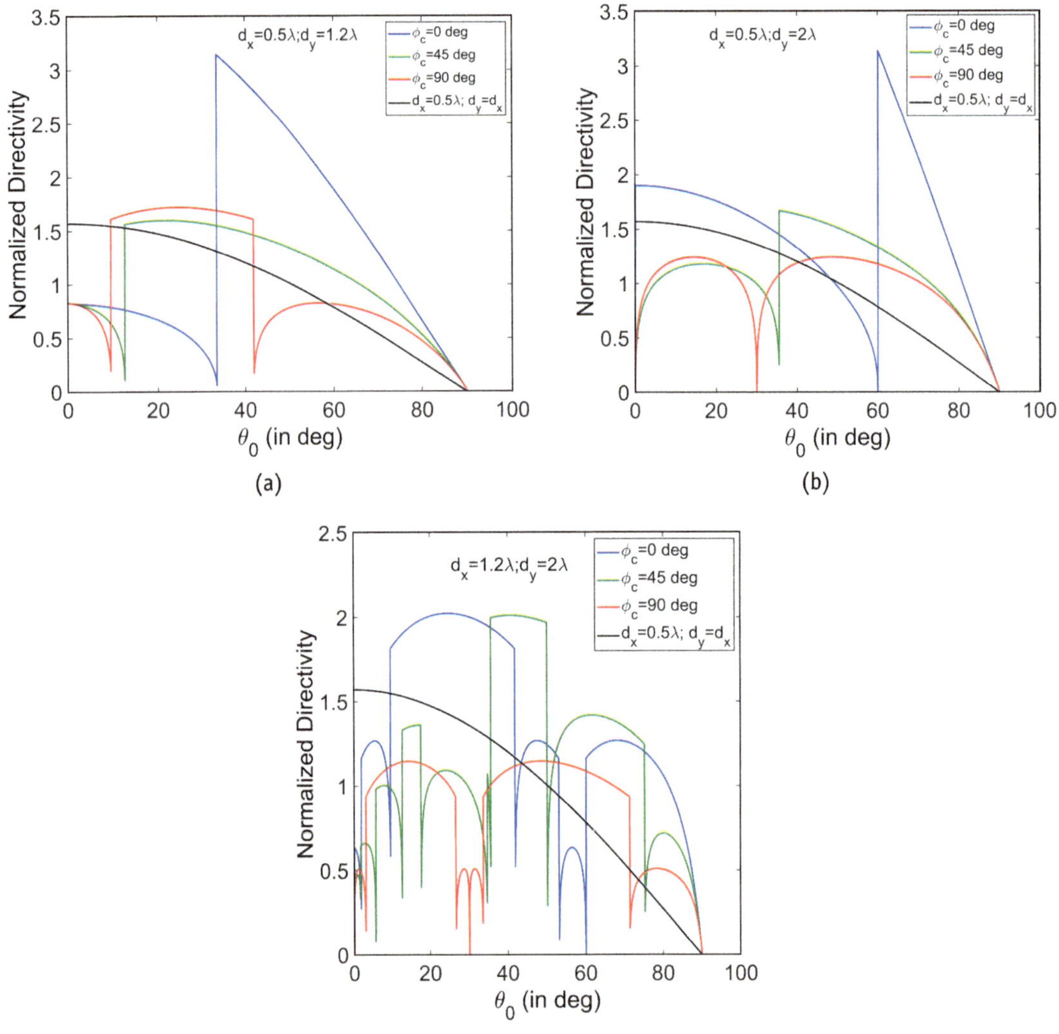

Figure 4.9
Normalized directivity versus the scan angle for a rectangular antenna array.
(© 2019 IEEE. Reprinted with permission from [21].)

to be presented here and their directive computations are carried out using (4.14) showing versatility of the expression with respect to any arbitrary shape of the antenna array. The usage of uniformly fed arrays also ensures maximum efficiency (ideally 100%) as all elements radiate or receive the maximum RF power. It is shown that the tapered amplitude distribution can be achieved via isophoric CRA array by spatially tapering its rings' radii [32–39]. This topic will be taken up in Chapter 6 on concentric ring sparse arrays (CRSAs). Here the focus is solely upon the utilization of the exact expression of directivity in predicating the characteristics of these arrays. The SLL of a CRA array can be controlled by controlling:

- The number of rings;
- The radii of the rings;
- The number of elements per ring.

Figure 4.10 shows the geometry of CRA comprising M-rings and the associated coordinate system considered for the analysis. Each ring consists of n_M elements and the total number of elements in an array is N $(= n_1 + n_2 + n_3 + \dots + n_M)$. The location of mn^{th} element in the CRA array is given by (x_{mn}, y_{mn}) or (a_m, φ_{mn}) in Cartesian or spherical coordinates, respectively, where a_m is the radii of the mth ring ($m = 1, 2, \dots, M$) in which the particular element is located and

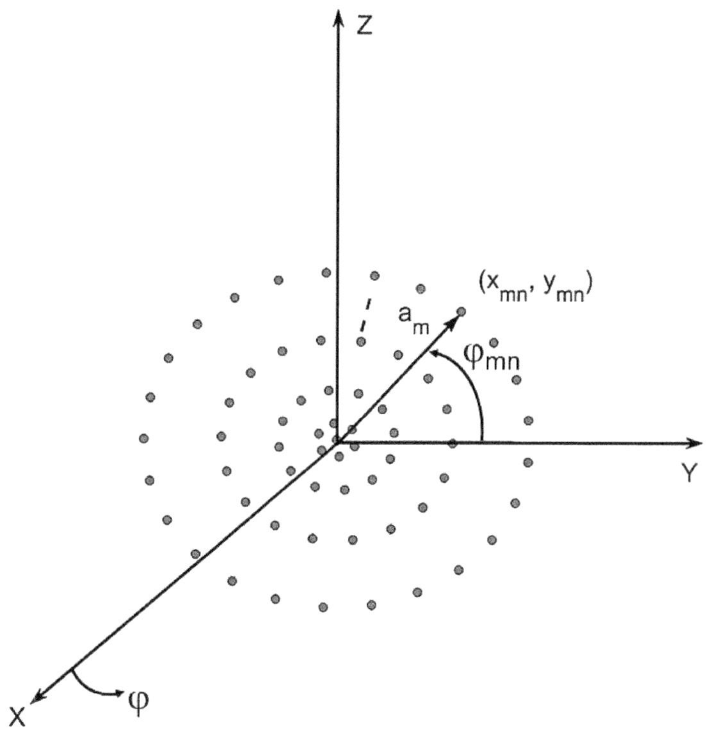

Figure 4.10
General schematic of a concentric ring array consisting of *N* rings.

φ_{mn} ($m = 1, 2,\ldots, M$; $n = 1, 2,\ldots, n_M$) is the angle which it makes with the y-axis. The relation between polar and rectangular coordinates may be given as:

$$x_{mn} = a_m \cos(\varphi_{mn});\ y_{mn} = a_m \sin(\varphi_{mn}) \quad (4.31)$$

Hence, (4.31) is utilized to compute the respective locations of the elements of CRA based on the chosen rings' radii and the corresponding number of elements therein the ring. Further, using the computed elements' locations, directivity is computed using (4.15).

First of all, a dense CRA (DCRA) with $M = 9$ rings is considered with uniformly-distributed elements (interelement separation of 0.5λ) and having one element at the center. The separation between radii of the rings, d_R is chosen equal to 0.5λ. The number of elements in each ring is estimated as:

$$N_m = \frac{2 \bullet \pi \bullet a_m}{d_{mn}} \quad (4.32)$$

In (4.32), d_{mn} is the interelement spacing between n^{th} and $(n-1)^{\text{th}}$ elements of the mth ring. Since CRA consists of uniformly distributed elements, d_{mn} is fixed for all elements in each ring of the CRA ($d_{mn} = 0.5\lambda$). The chosen radii and the number of elements in each ring are given in Table 4.3. Using these tabulated values and (4.31), the planar location of each of the elements is obtained. Figures 4.11(a, b) show the computed directivity with respect to the scan angle and AF plots, respectively. Figure 4.11(a) clearly illustrates the cos behavior followed by the DCRA. Figure 4.11(b) shows the array factor plot at different phi- (φ) cuts showing a symmetrical pattern with respect to the phi variation. The observed values of directivity at boresight and peak SLL (PSLL) are 29.5 dB (see Figure 4.11(a)) and

Table 4.3
Uniformly Distributed CRA Design Parameters

Uniform Ring Array (M = 9)									
m	1	2	3	4	5	6	7	8	9
a_m (in λ)	0.5	1.0	1.5	2.0	2.5	3.0	3.5	4.0	4.5
N_m	6	12	18	25	31	37	43	50	56

–19.5 dB (see Figure 4.11(b)), respectively. However, far SLL (FSLL) rises beyond 70° (<–19 dB), which may deteriorate the average SLL (ASLL). These observations enunciate the fact that comparatively lower SLLs can be attained with uniformly fed CRA arrays as compared to the rectangular arrays.

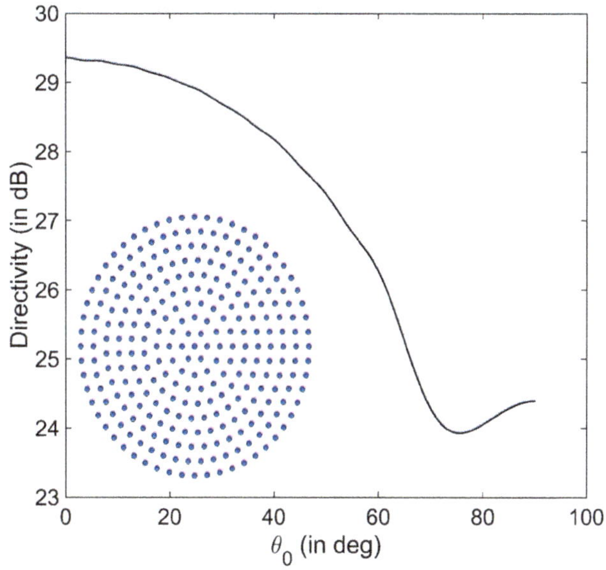

(a)

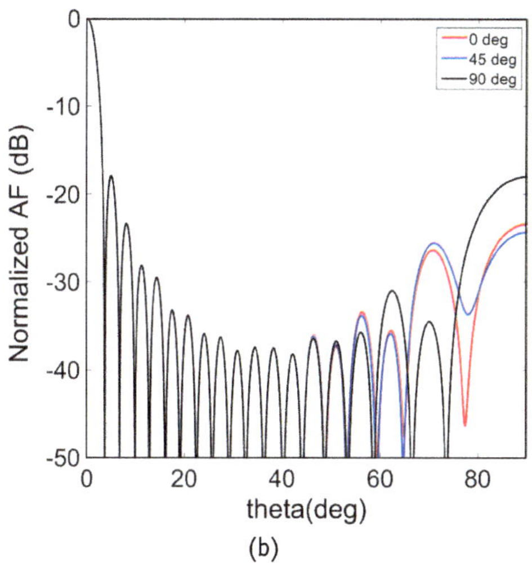

(b)

Figure 4.11
Uniformly distributed concentric ring array. (a) Computed directivity versus the scan angle. (b) Normalized array factor (in dB) plot in azimuth plane at different phi-cuts.

Table 4.4
Spatially Tapered CRA Design Parameters

Spatially Tapered Concentric Ring Array (M = 10)										
M	1	2	3	4	5	6	7	8	9	10
a_m (in λ)	17	13.85	11.55	9.51	7.70	6.13	4.74	3.36	2.23	1.11
$diff\_a_m$	—	3.15	2.3	2.04	1.81	1.57	1.39	1.38	1.13	1.12
N_m	104	122	116	110	91	73	59	42	28	14
d_{mn} (in λ)	1.027	0.713	0.625	0.543	0.531	0.527	0.505	0.503	0.5	0.498

Next, a 759-element space tapered CRA consisting of 10 rings (M = 10) is considered. Table 4.4 gives the details of the radii of the rings, a_m (in λ); interring separation (in λ), $diff\_a_m$; the number of elements in each ring, N_m; and the interelement spacing between the elements in a ring, d_{mn} (in λ), respectively. In this array, the interring separations, $diff\_a_m$, are spatially tapered, but the interelement spacing are within the limits of a dense array ($\lambda/d_{mn} \geq 2$). This can be considered as a pseudo-sparse array.

Figures 4.12(a, b) show the respective plots of the computed directivity with respect to the scan angle and the array factor with respect to the angle. It is observed that the directivity is almost constant up to 60° and then it falls gradually (Figure 4.12(a)), which clearly establishes the enhanced scan performance of CRA up to 70°. Figure 4.12(b) shows the array factor plots for different phi-cuts, illustrating the φ-symmetry of CRAs. The boresight gain observed is 32.5 dB (Figure 4.12(a)), PSLL < –34.5 dB, and the worst SLL < –20 dB (see Figure 4.12(b)).

It is observed that by utilizing the space-tapering technique while designing CRAs, ASLL can be improved. However, still the scanning performance as observed is limited. Hence, some additional techniques need to be explored that would further enhance the scanning performance. This is achieved by utilizing the sparse array techniques and will be discussed in subsequent chapters.

4.4.1 Concentric Ring Sparse Array (CRSA)

A CRA with 3 rings (M = 3) is considered as the first test case. The number of elements, N_m, considered in each ring is 13, 23, and 39, respectively, leading to a total number of 75 elements (= $N = N_1 +$

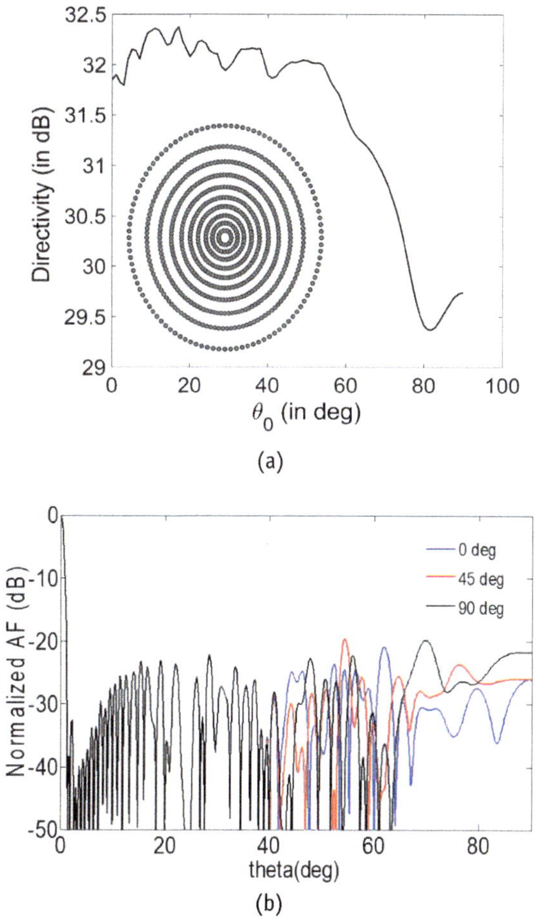

Figure 4.12
Spatially tapered concentric ring antenna array. (a) Computed directivity versus the scan angle. (b) Array factor with respect to the angle in the azimuth plane at different phi-cuts.

$N_2 + N_3 = 75$). The average interelement spacing, $\mathbf{d}_{\mathbf{mn,avg}}$, considered between the successive elements in each ring is 2.4λ and 0.42λ representing a CRSA and DCRA, respectively. The respective ring radii, a_m ($m = 1, 2$, and 3) for CRSA and DCRA arrays are chosen as (5λ, 9λ, and 11.25λ) and (1λ, 1.8λ, and 2.24λ), respectively.

Based on the chosen parameters, the scan performance of each array is plotted in terms of the variation of the normalized directivity with respect to the scan angle in Figure 4.13. The solid curve and the dotted curves show the scan performance of CRSA and DCRA, respectively. It is apparent from the figure that the directivity falls

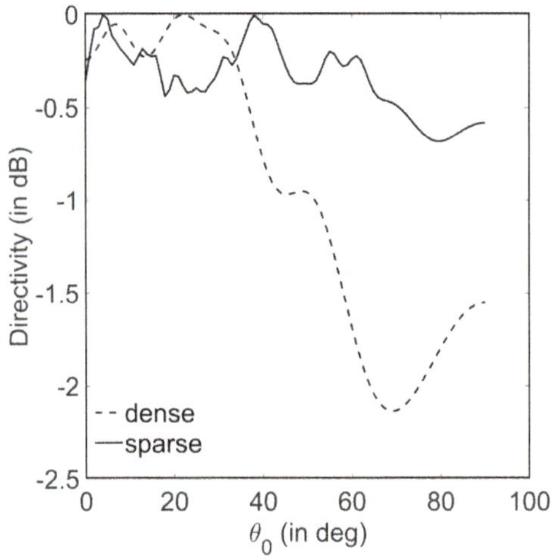

Figure 4.13
Comparison of the scan performance between 3-ring sparse and dense CRAs.

rapidly in the denser case as the scan angle approaches 60°. On the contrary, the directivity remains almost constant with respect to the variation in the scan angle in the case of CRSA even beyond the scan angle of 80°, closer to the horizon, which is a highly desirable feature for wide scanning radars. Figure 4.14 shows the AF plots of CRSA across different phi-cuts showing an almost similar performance, a desirable feature. It is observed that the far SLLs rise in these types of arrays, which is not desirable as it leads to higher average SLL. In the further sections, it is shown that, by employing the rotation of individual rings by small angles, this problem can also be tackled.

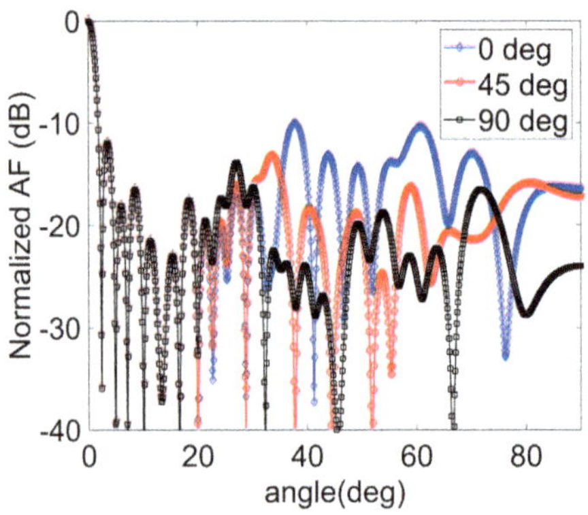

Figure 4.14
Array factor plot of a 3-ring sparse CRA across different φ-cuts.

A study is undertaken to exemplify the effect of $\mathbf{d}_{mn,avg}$, when it is varied from 0.5λ to 6λ, on the performance of a CRA. A 10-ring (M = 10) CRA is considered as a test case. Table 4.5 shows the various cases being analyzed. The radii of rings, a_m, are considered to be same in all the cases to keep the overall aperture size fixed. Using (4.30), the number of elements in each ring, N_m, is obtained and listed in the table.

It is observed from the table that almost the same PSLL and directivity is obtained in all the cases, although the total number of elements reduces from 2,747 to 225, showing a significant reduction

Table 4.5
Different Test Cases of 10-Element CRSA (M = 10)

$d_{mn,avg}$	N_m	a_m (in λ)			PSLL (dB)	Total N	D_o (dB)
0.5	62 124 182 236	5.0000	9.8769	14.5106	−18.19	2,747	20
	284 324 357 381	18.7869	22.6007	25.8580			
	396 401	28.4786	30.3979	31.5688			
		31.9623					
1.0	31 62 91 118	5.0000	9.8769	14.5106	−18.19	1,372	20
	142 162 178 190	18.7869	22.6007	25.8580			
	198 200	28.4786	30.3979	31.5688			
		31.9623					
2.0	15 31 45 59 71	5.0000	9.8769	14.5106	−18.19	685	20
	81 89 95 99	18.7869	22.6007	25.8580			
	100	28.4786	30.3979	31.5688			
		31.9623					
3.0	10 20 30 39 47	5.0000	9.8769	14.5106	−18.19	454	20
	54 59 63 66 66	18.7869	22.6007	25.8580			
		28.4786	30.3979	31.5688			
		31.9623					
4.0	7 15 22 29 35	5.0000	9.8769	14.5106	−18.19	338	20
	40 44 47 49 50	18.7869	22.6007	25.8580			
		28.4786	30.3979	31.5688			
		31.9623					
5.0	6 12 18 23 28	5.0000	9.8769	14.5106	−18.19	271	20
	32 35 38 39 40	18.7869	22.6007	25.8580			
		28.4786	30.3979	31.5688			
		31.9623					
6.0	5 10 15 19 23	5.0000	9.8769	14.5106	−18.19	225	20
	27 29 31 33 33	18.7869	22.6007	25.8580			
		28.4786	30.3979	31.5688			
		31.9623					

in the number of elements and, henceforth, the cost of the array. Also, it is observed from Figure 4.15(b) that the scan volume of array grows with an increase in the sparseness of the elements. It has been observed that for a sparse CRA array with $d_{mn,\text{avg}} > 4\lambda$,

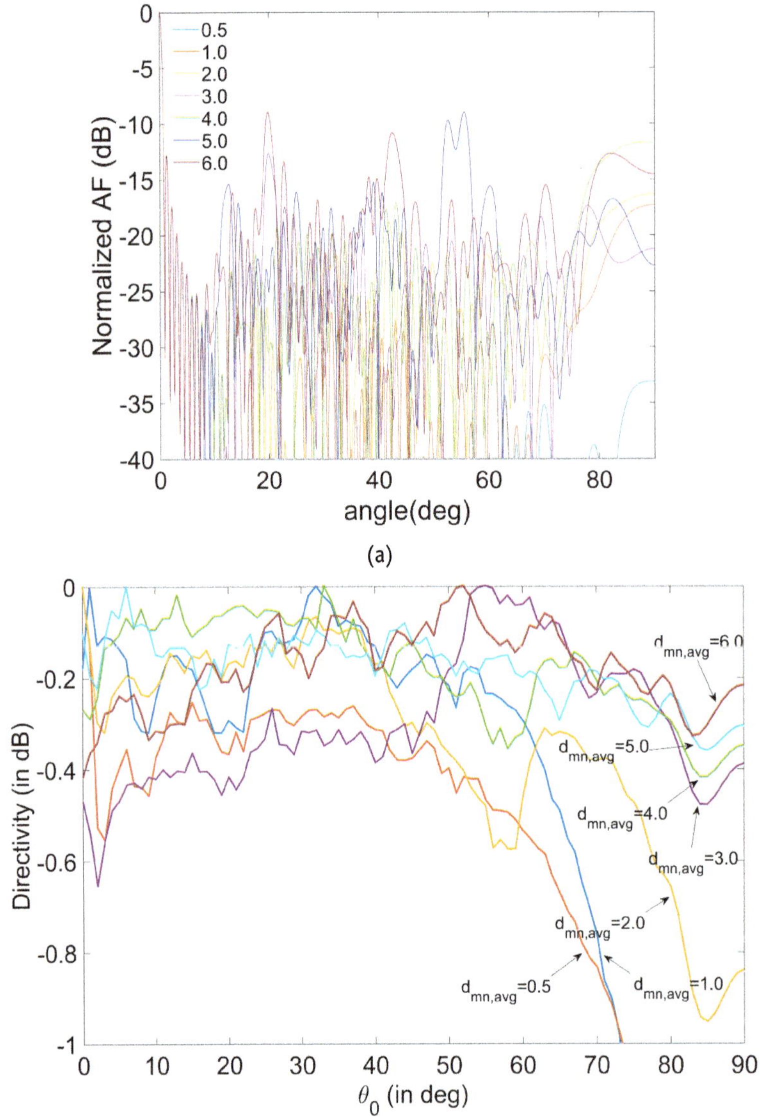

(a)

(b)

Figure 4.15

The performance of CRAs at values of $d_{mn,\text{avg avg}}$ (in λ). (a) Normalized array factor plot of a 10-ring CRA array. (b) Computed directivity curves with respect to the scan angle for a 10-ring CRA array.

scanning can be achieved even up to the horizon (i.e., nearly hemispherical coverage). This is a highly desirable feature for very long-range radars such as the surface wave over-the-horizon (SWOTH) radars. The only disadvantage seen is the rise in far SLL leading to deterioration in average SLL. This problem can be tackled by the usage of nonperiodic CRSAs and by rotating the individual rings by small amounts.

Another study is undertaken in which $d_{mn,\text{avg}}$ is fixed at 3λ and CRA performance is studied by varying interring separation, $d_{\text{avg, ring}}$. Table 4.6 lists the combinations studied and the resulting curves for the directivity computed are plotted in Figure 4.16. A nearly hemispherical scan coverage is seen for $d_{\text{avg, ring}} = 6\lambda$, which depicts a highly sparsed CRA and is explicitly shown in Figure 4.17. On the top of Figure 4.17, the optimized radii of 10 rings are shown, which yield the computed directivity curve. A very negligible variation in directivity is observed over the scan range.

4.4.2 Effect of Rotation on CRSA

Rotation is important to achieve randomization of the elements location destroying the periodicity in the array leading to the

Table 4.6
A 10-Ring Array with $d_{mn\_\text{avg}} = 3\lambda$

$D_{\text{avg,ring}}$	N_m	a_m (in λ)	Total N
0.5	1 2 3 3 4 5 5 6 6 6	0.5000 0.9877 1.4511 1.8787 2.2601 2.5858 2.8479 3.0398 3.1569 3.1962	41
1.0	2 4 6 7 9 10 11 12 13 13	1.0000 1.9754 2.9021 3.7574 4.5201 5.1716 5.6957 6.0796 6.3138 6.3925	87
2.0	4 8 12 15 18 21 23 25 26 26	2.0000 3.9508 5.8042 7.5148 9.0403 10.3432 11.3914 12.1592 12.6275 12.7849	178
3.0	6 12 18 23 28 32 35 38 39 40	3.0000 5.9261 8.7063 11.2722 13.5604 15.5148 17.0872 18.2388 18.9413 19.1774	271
4.0	8 16 24 31 37 43 47 50 52 53	4.0000 7.9015 11.6085 15.0296 18.0806 20.6864 22.7829 24.3183 25.2550 25.5698	361
5.0	10 20 30 39 47 54 59 63 66 66	5.0000 9.8769 14.5106 18.7869 22.6007 25.8580 28.4786 30.3979 31.5688 31.9623	454
6.0	12 24 36 47 56 64 71 76 79 80	6.0000 11.8523 17.4127 22.5443 27.1209 31.0296 34.1743 36.4775 37.5000 38.5000	545

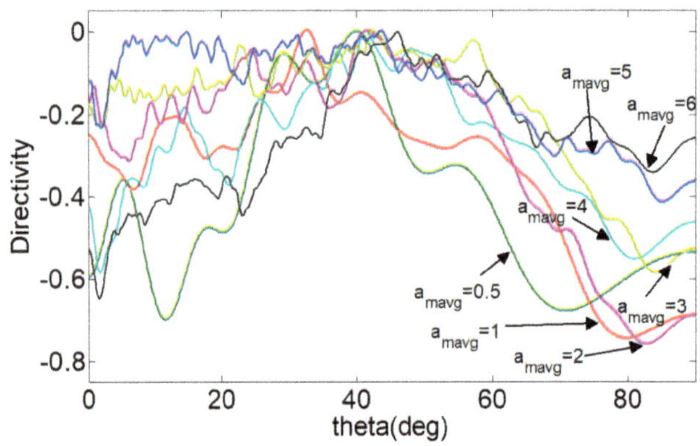

Figure 4.16
Computed directivity with respect to the scan angle of a 10-ring CRA array for different $a_{m,\text{avg}}$.

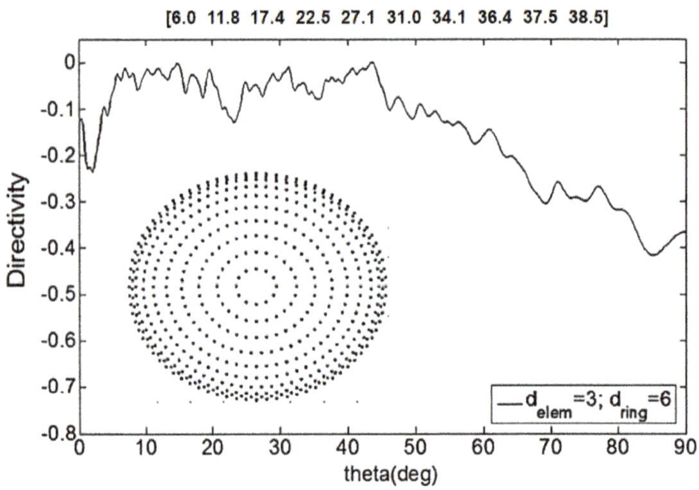

Figure 4.17
Computed directivity plot of 10-ring CRAs.

nonappearance of grating lobes, which appear due to the mutual coupling phenomenon in periodic arrays. The angular spacing between the two elements in the mth ring is given as $2\pi/N_m$. A small rotation of a ring by an angle, $m_{m,\text{rot}}$ (= 0,1,2,3,...) means a small step in φ_{mn} being given as:

$$\varphi_{mn} = n \cdot \frac{2\pi}{N_m} + m_{m,\text{rot}} \cdot \frac{2\pi}{N_m} \tag{4.33}$$

Herein, only 5 small rotations are considered (m_{rot} = 0,1,2,3,4) and its effect on the performance of 5-ring CRSA in terms of directivity is studied. During analysis, various combinations of rotations in rings are considered to optimize the scan performance and reduce the far SLL as will be shown in Chapter 6. Figure 4.18(a) shows the effect of the combinations of rotations in the constituent rings of CRSA on its scan performance. It is inferred that the combination of various rotations in different rings leads to an improvement in the scan range and the reduction in far SLL (Figure 4.18(b)).

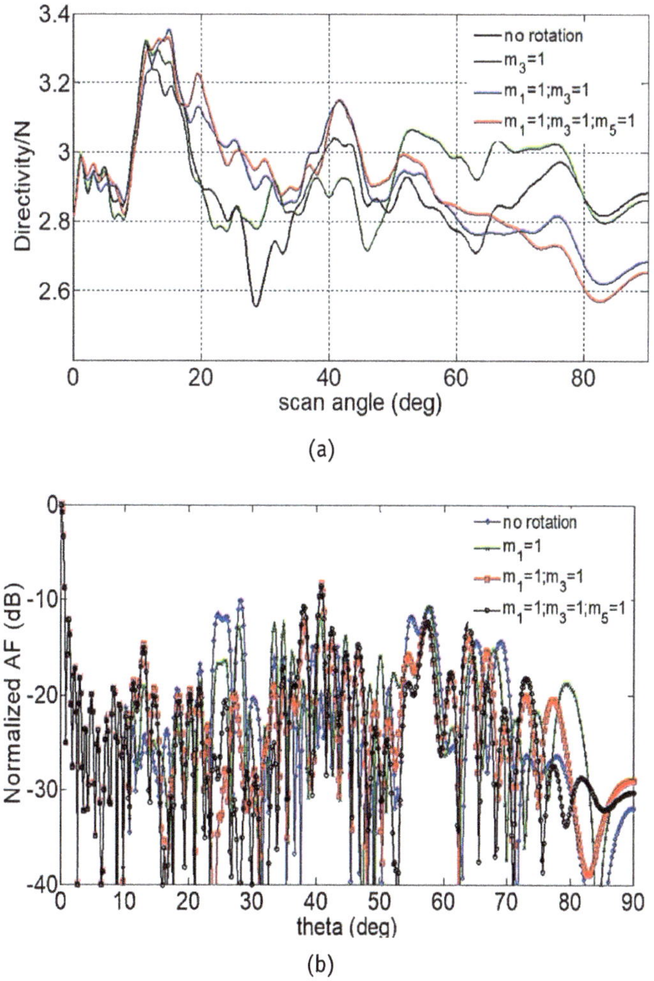

Figure 4.18
Effect of rotation of constituent rings of CRA array on its performance. (a) Computed directivity versus scan angle. (b) Normalized array factor plot.

The study shown in figures is just an indication of the possibilities as a result of rotating the individual rings in a controlled manner and may be further combined with varied ring radii, interring separation and the number of elements during optimization to arrive at the desired performance of CRSA by making use of some deterministic or evolutionary optimization techniques as will be discussed in Chapter 6.

4.5 Conclusions

This chapter presents a generalized expression for the computation of the directivity of antenna arrays, namely, an exact expression for finite arrays and an approximate expression for infinite arrays. These expressions are valid for dense and sparse antenna arrays. The approximate expression can run as a forerunner to all designs before the final computation using the exact expression to save computation time. The ease and effectiveness of the expression in analyzing the antenna arrays with arbitrary shapes are demonstrated with the help of case studies. Uniformly spaced CRAs are studied in this chapter for their characteristics and it is shown that the spatial tapering of these leads to the lowering of SLL. Later, Chapter 6 presents stress upon the utilization of aperiodicity to improve the scan performance of CRAs. The exact expression forms the basis of all directivity computations in this book.

References

[1] Balanis, C. A., *Antenna Theory Analysis and Design*, New York: John Wiley & Sons, 2005.

[2] Forman, B. J., "A Novel Directivity Expression for Planar Arrays," *Radio Science*, Vol. 5, No. 7, 1970, pp. 1077–1083.

[3] Tai, C., "The Optimum Directivity of Uniformly Spaced Broadside Arrays of Dipoles," *IEEE Transactions on Antennas and Propagation*, Vol. AP-12, No. 4, 1964, pp. 447–454.

[4] Bach, H., "Directivity of Basic Linear Arrays," *IEEE Transactions on Antennas and Propagation*, Vol. AP-18, No. 1, 1970, pp. 107–110.

[5] Langston, L. J., "Scanned Directivity of Linear Arrays," *IEEE Transactions on Antennas and Propagation*, Vol. AP-19, No. 2, 1971, pp. 282–284.

[6] Hansen, R. C., "Comparison of Square Array Directivity Formulas," *IEEE Transactions on Antennas and Propagation*, Vol. AP-20, No. 1, 1972, pp. 100–102.

[7] Daniel, J. P., "Directivity of Linear Microstrip Arrays," *Electronics Letters*, Vol. 23, No. 17, 1987, pp. 897–899.

[8] Safaai-Jazi, A., "Directivity of Chebyshev Arrays with Arbitrary Element Spacing," *Electronics Letters*, Vol. 31, No. 10, 1995, pp. 772–774.

[9] Hansen, R. C., "Dolph-Chebyshev Array Directivity Versus Spacing," *Electronics Letters*, Vol. 32, No. 12, 1996, pp. 1050–1051.

[10] Werner, D. H., D. Mulyantini, and P. L. Werner, "Closed-Form Representation for Directivity of Nonuniformly Spaced Linear Arrays with Arbitrary Element Patterns," *Electronics Letters*, Vol. 35, No. 25, 1999, pp. 2155–2157.

[11] Mahony, J. D., "An Approximate Expression for the Directivity of a Tapered-Cosine Distribution on a Circular Aperture," *IEEE Antennas and Propagation Magazine*, Vol. 52, No. 2, 2010, pp. 126–128.

[12] Lee, M. J., et al., "Evaluation of Directivity for Planar Antenna Arrays," *IEEE Antennas and Propagation Magazine*, Vol. 42, No. 3, 2000, pp. 64–67.

[13] Nuttall, A. H., and B. A. Cray, "Approximations to Directivity for Linear, Planar, and Volumetric Apertures and Arrays," *IEEE Journal of Oceanic Engineering*, Vol. 26, No. 3, 2001, pp. 383–398.

[14] Forman, B. J., "Directivity Characteristics of Scannable Planar Arrays," *IEEE Transactions on Antennas and Propagation*, Vol. AP-20, No. 3, 1972, pp. 245–252.

[15] Ligthart, L.P., "A New Numerical Approach to Planar Phased Arrays," *Proc. IEEE AP-S*, September 1975, pp. 453–457.

[16] Das, S., et al., "A Generalized Closed Form Expression of Directivity of Arbitrary Planar Antenna Arrays," *IEEE Transactions on Antennas and Propagation*, Vol. 61, No. 7, 2013, pp. 3909–3911.

[17] Mailloux, R. J., *Phased Array Antenna Handbook*, Second Edition, Norwood, MA: Artech House, 2005.

[18] Kahn, W. K. "Element Efficiency: A Unifying Concept for Array Antennas," *IEEE Antennas and Propagation Magazine*, Vol. 49, No. 4, 2007, pp. 48–56.

[19] Kildal, P. S., S. Maci, and A. Vosoogh, "Fundamental Directivity Limitations of Dense Array Antennas: A Numerical Study Using Hannan's Embedded Element Efficiency," *IEEE Antennas and Wireless Propagation Letters*, Vol. 15, 2015, pp. 766–769.

[20] Pozar, D. M., "A Relation Between the Active Input Impedance and the Active Element Pattern of a Phased Array," *IEEE Antennas and Propagation Magazine*, Vol. 51, No. 9, 2003, pp. 2486–2489.

[21] Kedar, A., and L. P. Ligthart, "Wide Scanning Characteristics of Sparse Phased Array Antennas Using an Analytical Expression for Directivity," *IEEE Transactions on Antennas and Propagation*, Vol. 67, No. 2, 2019, pp. 905–915.

[22] Abramowitz, M., and I. A. Stegen, *Handbook of Mathematical Functions: with Formulas, Graphs, and Mathematical Tables*, Washington, D.C.: National Bureau of Standards, 1972.

[23] Watson, G. N., *A Treatise on the Theory of Bessel Functions*, Second Edition, Cambridge, U.K.: Cambridge University Press, 1995.

[24] Amitay, N., V. Galindo, and C. P. Wu, *Theory and Analysis of Phased Array Antennas*, New York: Wiley-Interscience, 1972.

[25] Makarov, S. N., A. Puzella, and V. Iyer, "Scan Impedance for an Infinite Dipole Array: Accurate Theoretical Model Compared to Numerical Software," *IEEE Antennas and Propagation Magazine*, Vol. 50, No. 6, 2008, pp. 132–149.

[26] Chiavacci, P. A., "Limited Scan Antenna Systems Using Phased-Array Feeds (Versus Direct Radiating Array Apertures) Require Fewer Array Elements; Or Do They?" *Proceedings 2000 IEEE International Conference on Phased Array Systems and Technology*, Dana Point, CA, 2000, pp. 291–297.

[27] Skobelev, S. P., "One More Look at the Reduction of the Number of Controlled Elements in Limited-Scan Phased Array Antennas," *The 8th European Conference on Antennas and Propagation (EuCAP 2014)*, The Hague, 2014, pp. 3625–3628.

[28] Skobelev, S. P., and A. V. Shishlov, "On the Use of Yagi-Uda Elements in Array Antennas for Limited Scan Applications," *Proceedings 2000 IEEE International Conference on Phased Array Systems and Technology*, Dana Point, CA, 2000, pp. 299–302.

[29] Iqbal, Z., and M. Pour, "Grating Lobe Mitigation in Scanning Planar Phased Array Antennas," *2019 IEEE International Symposium on Phased Array System & Technology (PAST)*, Waltham, MA, 2019, pp. 1–3.

[30] Hansen, R. C., *Phased Array Antennas*, New York: Wiley-Interscience, 2009.

[31] Pozar, D. M., *Microwave Engineering*, New York: Wiley, 2005.

[32] Stearns, C., and A. Stewart, "An Investigation of Concentric Ring Antennas with Low Sidelobes," *IEEE Transactions on Antennas and Propagation*, Vol. 13, No. 6, 1965, pp. 856–863.

[33] El-Kamchouchi, H., "Ultra-Low-Sidelobe-Level Concentric-Ring-Array Pattern Synthesis Using Bessel Neural Networks [Antenna Designer's Notebook]," *IEEE Antennas and Propagation Magazine*, Vol. 52, No. 4, 2010, pp. 102–105.

[34] Kumar, B. P., and G. R. Branner, "Design of Low Sidelobe Circular Ring Array by Element Radius Optimization," *IEEE Antennas and Propagation Society International Symposium*, Orlando, FL, July 11–16, 1999, pp. 2032–2035.

[35] Das, R., "Broadbanding of Concentric Planar Ring Arrays by Space Tapering," *Radio and Electronic Engineer*, Vol. 33, No. 4, 1967, pp. 211–222.

[36] Milligan, T. A., "Space-Tapered Circular (Ring) Array," *IEEE Antennas and Propagation Magazine*, Vol. 46, No. 3, 2004, pp. 70–73.

[37] Haupt, R. L., "Optimized Element Spacing for Low Sidelobe Concentric Ring Arrays," *IEEE Transactions on Antennas and Propagation*, Vol. 56, No. 1, 2008, pp. 266–268.

[38] Jiang, Y., et al., "Synthesis of Uniformly Excited Concentric Ring Arrays Using the Improved Integer GA," *IEEE Antennas and Wireless Propagation Letters*, Vol. 15, 2016, pp. 1124–1127.

[39] Guo, Q., C. Chen, and Y. Jiang, "An Effective Approach for the Synthesis of Uniform Amplitude Concentric Ring Arrays," *IEEE Antennas and Wireless Propagation Letters*, Vol. 16, 2017 pp. 2558–2561.

APPENDIX 4A

Derivation of (4.14)

$I_n = I_{on}$ (the maximum value of I_n at $(\theta_0, \varphi_0) = (0, 0)$) is computed as

$$\left| \iint_s I_0(\xi,\eta)\,d\xi\,d\eta \right|^2 = \sum_{m=1}^{N} I_{om}^2 \tag{4A.1}$$

Referring to the denominator of (4.10),

$$\iint \chi_0(\xi,\eta)\cos\left[k\sin\theta_0\left(\xi\cos\varphi_0 + \eta\sin\varphi_0\right)\right] \frac{\sin k\sqrt{\xi^2 + \eta^2}}{k\sqrt{\xi^2 + \eta^2}}\,d\xi d\eta \tag{4A.2}$$

From (4.10),

$$\chi_0(\xi,\eta) = \iint_s I_0(x,y) I_0^*(x - \xi, y - \eta)\,dx\,dy \tag{4A.3}$$

Now, for a planar phased array antenna,

$$I_o(x,y) = \sum_{m=1}^{N} I_m \delta(x - x_m)\delta(y - y_m) \tag{4A.4}$$

$$I_o(x-\xi, y-\eta) = \sum_{n=1}^{N} I_n \delta(x-\xi-\xi_n)\delta(y-\eta-\eta_n) \tag{4A.5}$$

$$\begin{aligned}
\chi_0 &= \sum_{m=1}^{N}\left\{ I_m(x = x_m, y = y_m)\sum_{n=1}^{N} I_n^*(\xi = x_m - \xi_n, \eta = y_m - \eta_n)\right\} \\
&= \sum_{m=1}^{N}\left\{\begin{aligned} & I_{0m}\exp\left[-ik\sin\theta_0\left(x_m\cos\varphi_0 + y_m\sin\varphi_0\right)\right] \\ & \bullet \sum_{n=1}^{N} I_{on}^*\exp\left[ik\sin\theta_0\left((x_m-\xi_n)\cos\varphi_0 + (y_m-\eta_n)\sin\varphi_0\right)\right]\end{aligned}\right\} \\
&= \sum_{m,n=1}^{N} I_{0m}I_{on}^*\exp\left[-ik\sin\theta_0\left(\xi_n\cos\varphi_0 + \eta_n\sin\varphi_0\right)\right] \\
&= \chi_0(\xi_n, \eta_n)\bullet\cos\left[k\sin\theta_0\left(\xi_n\cos\varphi_0 + \eta_n\sin\varphi_0\right)\right]
\end{aligned} \tag{4A.6}$$

∴ Denominator of (4.10) becomes:

$$\sum_{m,n=1}^{N} I_{om}I_{on}^*\cos\left[k\sin\theta_0\left\{(x_m-\xi_n)\cos\varphi_0 + (y_m-\eta_n)\sin\varphi_0\right\}\right]$$

$$\bullet\frac{\sin k\sqrt{(x_m-\xi_n)^2 + (y_m-\eta_n)^2}}{k\sqrt{(x_m-\xi_n)^2 + (y_m-\eta_n)^2}}$$

Putting, $\xi_n = x_n$; $\eta_n = y_n$;
Denominator =

$$\begin{aligned}
&\sum_{m,n=1}^{N} I_{om}I_{on}^*\bullet\cos\left[kr_{nm}\bullet\sin\theta_0\left\{\cos\alpha_{nm}\cos\varphi_0 + \sin\alpha_{nm}\sin\varphi_0\right\}\right]\bullet\frac{\sin kr_{nm}}{kr_{nm}} \\
&r_{nm} = \sqrt{(x_n-x_m)^2 + (y_n-y_m)^2}; \\
&\cos\alpha_{nm} = \frac{x_n - x_m}{r_{nm}};\ \sin\alpha_{nm} = \frac{y_n - y_m}{r_{nm}}
\end{aligned} \tag{4A.7}$$

Hence, we get the expression:

$$D = \frac{\left(\sum_{m=1}^{N} I_{0m}\right)^2}{\sum_{m,n=1}^{N} I_{0m}I_{0n}^*\cos\left[kr_{nm}\sin\theta_0\cos(\varphi_0-\alpha_{nm})\right]\frac{\sin kr_{nm}}{kr_{nm}}} \tag{4A.8}$$

APPENDIX 4B

Derivation of (4.21)

It is possible to rewrite (4.10) as:

$$D = \frac{\left(\sum_{m=1}^{N} I_{0m}\right)^2}{\sum_{m,n=1}^{N} I_{0m} I_{0n}^{*} \cos\left[kr_{nm} \sin\theta_0 \cos\left(\varphi_0 - \alpha_{nm}\right)\right] \frac{\sin kr_{nm}}{kr_{nm}}} \tag{4B.1}$$

$$I_n = I_{on} \exp\left(-ikx_n \sin\theta_0\right) \ \left(\because \varphi_0 = 0\right) \tag{4B.2}$$

$$I(x,y) = \sum_{n=1}^{N} I_n \delta\left(x - x_n\right) \delta\left(y - y_n\right) \tag{4B.3}$$

$$\sum_{m=1}^{N} I_{om}^2 = N^2 \left(= M_x^2 M_y^2\right) \tag{4B.4}$$

Equation (4B.1) may be rewritten as:

$$D = \frac{N^2 \left(= M_x^2 M_y^2\right)}{\sum_{m,n=1}^{N} \cos\left[kr_{nm} \sin\theta_0 \cos\alpha_{nm}\right] \cdot \frac{\sin kr_{nm}}{kr_{nm}}} \tag{4B.5}$$

$$r_{nm} = r_{m_1,m_2,n_1,n_2} = \sqrt{d_x^2\left(m_1 - m_2\right)^2 + d_y^2\left(n_1 - n_2\right)^2} \tag{4B.6}$$

It is known that

$$-M_x + 1 < m_1 - m_2 < M_x - 1 \text{ and } -M_y + 1 < n_1 - n_2 < M_y - 1$$

Introducing new variables $m = m_1 - m_2$ and $n = n_1 - n_2$, the number of equal r_{nm} terms for given m and n are $M_x \bullet (1 - |m|/M_x)$ and $M_y \bullet (1 - |n|/M_y)$, respectively.

Hence, denominator of (4B.5) becomes

$$M_x M_y \sum_{m=-(M_x-1)}^{M_x-1} \sum_{n=-(M_y-1)}^{M_y-1} \left(1 - \frac{|m|}{M_x}\right)\left(1 - \frac{|n|}{M_y}\right) \frac{\sin\left(k\sqrt{d_x^2 m^2 + d_y^2 n^2}\right)}{k\sqrt{d_x^2 m^2 + d_y^2 n^2}} \cos\left(kd_x m \sin\theta_0\right) \tag{4B.7}$$

For large arrays and considering the oscillating character of $\cos(kd_x m\sin\theta_0)$, it may be assumed that

$$\left(1 - \frac{|m|}{M_x}\right)\left(1 - \frac{|n|}{M_y}\right) \approx 1 \tag{4B.8}$$

$\therefore$, denominator in (4B.5) becomes

$$\sum_{m=-(M_x-1)}^{M_x-1} \left\{ \sum_{n=-(M_y-1)}^{M_y-1} \frac{\sin\left(k\sqrt{d_x^2 m^2 + d_y^2 n^2}\right)}{k\sqrt{d_x^2 m^2 + d_y^2 n^2}} \right\} \cos\left(kd_x m \sin\theta_0\right) \tag{4B.9}$$

Let

$$md_x = x \text{ and } nd_y = y, \Rightarrow dx = \frac{m \bullet d_x}{M_x} \Rightarrow \Delta m = 1 = \frac{M_x}{d_x} \bullet dx \tag{4B.10}$$

Similarly,

$$\Delta n = 1 = \frac{M_y}{d_y} \bullet dy$$

$$\Rightarrow \sum_{n=-(M_y-1)}^{M_y-1} \rightarrow \frac{M_y}{d_y} \bullet \int_{-\infty}^{\infty} \frac{\sin\left(k\sqrt{x^2+y^2}\right)}{k\sqrt{x^2+y^2})} dy(-\infty \le y \le \infty) \tag{4B.11}$$

Thus, the denominator of (4B.5) may be rewritten as:

$$\sum_{m=-(M_x-1)}^{M_x-1} \left\{ \frac{M_y}{d_y} \bullet \int_{-\infty}^{\infty} \frac{\sin\left(k\sqrt{x^2+y^2}\right)}{k\sqrt{x^2+y^2})} \right\} \cos(kx\sin\theta_0)dy$$

$$= \frac{M_x M_y}{d_x d_y} \int_{x=-\infty}^{\infty} \int_{y=-\infty}^{\infty} \frac{\sin\left(k\sqrt{x^2+y^2}\right)}{k\sqrt{x^2+y^2})} dy \cos(kx\sin\theta_0)\, dx(-\infty \le x \le \infty) \tag{4B.12}$$

Alternately, the expression in (4B.11) may be approximated as given below (making use of Poisson's formula and properties of modified Bessel functions [21]), for those values of l and p integers where $\sqrt{\bullet} = 0$.

$$\approx \lambda \frac{M_x M_y}{kd_x d_y} \sum_{l,p} \left[1 - \left(\frac{\lambda}{d_x} l - \sin\theta_0 \right)^2 - \left(\frac{\lambda}{d_y} p \right)^2 \right]^{-\frac{1}{2}} \tag{4B.13}$$

The qualifying (l, p) set gives an effect on the maximum D_{norm} as function of θ_0.

$$l = 0,\ p = 0 \Rightarrow D_{\text{norm}} = \frac{D}{M_x M_y} = \frac{2\pi d_x d_y}{\lambda^2\ 1/\cos\theta_0} = \frac{2\pi d_x d_y}{\lambda^2} \cos\theta_0 \tag{4B.14}$$

Equation (4B.13) is θ_0 dependent and it is valid for a dense array, depicting the angular dependence of D_{norm}. The condition for D_{norm} in the dense array is:

$$\cos\theta_0 > \sqrt{\frac{\lambda}{M_x d_x}} \tag{4B.15}$$

where $M_x d_x$ is the length of the array (for scanning in the xz-plane).

With respect to (4B.11), integral 1 in (4B.11) becomes:

$$\mathrm{Int}_1 = \int_{y=0}^{\infty} \frac{\sin\left(k\sqrt{x^2+y^2}\right)}{k\sqrt{x^2+y^2}}dy = \int_{0}^{\infty} \frac{\sin\left(k\sqrt{x^2+y^2}\right)}{k\sqrt{x^2+y^2}}dy \tag{4B.16}$$

In order to solve the integral, let

$$\begin{aligned} &z = k\sqrt{x^2+y^2} \Rightarrow dz = \frac{ky}{\sqrt{x^2+y^2}}dy \\ &k^2\left(x^2+y^2\right) = z^2 \Rightarrow y = \sqrt{\frac{z^2}{k^2}-x^2} \qquad y>0;\ z>kx \\ &\frac{dy}{k\sqrt{x^2+y^2}} = \frac{1}{k^2 y}dz = \frac{1}{k\sqrt{z^2-k^2x^2}}dz \\ &\therefore \mathrm{Int}_1 = \int_{z=kx}^{\infty} \frac{\sin(z)}{k\sqrt{z^2-k^2x^2}}dz \end{aligned} \tag{4B.17}$$

Introducing [21, Eq. 11.4.37]:

$$t = \frac{z}{kx} \Rightarrow dt \bullet kx = dz$$

$$\therefore \mathrm{Int}_1 = \int_{t=1}^{\infty} \frac{\sin kxt}{k\sqrt{t^2-1}}dt = \frac{2}{\sqrt{\pi}\ \Gamma(1/2)} \bullet \frac{1}{k}J_0(kx)$$

Since

$$\Gamma\left(\frac{1}{2}\right) = \sqrt{\pi} \Rightarrow \mathrm{Int}_1 = \frac{2}{\pi} \bullet \frac{1}{k}J_0(kx) \tag{4B.18}$$

Now [21, Eq. 11.4.37]:

$$\begin{aligned} \mathrm{Int}_2 &= \int_{x=0}^{\infty} \frac{2}{\pi} \bullet \frac{1}{k}J_0(kx)\cos\left(kx \sin\theta_0\right)dx \\ &= \frac{2}{\pi k}\frac{1}{\sqrt{k^2-k^2\sin\theta_0}} = \frac{2}{\pi k^2\cos\theta_0} \end{aligned} \tag{4B.19}$$

$$\therefore D_{\text{norm}} = \frac{M_x^2 M_y^2}{\dfrac{M_x M_y}{d_x d_y} \bullet \dfrac{2}{\pi k^2 \cos\theta_0}} = M_x d_x M_y d_y \frac{2\pi}{\lambda^2} \cos\theta_0 \tag{4B.20}$$

Comparing expressions (4B.19) and (4B.13), the dependence on the number of elements is an extra factor as shown. Further, we may write in case that no GL are visible ($d_x < \lambda/2$ and $d_y < \lambda$),

$$\text{Int}_2 = \frac{2}{\pi k^2 \sqrt{1 - \sin^2\theta_0}} \tag{4B.21}$$

On the contrary, in the case of visible GLs, for the case of a non-dense array, where $d_x > \lambda/2$, we may use the undermentioned treatment to assess the effect on the directivity. The array factor in the xz-plane [1] may be written in the alternate form as:

$$\frac{\sin\left(\dfrac{N}{2}\psi\right)}{\sin\left(\dfrac{\psi}{2}\right)} \quad \text{where } \psi = kd_x \sin\theta + \beta; \tag{4B.22}$$

In (4B.21), $\beta = kd_x \sin\theta_0$ is the phase difference between the successive elements, and θ_0 is the beam-pointing or scan direction, θ_0. Now, the GL appear when

$$\frac{kd_x}{2}\left(\sin\theta - \sin\theta_0\right) = 0, \pm\pi, \pm 2\pi, \ldots \ldots, m\pi$$

$$\therefore \left(\sin\theta - \sin\theta_0\right) = 0, \pm\frac{\lambda}{d_x}, \pm\frac{2\lambda}{d_x}, \ldots \ldots, \pm\frac{m\lambda}{d_x}$$

The value in the above expression in parentheses is equal to zero, if $d_x < \lambda/2$, representing the main lobe (desired peak direction). The undesired grating lobes occur for

$$\left(\sin\theta - \left(\sin\theta_0 \pm \frac{m\lambda}{d_x}\right)\right) = 0 \tag{4B.23}$$

Using (4B.2) and (4B.22), (4B.21) becomes

$$\text{Int}_2 = \frac{2}{\pi k^2 \sum_m \sqrt{1 - \left(\dfrac{m\lambda}{d_x} - \sin\theta_0\right)^2}} \text{ when } d_y < \lambda \tag{4B.24}$$

In a similar manner, the dependency for $d_y > \lambda$ is derived, considering GLs for $\sin\theta_0 \pm m\lambda/d_x$ in the xz-plane and $\pm n\lambda/d_y$ in the yz-plane (no φ_o dependence, in which scanning is only in the xz-plane).

Now,

$$\text{Int}_2 = \frac{2}{\pi k^2 \sum_{m,n} \sqrt{1 - \left(\dfrac{m\lambda}{d_x} - \sin\theta_0\right)^2 - \left(\dfrac{n\lambda}{d_y}\right)^2}} \tag{4B.25}$$

that is, summation over all (m, n) values for which

$$1 - \left(\frac{m\lambda}{d_x} - \sin\theta_0\right)^2 - \left(\frac{n\lambda}{d_y}\right)^2 \geq 0$$

Hence, the expression of normalized directivity becomes

$$\therefore D_{\text{norm}} = \frac{D}{M_x M_y}$$

$$= \frac{1}{M_x M_y} \cdot \frac{\left(M_x M_y\right)^2}{\dfrac{M_x M_y}{d_x d_y} \dfrac{2}{\pi k^2 \left[1 - \left(\dfrac{\lambda}{d_x} m - \sin\theta_0\right)^2 - \left(\dfrac{\lambda}{d_y} n\right)^2\right]^{1/2}}} \tag{4B.26}$$

$$= \frac{2\pi d_x d_y}{\lambda^2 \sum_{m,n} \left[1 - \left(\dfrac{\lambda}{d_x} m - \sin\theta_0\right)^2 - \left(\dfrac{\lambda}{d_y} n\right)^2\right]^{-1/2}}$$

5

Synthesis of Linear Sparse Array Antennas

5.1 Introduction

Linear antenna arrays (LAAs) have numerous applications in many arenas such as direction of arrival (DoA) estimation [1, 2], massive MIMOs (M-MIMOs) [3, 4], instrument landing system (ILS) [5], identification friend or foe (IFF) antennas [6], and Internet of Things (IoT) [7]. They have highly directional patterns along one of the φ planes (azimuth plane) as already discussed in Chapter 3. Furthermore, any planar antenna array (PAA) can be assumed to be an amalgamation of multiple LAAs arranged along different φ-cuts [8]. Hence, the performance characterization of PAA along a φ_c-cut in azimuth plane may be equivalently analyzed by analyzing an equivalent collapsed LAA (CLAA) along the respective cut. The concept of CLAA will be elaborated in detail in Chapter 7 while studying PAAs. Thus, a study of the characteristics and synthesis techniques for LAA is very important to understand the behavior of PAAs as well. This chapter focuses upon the LAAs and their characteristics and finally discusses the sparse LAAs design techniques and their advantages over dense LAA.

5.2 LAAs

The primary specifications or qualifying parameters of an antenna array are gain or directivity, D_o, −3 dB or half-power beamwidth

(HPBW), and sidelobe level (SLL) envelope. D_o and HPBW are inversely proportional to each other, and both of these depend upon the electrical dimensions of the array aperture [9–11]. Typically, for a uniformly fed (isophoric) array, $\mathbf{D_o} \propto \mathbf{N_{tot}}$ and HPBW ~ $50\lambda/\mathbf{D_o}$ (where $\mathbf{N_{tot}}$ is the total number of elements in an array and λ is the operating wavelength), provided that the practical constraints such as mutual coupling (MC) and finite array edge effects [12–19] are ignored. Thus, $\mathbf{D_o}$ and HPBW are achieved by the choice of the aperture dimensions. The next challenging design task is achieving a low SLL envelope over the hemisphere. This is the prime focus of ongoing research in this specific area up to today, and many attempts have been made by using different techniques belonging to analytical, metaheuristic, and deterministic technique categories [9–11]. The lowering of SLL with uniform interelement spacing in LAA is achieved by employing various analytical amplitude tapering distributions such as Taylor, Chebyshev, Villeneuve, or Bayliss [11] or stochastic techniques such as genetic algorithms (GA) and particle swarm optimizer (PSO) [12–17]. The synthesized weight coefficients are fed to the individual elements in an array either using solely RF power feed networks or using RF feed networks and beam-steering networks (BSN) together such as in phased array antennas (PhAAs) [6, 14]. However, the SLL lowering is always accompanied with an increase in the HPBW, thereby leading to a loss in the angular resolution. The unequal power division (sometimes with a high power ratio) required as per the applied amplitude taper leads to the realization of the complicated RF feed networks' design as the transmission line dimensions are proportional to the line impedances and the frequency of operation [18]. In the case of digital beam-forming (DBF)-based PhAAs, the computational burden of the digital BSN system is further increased. The size and complexity increase proportionately in order with the increase in the size of array ($\propto\mathbf{N_{tot}}$). Further, the dense LAAs (DLAAs) suffer from grating lobes (GLs), if interelement spacing $d > \lambda/2$, and by scan blindness due to tighter MC, if $d \leq \lambda/2$. Hence, in either case, DLAAs have inherent operational constraints that need a careful redressal while designing these antennas.

Linear sparse array antennas (LSAAs) provide the solution to the above-cited issues by reducing $\mathbf{N_{tot}}$ and applying equal excitations to all of the array elements. LSAAs achieve HPBW values similar to the dense antenna arrays (DAA) within a reasonable agreement and at the same time much reduced SLL in comparison. LSAAs can be classified into two main categories: (1) thinned arrays, where a few elements are removed from the aperture by different array thinning techniques, and (2) employing nonuniform spacing between the array elements, thereby creating nonuniform or aperiodic LSAAs. Using either of these two categories, the average interelement spacing, d_{avg}, is made much greater than $\lambda/2$ to decrease MC and destroy the periodicity in the array grid. The disrupted periodicity alleviates the GLs' onset and reduced MC eliminates the scan blindness. This chapter dwells upon LSAAs and various efficient design techniques for these antennas.

5.2.1 A Brief History of the Synthesis of LSAAs

The idea of nonuniform arrays consisting of arbitrarily located elements was originated by Unz [19], and later King et al. [20] demonstrated a 21-element nonuniformly spaced LAA with d_{avg} of 4λ and SLL ~ –5.2 dB. Andreason [21] designed a nonuniform LAA with d_{avg} of 2λ and 3λ and demonstrated through several examples the advantages of these arrays over equivalent DAAs. Harrington [22] used a perturbation technique for lowering the SLL. He demonstrated that an SLL ~ $2/N_{tot}$ is achievable without altering HPBW significantly with significant SLL reduction in large arrays beyond 22 dB down from main lobe. Ishimaru [23] used Poisson's sum formula and defined a new function, the source position function, to design LSAAs. The original radiation pattern was converted into a series of integrals, each considered as a radiation from continuous source distribution. The amplitude and phase of these sources can lead to synthesis of unequal positions of the elements in LSAA. A use of combinatorial mathematics was shown by Leeper [24] to synthesize aperiodic antenna arrays in a noniterative manner using combined difference sets (CDS). A use of inversion algorithm and grid search-type inspection-based placement techniques to synthesize nonuniform LAAs is also shown by Kumar and Branner [25] and Dickhoff et al. [26]. A graphical procedure based on quasi-analytical null matching technique for the synthesis of linear LSAAs was shown by Toso et al. [27]. A deterministic technique for the synthesis of large aperiodic antenna arrays was proposed by Skolnik based on statistical density tapering (SDT) technique [28]. The use of metaheuristic techniques is also made extensively for the synthesis of LSAAs by various researchers [29–33]. Hooker and Arora [33] gave an optimal thinning level in LSAAs for different size using PSO. They have shown that 75% to 85% filled arrays may yield a minimum PSLL of –21.18 dB and –22.56 dB typically for the fill ratio of 81.1% and 76.1%, respectively. They have also shown that the 40% to 50% filled arrays might yield PSLL in the range of –15 to –12 dB. Recently, a few hybrid procedures utilizing stochastic and analytical methods in a combined manner were also published [34–39]. Exhaustive literature [37] is available based on the various techniques on the synthesis of LSAAs and readers are encouraged to explore these (see the summarized flowchart in Figure 5.1). The choice of a particular technique for the synthesis of arrays, whether deterministic, metaheuristic, or hybrid, depends upon the availability of the computational resources, the desired accuracy, and the desired size of an array (N_{tot} and physical aperture). Usually, the metaheuristic techniques become computationally expensive with the increase in the size of an array and may lead to suboptimal solutions. At the same time, the deterministic techniques are very helpful in synthesizing any size of an array and also in understanding the physics of their operation; however, these

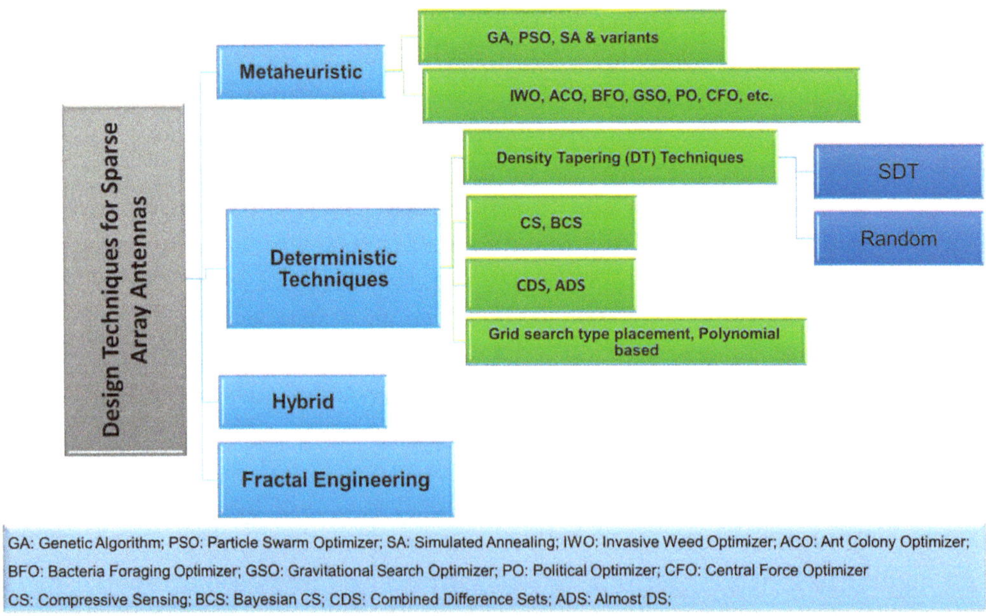

Figure 5.1
Overview of the synthesis techniques.

may not always give an optimal solution. For example, SDT or any probabilistic technique may give a different solution for a different number of iterations. Thus, numerous trials need to be run for such techniques to gain optimality by tailoring the statistical parameters to obtain a statistically optimal design. Few hybrid techniques have been proposed to take care of the accuracy, size, and attaining optimality [35–39]. This chapter will not dwell upon the elaborative details of these techniques, and readers are encouraged to refer to the mentioned literature. However, the author discusses the characteristics of sparse LAAs using few of the popular techniques selected from among these and using few novel techniques derived from his research work.

5.2.2 Sparse Versus Dense Linear Array Antennas

This section briefly discusses the comparison of DLSAAs and LSAAs before proceeding to a detailed discussion on the characteristics of LSAAs. LSAAs when compared to corresponding DAA counterparts have the following advantages:

1. In LSAAs, a significant increase in angular resolution (narrower HPBW) in contrast to amplitude tapered DAAs is achievable without introducing any GLs even with an increased average interelement spacing, d_{avg}.
2. In some instances, where transmitting arrays with maximum power and lowest possible SLL are required, LSAAs are potential candidates in comparison to the amplitude-tapered DAAs. LSAAs can control SLL lowering through spatial tapering with equally fed antenna elements instead of unequally fed elements as in DAAs. Thus, in LSAA, all constituent elements may operate with a maximum power output and SLL is lowered by the tailoring the element density distribution on the array aperture.
3. LSAAs are attractive for realizing conformal antenna arrays as the minimum spacing requirement of $\lambda/2$ is not required in these arrays and the antenna elements may be spread across the conformably shaped aperture.
4. LSAAs help in reducing size, weight, aperture, thermal, DC power, and cost (i.e., SWAP-C) resources, which is highly desirable in applications such as active phased array antennas [6].

5.3 Numerical Studies on LSAAs

This section gives a visualization on the various characteristics of LSAAs with the aid of a few numerical test cases. Very efficient designs of LSAAs are presented, clearly bringing out their advantages in comparison to dense counterparts.

5.3.1 An LSAA with 60 Grid Locations ($d = \lambda/2$)

This section presents the design of a thinned LSAA using metaheuristic techniques. A DAA with a uniform grid (antenna elements locations spaced 0.5λ apart) is considered. During the thinning process, the element locations are not varied (fixed); however, these elements are fed in a binary mode with 0 (low) and 1 (high) excitations. An LAA with 60 (= N_{tot}) locations is chosen for the study. Array thinning is carried out using a PSO-based synthesis algorithm [40–43] developed in the MATLAB environment. The number of iterations considered is 1,000 with a swarm population size of 150. A mask (shown as a red dotted line in Figure 5.2(a)) was defined as a fitness goal for the desired SLL envelope with upper and lower limits of –20 dB and −60 dB, excluding the main lobe region with approximated HPBW of a 30λ-long isophoric DAA. Figure 5.2(a) shows the radiation pattern of an LSAA (element gain, $D_{o,iso}$ = 3 dBi [9]) with radiation only in the upper hemisphere (i.e., half isotropic radiator, that is, element with a metallic back plane). Table 5.1 presents a comparative summary of the LSAA's performance with a DAA counterpart. The designed LSAA geometry is shown as a top inset in Figure 5.2(a).

The observations as inferred from the study are:

- N_{tot} is proportional to the SLL and SLL is almost independent of the aperture dimension.

Table 5.1
Comparison of Dense and Sparse LAAs

LAA Type	D_o (dBi)	HPBW (deg)	PSLL (dB)	ASLL (dB)	d_{avg} (λ)	A_{eff} (λ^2)	f (%)	L (λ)
Dense	20.8	1.6	−13.2	−19.5	0.5	1.65	100	30
Sparse	17.5	3.3	−18.8	−22.5	0.821	1.39	50	30

- The resolution (or HPBW) depends mainly on the aperture dimension.
- D_o is proportional to N_{tot} if d_{avg} is large. A large d_{avg} suggests that the total number of elements is considerably lesser than DAA.
- Starting with a fixed number of elements and aperture size, it is possible to improve the angular resolution by a factor of 10, 100, or more by spreading these elements over a larger aperture with little risk in obtaining a much higher sidelobe level and a lower directive gain using LSAAs.
- Figure 5.2(a) and Table 5.1 show that, while keeping the overall length fixed, a much lower SLL (i.e., PSLL < –18.8 dB) is obtained. A fill factor f of 53.3% (see (5.1)) together with a d_{avg} of 0.82λ is achieved showing a minimal effect of MC and a high sparsity in the LSAA. The decrease in the effective aperture area, A_{eff}, compared to DAA, defined by (5.2), validates the increase in HPBW and the reduction in D_o.

$$f = 1 - \frac{N_{sp}}{N_{tot}} \tag{5.1}$$

$$A_{eff} = \frac{D_o}{4\pi}\left(\lambda^2\right) \tag{5.2}$$

- Here, $D_o \propto N_{tot} + D_{o,iso}$, where N_{tot} is equal to N_{sp}, the number of active elements in the LSAA. In this case, N_{sp}= 30 and $D_{o,iso}$ = 3 dBi (directivity of isotropic radiator backed with a ground plane or half-isotropic radiator), leading to D_o = 17.7 dBi, a value closer to the one obtained with simulations based on (4.14) in Chapter 2, as shown in Table 5.1. Thus, an important characteristic of the LSAA is inferred that the directivity of the LSAA is proportional to the total number of elements in an array.
- The achieved PSLL is proportional to the fill factor, f.
- Further, it is observed that the elements' concentration near the center of the array is responsible for the lowering of SLL, Henceforth. the PSO algorithm tailors the elements' density and reduces their concentration from the array's center

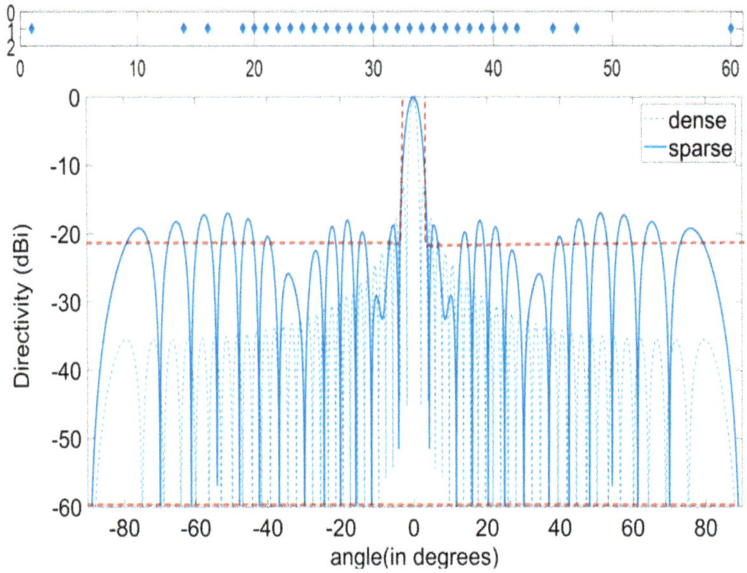

(a)

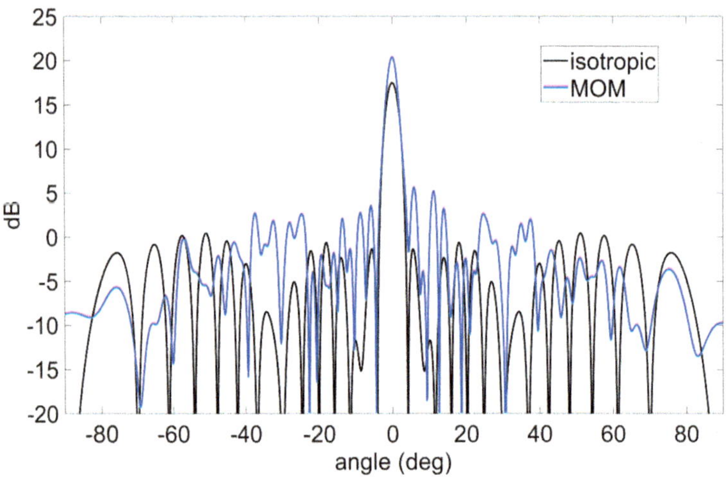

(b)

Figure 5.2

Radiation pattern of a 60-element LSAA: (a) MATLAB simulations, (b) FEKO simulations, (c) HFSS simulations, and (d) 3-D plot using HFSS.

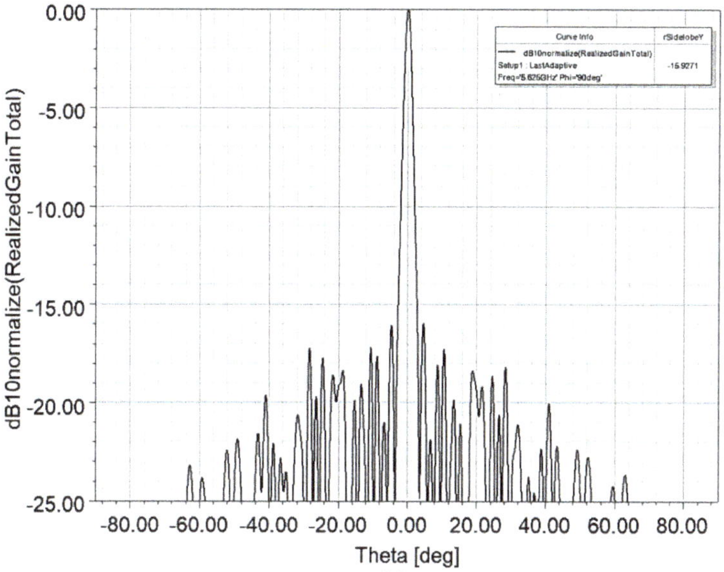

(c)

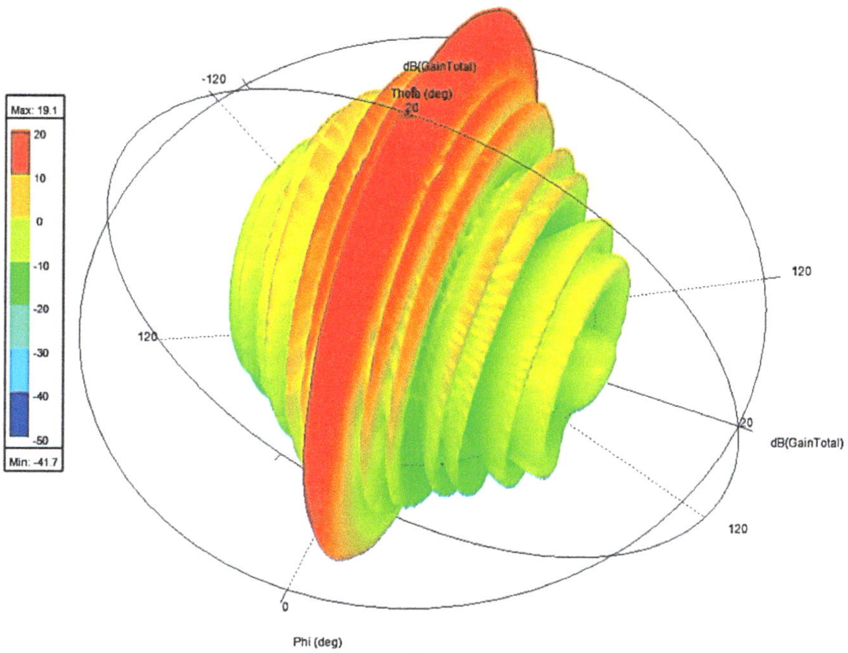

(d)

Figure 5.2
(Continued)

towards the edges (see inset in Figure 5.1(a)). However, it creates an appreciable MC in the central portion of an LSAA. Thus, the behavior of the elements in the center and near the array's edges is different. This problem is also moderated by increasing the d_{avg} as shown in the following case studies.

5.3.2 The Effect of Mutual Coupling and the Choice of the Array Element

Figure 5.2(a) shows the simulations carried out using MATLAB assuming a half-isotropic radiator. However, the choice of element type and the study of inherent MC between array elements are necessary for the realization of the LSAA practically. Figure 5.2(a) shows the simulated radiation pattern comparison of LSAA with dense counterpart. The inset in Figure 5.2(a) shows the LSAA geometry as optimized. A cavity-backed microstrip patch antenna (CB-MPA) element [44] (with gain~5–6 dBi) is chosen to model the LSAA. CB-MPA has nearly a 1-GHz bandwidth in C-band and has HPBW > 120°/90° in E- and H-planes, respectively. The LSAA is modeled using the method of moments (MoM)-based electromagnetic (em) solver, Altair FEKO [45]. Figure 5.2(b) shows the comparison of radiation pattern obtained using the FEKO solver compared with MATLAB simulations. The observed array parameters using FEKO are $D_o \sim 20.3$ dBi and HPBW ~ 3.34°. D_o is higher as compared to the MATLAB simulations due to the gain of CB-MPA (gain ~5–6 dBi) higher than 2–3 dB compared to $D_{o,iso}$. Further, a slight asymmetry in the SLL envelope while comparing PSLL on the right side (–14.98 dB) and the left side (–16.98 dB) is observed due to stronger MC in the center of the LSAA and reflections from finite array edges. Figure 5.2(c) shows the modeled performance as obtained from an FEM-based em solver, Ansys HFSS [46]. It shows a similar performance to FEKO with $D_o = 22$ dBi and a symmetric PSLL of –15.3 dB. Again, the effect of MC is observed in HFSS as well due to more elements in the middle of the array. In general, a good agreement between MATLAB and commercial em solvers is observed that validates the LSAA's design. Thus, in order to avoid large computation time involved in simulations using em solvers, (4.14) is preferred owing to its accuracy and will suffice and is used to study SAAs' characteristics with sufficient accuracy throughout this book.

5.3.3 The Relationship Between SLL and the Number of Elements

Table 5.2 shows a few simulations carried out for the LSAA with different size using PSO to derive a relationship between SLL and array size. In general, it is observed that:

- PSLL increase with increase in the sparsity, that is, a lower PSLL value is obtained for minimal sparsity as shown for a 100-element array, with $f = 80\%$, PSLL ~ –20.7 dB is achieved, but as the f decreases to 63, PSLL achieved is –16.7 dB. Hence, an optimality criterion is observed stating that a lowest PSLL is obtained if $f \geq 80\%$ and for heavy thinning, $f \leq 80\%$, a PSLL $< (2 - 5)/N_{tot}$ is achieved.
- PSLL lowers with an increase in the number of elements. It is observed that for a 500-element LSAA, a PSLL of –25.1 dB is achieved with $f \leq 55\%$ (i.e., proportional to $2 - 5/N_{tot}$).
- ASLL lowers as the number of elements increases (i.e., ASLL $\propto 1/N_{tot}$).
- The d_{avg} increases with the sparsity, thereby reducing the effects of MC.
- The results show that the LSAA is able to demonstrate a low SLL, reduction in N_{sp} while maintaining reasonable D_o and HPBW comparable to their DAA counterparts.

5.3.4 Relationship Between the Thinning Ratio (Fill Factor) and the Number of Elements

A 200-element LAA with $d = \lambda/2$ is considered a demonstrator to establish a relationship between f and N_{tot}. An iterative Fourier technique (IFT) is used to design the various designs with different fill

Table 5.2
LSAA Configurations

Array Grid Size	D_o (dBi)	HPBW (deg)	PSLL (dB)	ASLL (dB)	d_{avg} (λ)	A_{eff} (λ^2)	f (%)	L (λ)
60	17.5	3.3	−18.8	−22.5	0.821	1.39	50	30
100	22.0	1.0	−20.7	−26.4	0.626	1.75	80	50
100	21.4	1.0	−16.7	−21.4	0.798	1.67	63	50
500	27.4	0.2	−25.1	−28.9	0.904	2.18	55	250

factors [47, 48]. The IFT technique is computationally much faster as compared to PSO specifically for larger array size and, henceforth, is used to demonstrate the above-said relationship here. The IFT technique is based on the fact that an inverse discrete Fourier transform (DFT) relationship holds true between the array factor (AF) of an array and the complex excitations applied to the array elements. It follows an iterative procedure comprising an application for FFT and inverse FFT sequentially to obtain a suitable LSAA distribution comprising 1s and 0s to get the desired PSLL. Readers are encouraged to read [47] for details of the technique and the algorithm followed here for the simulations [47, 48] implemented here in MATLAB environment for the respective simulations.

Table 5.3 summarizes the various LSAAs' designs for different fill factors, f, for a 200-element LAA with the targeted PSLL < –20 dB using the IFT. It is clearly apparent that PSLL increases with f (i.e., in proportion to the leftover elements in an array). In comparison to isophoric DAA, there is more than a 3-dB decrease in D_o (computed using (4.14)), a 15% increase in HPBW, and a 3.3 to 9.6-dB improvement in PSLL, and, furthermore, there is more than 50% savings in the N_{TOT}. Thus, except for the decrease in D_o, there is a positive improvement in all other aspects of the LSAA. These factors lead to a reduction in the cost and power and increase the array efficiency. Further, the decrease in D_o is compensated with the decrease in the scan loss. This is shown in the following section.

5.3.5 Scan Performance

It is a well-known fact that that the DAAs suffer from the effects of MC leading to the scan blindness and GLs after certain scan angle

Table 5.3
LSAA Configurations Using IFT

N_{tot}	D_o (dBi)	HPBW (deg)	PSLL (dB)	ASLL (dB)	d_{avg} (λ)	A_{eff} (λ^2)	f (%)	L (λ)
200	26.0	0.4	−13.3	−35.0	0.5	2.07	100	100
200	23.8	0.6	−22.56	−28.5	0.794	1.89	60	95
200	24.2	0.6	−22.9	−28.7	0.759	1.92	66	100
200	22.9	0.8	−20.8	−27.6	0.665	1.82	49	65
200	22.9	0.52	−17.2	−22.6	1.02	1.82	49	100
200	21.9	0.6	−16.4	−22.2	1.29	1.74	39	100

due to the inherent periodicity of the array grid. The frequency-dependent array grid structure also leads to the issue of beam squint observed during the electronically scanning of phased array antennas. Thus, these factors lead to a limited frequency bandwidth (FBW) and limited scan volume. It has been shown here that LSAAs serve a riddance from these issues by destroying the periodicity in the array grid and increasing d_{avg}. The decrease in D_o in LSAAs due to a decrease in the number of elements gets compensated by the decrease in the scan loss with respect to the variation in the scan angle.

The diminishing effect of MC in LSAAs leads to a wide scan performance as is evident from Figure 5.3, which plots a scanned radiation pattern from boresight until 60° using an LSAA modeled with a grooved tapered slot antenna (GTSA) [49] using Ansys HFSS (shown as an inset in Figure 5.3). It has a wide HPBW (>120° and 90° in E- and H-planes, respectively) and D_o ~ 3–4 dBi. The dotted curve in Figure 5.3 shows the element pattern (EP) in the concerned φ-plane superimposed upon the array pattern. GTSA demonstrates a variation D_o within ±1 dB over ±60° scan volume. Thus, as per $\cos\theta$ law [50], AF demonstrates only 3–4-dB variation and, henceforth, in total, a 4–5-dB combined variation due to EP and AF is shown in D_o by a modeled LSAA. Figure 5.3 and Table 5.4 show

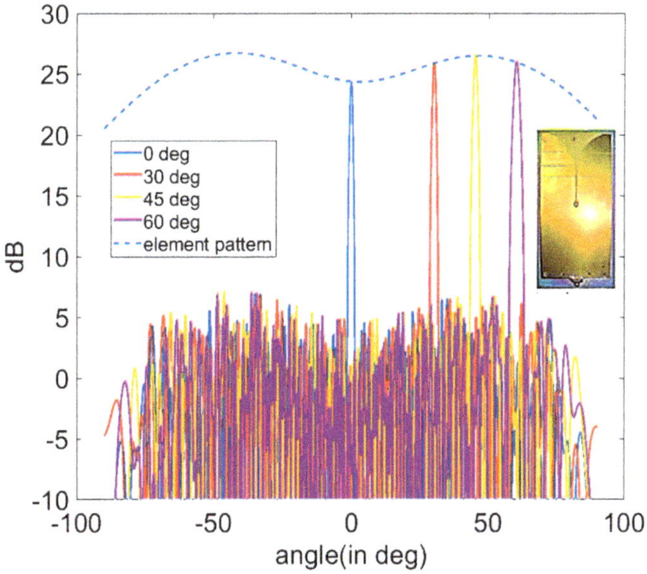

Figure 5.3
Radiation pattern of a 200-element LSAA with a superimposed element pattern.

Table 5.4
Scanned Performance of 200-Element LSAA with $f = 50\%$

Scan Angle (deg)	D_o (dBi)	HPBW (deg)	PSLL (dB)
0	24.32	1.08	17.4
30	25.82	1.25	18.9
45	26.46	1.53	19.5
60	25.9	2.15	18.9

the wide scan capability of the LSAA with 50% array thinning without occurrence of any GLs in the visible region. A variation in D_o, within 2.16 dB (maximum) over the scan volume with a maximum and minimum gain of 26.46 dB and 24.32 dB at 45° and 0°, respectively, is shown. HPBW varies in proportion to the cos rule and doubles at 60° scan angle and the variation in PSLL is within 1–2 dB. Thus, the proper choice of the elementary radiator is necessary to completely utilize the advantages of the sparse antenna arrays.

Henceforth, the inferences drawn from the numerical experiment are:

- LSAAs demonstrate a wide scan volume with minimal scan loss as compared to DAA. The advantage in terms of the lesser deterioration in D_o compensates the loss in D_o due to the decrease in the number of active (on) elements in the LSAA.
- A larger d_{avg} that minimizes the effect of the mutual coupling on the LSAA performance.
- PSLL is inversely proportional to the aperture length and the number of elements.
- The SLL envelope is almost constant with respect to the variation in the scan angle.
- D_o is directly proportional to the number of active elements.

Thus, these case studies using PSO and IFT bring out the salient features and advantages of LSAAs. The next section focuses upon LSAAs with nonuniform array spacing but with an equal number of elements as in DAA counterparts. A new deterministic synthesis technique called the array dilation technique (ADT) [51] that I innovated is presented to further demonstrate and establish the capabilities of nonuniform linear sparse array antennas (NU-LSAAs).

5.4 ADT for the Synthesis of Nonuniform LSAA

This section deals with another class of sparse LAAs (i.e., arrays with nonuniformly placed elements [28, 52]). These NU-LSAAs oblige a specific pattern of element concentration on the array aperture following the amplitude taper in DAAs to achieve SLL lowering with equally excited array elements. Thus, an isophoric LSAA with nonuniform spacing may be chosen to design LSAAs with higher d_{avg} (i.e., sparsity) and low SLL with comparable performance to the dense (uniformly placed elements) counterparts.

In order to lower the SLL envelope in a DAA, it is required to apply some kind of tapering (see Chapter 3 and [11]) to the continuous aperture current distribution. In isophoric SAAs, the tapered continuous aperture current can be approximated by a nonuniform element placement across the aperture. This technique is called density tapering and was shown initially by Skolnik [28] and later was improvised by many researchers [27, 52–56]. Thus, the elements' positions act as a degree of freedom for the synthesis. Equally fed elements help in the operation of amplifiers feeding these elements in the same optimal conditions, thus gaining power added efficiency (PAE) in phased array antennas. The idea behind the proposed ADT is to approximate a given positive continuous aperture current distribution (taken as a reference) with an isophoric array with nonuniformly placed elements.

5.4.1 Usage of ADT for the Array Synthesis

Figure 5.4 shows the LAA and AF for it is written as:

$$F(\theta) = \sum_{n=-N_n}^{N_n} I_n e^{jkx_n\Delta u} \tag{5.3}$$

Further, in (5.3),

$$\Delta u = \sin\theta - \sin\theta_s \tag{5.4}$$

$$x_n = x_{n-1} + d_n \tag{5.5}$$

where θ_s is the steering angle along the x-axis and x_n represent the element locations.

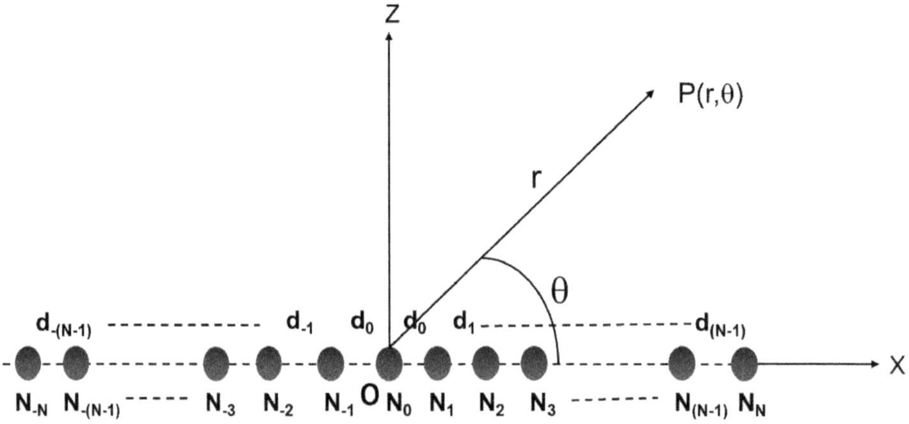

Figure 5.4
LAA geometry and the associated coordinate system. (© 2020 IEEE. Reprinted with permission from [51].)

Equation (4.15) can be rewritten in a simplified form as applicable to LAAs. It considers the effect of arbitrary array geometry (element locations), mutual coupling, scan angle, and complex excitations (amplitude and phase) on the array performance.

$$D_o(\theta_0) = \frac{\left(\sum_{m=1}^{K} I_m\right)^2}{\sum_{m,n=1}^{K} I_m I_n^* \cos\left[2\pi r_{nm} \sin\theta_0\right] \text{sinc}\left(k r_{nm}\right)} \tag{5.6}$$

In (5.6), $r_{nm} = (x_n - x_m)$ is the separation between the mth and n^{th} elements, where x_n and x_m are the respective locations (in terms of λ). Any nonuniformity in the separation manifests itself in the form of the variations in r_{nm}. Herein, the term sinc (kr_{nm}) is the coupling term and $I_m = I_n = 1$ for isophoric arrays. Here, $x_n = x_m$ represents self-coupling and $x_n \neq x_m$ represents mutual coupling terms, respectively. It is observed from the denominator that in the case of a uniform placement of antennas (i.e., $d_n \sim \lambda/2$ $(n = 1, 2, \ldots)$), it leads to sinc $(kr_{nm}) \sim 1$ and$D_o \rightarrow K$ (i.e., no scan dependence). However, if $d_n \neq n\lambda/2$, it leads to the variation in D_o with respect to a scan angle [57]. It was shown in Chapter 4 that uniform LSAAs exhibit very small variation in D_o (<1 dB) for $d_n \geq 0.7\lambda$, due to the decrease in the MC (negligible mag(sinc (·))). The term uniform implies that all of the array elements are spaced equally ($d_n = d_{no}$; d_{no} is fixed spacing in array grid) and the term "sparse" implies that $d_{no} > 0.5\lambda$. However, the uniform array grid, even in SAAs, leads to the onset of GLs on scanning, thereby limiting the scanning range. GLs may be alleviated by aperiodicity in element placement in an array by varying d_n between a pair of elements.

ADT stretches the array in analogy to an elastic membrane, henceforth creating an aperiodicity in the array. ADT considers uniformly fed and uniformly spaced ($d_n = d_{no}$) DAA grid as the base and applies the analytical expression (5.7) given below. ADT dilates DAA in an analogy to an elastic membrane with the different amount of dilation (or stretching) applied to its different parts, following (5.7).

$$d_n = d_{no}\left\{1 + \frac{\alpha_J}{K}\left(|n| - 1\right)\right\} \tag{5.7}$$

In (5.7), $n = 1, 2, \ldots, N_n$ and $d_{no} = m\lambda$ ($m = 0.5, 1, 2, \ldots$). Here, α_J is the dilation parameter and J represents the index of the section subjected to the dilation factor, α_J. The value of α_J thus decides the dilation in the interelement separation, d_n, nonuniformly in that particular region varies from 'd_{no}' to '$\alpha_J d_{no}$'. The value of α_J as chosen varies only from 0 to 4. However, $\alpha_J > 4$ is possible to choose, but will result in an excessive stretching of LSAAs leading to the large variations in the angular resolution with respect to DAA and high far SLL, which is not desirable. Further, the arrays are considered symmetrical, which saves computation time as only one-half of the array needs to be studied. One-half of the array is partitioned into four regions, although the number of regions may be increased further in the case of very large arrays (more than thousands of elements). Thus, herein totally 120 combinations of α_J are studied and presented to assess the array performance. Here, $\alpha_J = [0\ \ 0\ \ 0\ \ 0]$ represents a uniform isophoric DAA with interelement spacing, d_{no}. A simple code written in MATLAB executed on an Intel Corei7 processor with 8-GB RAM analyzes the combinations of α_J. The array performance monitors (APMs), namely, D_o, HPBW, first SLL, PSLL, ASLL, and d_{avg}, are computed and stored corresponding to each of the combinations of α_J.

5.4.2 Numerical Case Studies

This section presents a few design case studies to illustrate the ease of applicability of ADT and bring out the various important features of nonuniform LSAAs.

5.4.2.1 A 17-Element LSAA

A symmetric DAA with uniform grid, $d_{no} = 0.5\lambda$, is chosen and its one-half is divided into four regions to apply the dilation, α_J. These regions, namely, Region I, II, III, and IV, consists of 3, 2, 2, and 1 elements, respectively. The various combinations of α_J are studied and APMs computed at boresight and tabulated in Table 5.5. Further, $D_o = 15.3$ dBi is computed using (5.6), varying from 15.3 dBi (DAA) to 16.6 dBi (maximum), showing a variation of 1.3 dB, due to variation in the overall length of the array and d_n.

The design guidelines for the synthesis of LSAAs as inferred from Table 5.5 may be utilized for designing LSAA with low SLL using ADT, while realizing APMs close to the DAA counterpart. These are summarized here:

- The combination $\alpha_J = [0\ \ 0\ \ 0\ \ 0]$ is the same as DAA showing a PSLL = –13.28 dB and $D_o = 15.3$ dBi.
- The dilation of the outermost regions, Region III and IV, lowers the SLL envelope in the region near to the main beam of the array factor (shown by the first SLL). However, it leads to the rise in the SLL region away from the main beam (shown by PSLL). The near SLL (NSLL) region denotes 0° to 20°–30° away from the main lobe and the far SLL (FSLL) region denotes the region greater than 30° away from the main beam.
- The dilation of the middle region(s), Regions I and II, controls SLL in the far angular region.
- The PSLL degrades with the increase in the dilation of the array regions. However, the dilation leads to higher sparsity, η_{sp} ($d_{avg} >> 0.5\lambda$), in the array (i.e., increasing d_{avg}). Thus, PSLL and d_{avg} are inversely proportional to each other.
- An excessive stretching may lead to a finer angular resolution (smaller HPBW) due to an increase in the array's physical extent. HPBW varied from 5.8° (UF-USDA) to 4.0° (31% increase) for the maximum dilation corresponding to $\alpha_J = [4\ \ 4\ \ 4\ \ 4]$.

- ASLL is better than a statistical estimate of $1/K \sim -12.3$ dB [28], which is demonstrated in general for all combinations using ADT.
- The minimal PSLL is obtained for α_J [1 2 3 4], with PSLL = –21.02 dB; ASLL = –28.33 dB; and $d_{avg} = 0.66\lambda$ and η_{sp} (compared to DAA) = ((0.66 – 0.5)/0.5) • 100 = 32%, agreeing with [33].
- The highest sparsity, $d_{avg} = 0.82\lambda$, is shown for α_J = [4 4 4 4], demonstrating a PSLL = –14.65 dB and SLL < –16.59 dB in NSLL region, much better than statistical estimates of PSLL for sparse arrays given as $(2–5)/K \sim -9.2$ dB to –4.5 dB [28, 33]. This combination shows η_{sp} = ((0.82 – 0.5)/0.5) • 100 = 64%, much higher with low SLL compared to the observations given in [33] (i.e., –12 to –15 dB for 40% to 50% filled arrays).

ADT results are validated with a few similar designs presented in earlier works by other researchers [25, 35, 58]. They utilized the Legendre transformation (LT) [25], weighted density (WD), and two-step approach (TSA) [35], and improved genetic algorithm (IGA) [58]. Table 5.6 presents the summary of the comparison of the results obtained using these with those obtained using ADT.

ADT demonstrates comparatively lower PSLL ~ –21.02 dB for α_J [1 2 3 4] and $d_{no} = 0.5$, which is better compared to the references by the amounts shown within brackets in Table 5.6. The element locations (in wavelengths) are [–5.286, –4.235, –3.426, –2.676, –2.088, –1.529, –1, –0.5, 0, 0.5, 1, 1.529, 2.088, 2.676, 3.426, 4.235, 5.286]. The references do not present directive computations and any information on the sparsity. Here, the directive computations are done using (5.6) accounting for the mutual coupling effects assuming that the array consist of an isotropic radiator. The sidelobe envelope is studied in details in terms of near and far sidelobe levels. It is observed that an almost constant SLL envelope is obtained with ASLL < –28 dB. A medium sparsity is achieved that indicates that the mutual coupling is not totally ignorable but definitely less effective. Further, another case of $d_{no} = 0.4\lambda$ is considered that yielded even lower PSLL of –23.5 dB but at the cost of lesser $d_{avg} = 0.53\lambda$. Thus, ADT is an effective, direct, and deterministic synthesis technique that requires no iterations and can give low SLL with higher sparsity. The choice and effect of a practical radiator are

Table 5.5
Array Performance Parameters for 17-Element LSAA at Boresight

$[\alpha_1\alpha_2\alpha_3\alpha_4]$	D_o (dBi)	HPBW (deg)	PSLL (dB)	First SLL (dB)	ASLL (dB)	d_{avg} (λ)	L (λ)
0 0 0 0	15.3	5.8	−13.28	−13.28	−30.14	0.5	8.5
0 1 1 1	15.3	5.6	−16.52	−16.52	−31.40	0.56	9.34
0 2 2 2	15.9	5.2	−19.68	−21.3	−30.1	0.62	10.54
0 3 3 3	16.1	5.0	−17.70	−23.17	−26.38	0.67	11.39
0 4 4 4	15.5	4.6	−13.74	−20.61	−23.29	0.74	12.58
1 0 0 0	15.5	5.6	−13.59	−13.59	−29.35	0.52	8.84
2 0 0 0	15.6	5.4	−13.89	−13.89	−28.56	0.54	9.18
3 0 0 0	15.7	5.0	−13.05	−14.16	−27.79	0.57	9.69
4 0 0 0	15.7	4.8	−12.05	−14.42	−26.81	0.58	9.86
1 1 1 1	15.9	5.2	−16.66	−16.66	−31.24	0.58	9.86
1 2 2 2	16.1	5.0	−21.01	−21.12	−30.48	0.64	10.88
1 3 3 3	15.7	4.8	−18.43	−21.06	−26.12	0.69	11.73
1 4 4 4	15.6	4.6	−13.85	−19.61	−23.92	0.75	12.75
2 1 1 1	16.0	5.0	−16.80	−16.81	−31.00	0.60	10.20
2 2 2 2	16.3	4.8	−18.65	−20.97	−30.30	0.66	11.22
2 3 3 3	15.8	4.6	−19.22	−28.40	−26.61	0.72	12.24
2 4 4 4	15.7	4.4	−14.68	−18.58	−23.94	0.77	13.09
3 1 1 1	16.1	4.8	−15.15	−16.94	−30.16	0.62	10.54
3 2 2 2	16.4	4.6	−16.84	−20.85	−29.59	0.69	11.73
3 3 3 3	16.1	4.4	−17.68	−27.49	−26.84	0.74	12.58
3 4 4 4	15.7	4.2	−14.48	−17.55	−23.65	0.79	13.43
4 1 1 1	16.4	4.6	−13.84	−17.07	−29.73	0.64	10.88
4 2 2 2	16.6	4.4	−15.39	−20.75	−26.56	0.70	11.9
4 3 3 3	16.4	4.2	−16.35	−26.76	−26.34	0.76	12.92
4 4 4 4	15.7	4.0	−14.65	−16.59	−24.00	0.82	13.94
0 1 2 3	15.7	5.4	−17.77	−17.77	−30.94	0.58	9.86
1 1 3 4	15.7	5.2	−19.25	−19.25	−28.07	0.63	10.71
1 1 2 3	15.9	5.2	−17.84	−17.84	−30.32	0.61	10.37
1 2 3 4	15.9	5.0	−21.02	−23.11	−28.33	0.66	11.22

Table 5.6
Comparative Summary of 17-Element LSAA at 0°

Case	D_o (dBi)	HPBW (deg)	PSLL (dB)	First SLL (dB)	ASLL (dB)	d_{avg} (λ)
LT [25]	—	—	−19.50 (+1.52)	—	—	—
WD [35]	—	—	−19.55 (+1.47)	—	—	—
TSA [35]	—	—	−20.59 (+0.43)	—	—	—
IGA [58]	—	—	−20.30 (+0.72)	—	—	—
ADT (d_{no} = 0.5λ)	15.3	5	−21.02	−23.11	−28.33	0.66
ADT (d_{no} = 0.4λ)	15.3	6.2	−23.15	−23.15	−30.63	0.53

studied next to analyze the feasibility of the design obtained using ADT for realization.

5.4.2.2 A 17-Element Linear Cavity-Backed Microstrip Antenna Array

The previous section presented the design of a 17-element LSAA assuming an isotropic radiator, while considering the effect of the mutual coupling in (4.15). It further validates the design with the existing literature. However, there is no practical validation of the design. Nowadays, the commercial electromagnetic (em) simulators based on full-wave numerical techniques [59] such as MoM, finite element method (FEM), and finite-difference time-domain (FDTD) method are available that gives the results very close to the fabricated versions of the same. In order to save the fabrication challenges and save the time, the cost of the material, and testing, a judicious use of these em solvers can be made for validating the feasibility of the designs. This book makes use of two commercial em solvers, MoM-based Altair FEKO [45] and FEM-based Ansys HFSS [46], which have a proven legacy of good agreement with the measurements [44, 49].

A CB-MPA has been chosen as a demonstrator for the validation of the design. A CB-MPA [44, 60] is known for its wide bandwidth performance due to features such as surface-wave suppression and extended height of the substrate. A 17-element LSAA is realized using CB-MPA in Ansys HFSS based on the design presented previously (see Figure 5.5).

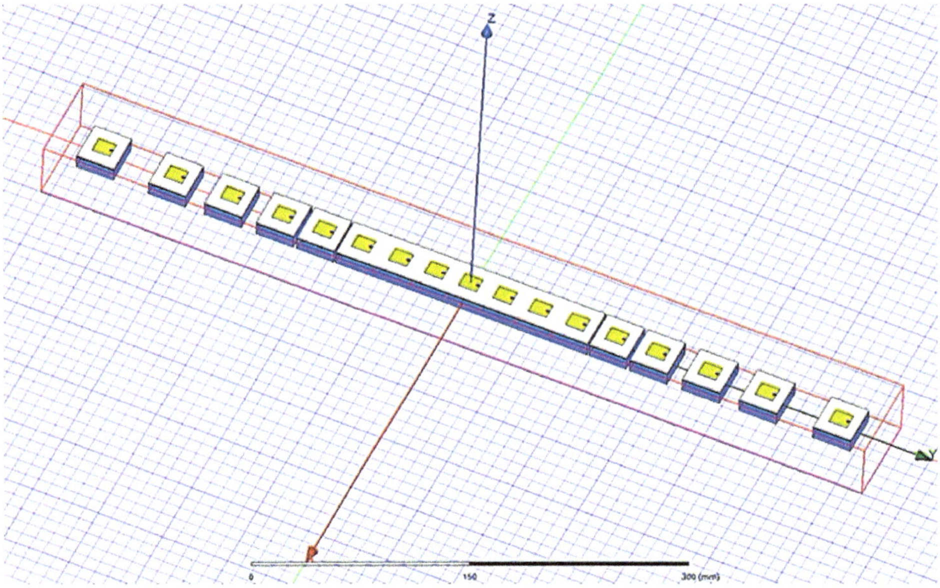

Figure 5.5
HFSS model of a 17-element CBMPA-based LSAA.

Figure 5.6 shows the comparison of the radiation pattern (in decibels) obtained using MATLAB code with the simulated results obtained from the HFSS model. It shows three respective traces, namely:

- *CB-MPA (finite):* This plot shows results from HFSS simulations considering the actual 3-D full-wave model. It considers the effect of the element pattern, mutual coupling, and the finite boundaries of the model.
- *CB-MPA (infinite):* It considers the single element radiation pattern obtained from HFSS simulations multiplied it to the array factor obtained from ADT code to get the radiation pattern. It considers the MC effect but ignores the finite boundaries of the array. The effect of embedded element pattern (EEP) is included in gain as [48]:

$$G_a(\theta,\varphi) = \left(1-\left|\Gamma_a\right|^2\right)\varepsilon_{\text{loss}} D_o(\theta,\varphi) \tag{5.8}$$

- *Isotropic (infinite):* It presents the MATLAB simulations assuming an isotropic radiator.

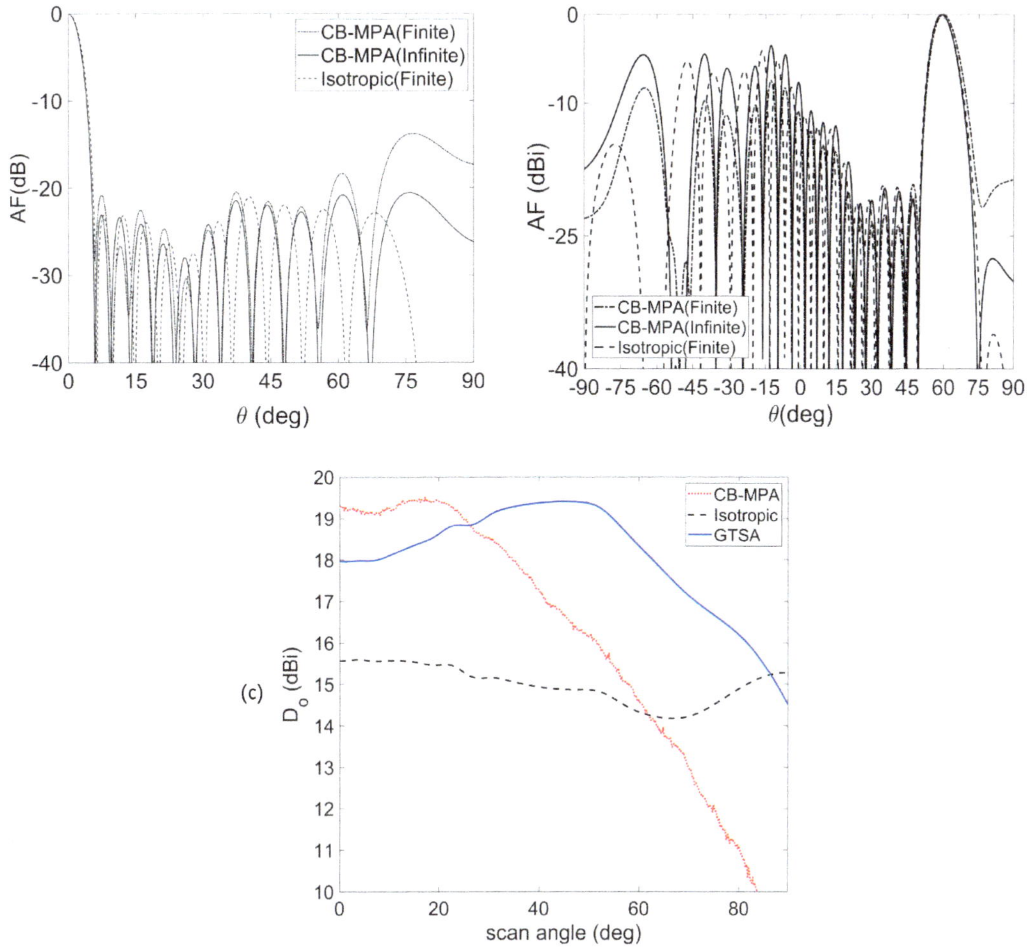

Figure 5.6
A 17-element LSAA performance. (a) At boresight. (b) At 60° scan angle. (c) Directivity versus scan angle. (© 2020 IEEE. Reprinted with permission from [51].)

It is clearly apparent from Figure 5.6(a, b) that CB-MPA (infinite) and isotropic (infinite) are in very good agreement. It justifies the choice of an isotropic radiator for the simulations. Further, both of these simulations are in agreement with the em simulations, CB-MPA (finite), within 2–3 dB over an angular extent up to 60° from the main beam. There is a discrepancy of more than 5 dB observed beyond a 70° scan angle. This can be attributed to the MC dominance in the central portion of array and the finite edges of the LSAA. These factors led to the rise of the GLs smeared over an angular extent. The scan performance is also good in the NSLL from the main beam up to 60° but shows an increase in the SLL in the far angular extent again due to smearing of grating lobes. Figure 5.6(c) shows the variation of the directivity with respect to the scan angle assuming isotropic and CB-MPA radiating elements. The variation in CB-MPA case is 4.7 dB up to 60°, solely due to the angular variation of the element pattern. The variation in the case of the isotropic element is only 1.2 dB up to 60°. HPBW varies from 5° to 10.5° from 0° to 60° following cosθ behavior. Thus, a wide scan performance is visible and it is suggested that the choice of radiating element is critical in ensuring wide scan volume. Figure 5.6(c) also shows the variation in directivity if another element GTSA [49] is chosen. In this case, the variation is only 2.5 dB up to 70°, showing an even wider scan performance. In general, the design is in good agreement with the full-wave FEM-based simulations. This validates the array dilation technique and its applicability in designing sparse antenna arrays.

The points to remember are that the correct choice of antenna element is necessary to make your design practically realizable.

5.4.2.3 A 37-Element LSAA

Another case study of a 37-element LSAA with low SLL and no restriction on the array length is considered (choosing $d_{no} = 0.5\lambda$). An array half is divided into four regions, I, II, III, and IV consisting of 8, 5, 4, and 1 elements and ADT is applied. Table 5.7 summarizes the computed APMs against the various combination of α_J as considered. The results conform to the design guidelines/observations presented previously. The lowest PSLL of –20.72 dB is obtained for $\alpha_J = [1\ \ 1\ \ 2\ \ 3]$ with $d_{avg} = 0.72\lambda$. Table 5.8 compares the APMs obtained from ADT with the results shown by LT [25], WD and TSA [35], and IGA [58]. ADT achieves PSLL = –23.27 dB for $\alpha_J = [0\ \ 1\ \ 2\ \ 3]$ (four regions consist of 7, 5, 3, and 3 elements) and is better by the amounts shown within brackets. Figure 5.7 shows the radiation pattern plots at 0° (left plot) and 60° (right plot) scan angles including the effect of the isotropic element pattern. A wide scan performance is evident from the figure. A low SLL envelope is maintained through the complete angular scan volume.

In these two case studies, it is observed that by not restricting the length of the array, similar angular resolution (HPBW), and lower

Table 5.7
A 37-Element Linear Array Optimization (d_{no} = 0.5λ) and APMs at 0°

Array Geometry $[\alpha_1\alpha_2\alpha_3\alpha_4]$	D_o (dBi)	HPBW (deg)	PSLL (dB)	First SLL (dB)	ASLL (dB)	d_{avg} (λ)
0 0 0 0	18.7	2.6	−13.27	−13.27	−34.38	0.5
0 1 1 1	19.4	2.2	−17.37	−17.37	−35.88	0.61
0 2 2 2	19.8	2.0	−17.44	−22.72	−34.48	0.71
0 3 3 3	19.2	1.8	−16.28	−16.28	−28.82	0.81
0 4 4 4	18.9	1.6	−14.84	−15.23	−26.7	0.91
1 0 0 0	18.7	2.6	−12.99	−12.99	−34.33	0.51
2 0 0 0	18.7	2.6	−12.74	−12.74	−34.25	0.51
3 0 0 0	18.8	2.6	−12.48	−12.48	−34.14	0.51
4 0 0 0	18.8	2.6	−12.26	−12.26	−34.01	0.52
1 1 1 1	19.5	2.2	−16.94	−16.94	−35.89	0.61
1 2 2 2	19.9	2.2	−17.66	−17.66	−34.51	0.72
1 3 3 3	19.2	1.8	−16.51	−30.13	−28.89	0.82
1 4 4 4	18.9	1.6	−14.74	−15.44	−26.68	0.92
2 1 1 1	19.5	2.2	−16.52	−16.52	−35.85	0.61
2 2 2 2	19.9	2.0	−17.87	−21.27	−34.46	0.72
2 3 3 3	19.2	1.8	−16.89	−16.89	−28.79	0.82
2 4 4 4	19.0	1.6	−14.74	−15.65	−26.64	0.92
3 1 1 1	19.5	2.2	−16.15	−16.15	−35.77	0.62
3 2 2 2	19.9	2.0	−18.06	−20.65	−34.59	0.72
3 3 3 3	19.3	1.8	−16.96	−27.24	−28.86	0.83
3 4 4 4	19.0	1.6	−14.83	−15.94	−26.66	0.93
4 1 1 1	19.6	2.2	−15.81	−15.81	−36.66	0.62
4 2 2 2	20.0	2.0	−18.23	−20.09	−34.35	0.72
4 3 3 3	19.3	1.8	−17.17	−26.14	−28.77	0.83
4 4 4 4	19.1	1.6	−14.99	−16.08	−26.79	0.94
0 1 2 3	19.2	2.0	−20.91	−27.37	−30.36	0.72
1 1 3 4	19.1	2.0	−17.13	−20.06	−28.14	0.80
1 1 2 3	19.3	2.0	−20.72	−26.05	−30.52	0.72
1 2 3 4	19.2	1.8	−17.57	−18.90	−28.49	0.83

Table 5.8
Comparative Summary of 37-Element Linear Array Performance at 0°

Case	D_o (dBi)	HPBW (deg)	PSLL (dB)	First SLL (dB)	ASLL (dB)	d_{avg} (λ)
LT [25]	—	—	−19.37 (+3.9)	—	—	—
WD [35]	—	—	−21.02 (+2.3)	—	—	—
TSA [35]	—	—	−22.68 (+0.6)	—	—	—
IGA [58]	—	—	−22.4 (+0.8)	—	—	—
ADT (d_{no} = 0.5λ)	19.2	2.2	−23.27	−23.57	−31.5	0.67

SLL performance is achievable with a higher sparsity. The effect of MC is also reduced resulting into a wider scan performance. Next, the effect of restricting the length of the LSAA same as the denser counterpart is studied.

5.4.2.4 LSAA with Fixed Aperture and Minimum Spacing

The case studies consider the synthesis of 17 and 37-element LSAA with restricted aperture size and minimum d_{no}. The edge elements' location is fixed at 4.872λ and 10.998λ, respectively. The simulations carried out using ADT are compared with similar geometries obtained using TSA [35], modified real GA (MRGA) [30], and differential evolution (DE) [31]. The results are summarized in Table 5.9.

The ADT simulations demonstrate a PSLL of −19.95 dB and −21.2 dB for 17 and 37 elements, respectively. The four regions considered consist of {3, 2, 2, 1} and {7, 4, 1, 6} elements and α_J = [0 1 2 3]

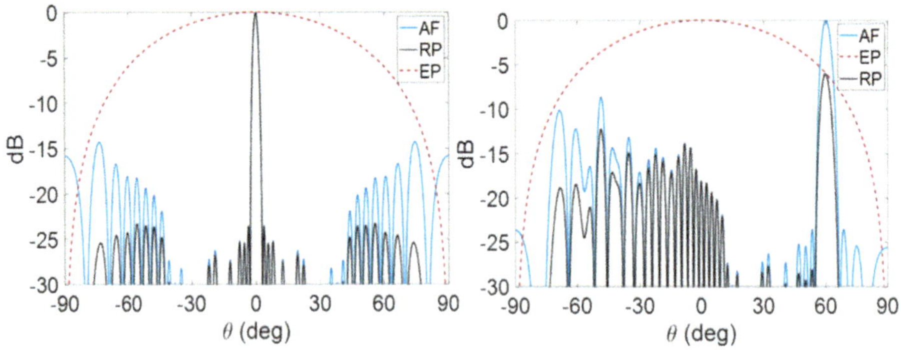

Figure 5.7
A 37-element LSAA performance (AF: array factor, RP: radiation pattern, EP: element pattern).

Table 5.9
A 37-Element Array Performance (Fixed Length and Spacing) at 0°

Case_17 Elements	PSLL (dB)	L (λ)	37 Elements PSLL (dB)	L (λ)
MRGA [30]	−19.79 (+0.2)	9.744	−20.562 (+0.6)	21.996
DE [31]	−19.89 (+0.1)	9.744	−20.942 (+0.3)	21.996
TSA [35]	−19.89 (+0.1)	9.744	−21.025 (+0.2)	21.996
ADT	−19.95	9.744	−21.2	21.996

and [0 2 3 4] in 17 and 37-element LSAAs, respectively. Table 5.10 summarizes APMs for 17 and 37-element A-LSAAs with $\eta_{sp} \sim 24\%$ and 36%, respectively.

5.4.3 Oversized Linear Sparse Array Antennas

In general, it is observed that the results obtained from metaheuristic and deterministic techniques are comparable and within acceptable limits. However, the real advantage of the deterministic techniques is realized in the case of very large arrays (VLAs) where the elements are thousands or more in quantity. In such cases, the heuristic techniques may either require huge number of iterations leading to unimaginable computation time or they may converge prematurely in the case of limited iterations. However, the deterministic techniques, being noniterative in nature, provide a quick solution that may be optimal as well. In case it is suboptimal, it can provide a cue to other numerical techniques for fast convergence to an optimal solution. This section presents a few designs of an oversized LSAA, bringing out the advantages of ADT as it does not require any huge CPU time and resources. Here, two designs are chosen, one with restricted minimum interelement separation, d_n,

Table 5.10
Summary of Array Performance (Fixed Length and Spacing) at 0°

K	D_o (dBi)	HPBW (deg)	PSLL (dB)	First SLL (dB)	ASLL (dB)	d_{avg} (λ)	L (λ)
17	15.8	5.2	−19.95	−27.37	−30.5	0.62	9.744
37	18.7	2.4	−21.2	−22.5	−28.8	0.68	21.996

and another with a restricted aperture size and d_n. Two designs, one with only restricted minimum d_n and another with both restricted aperture and d_n, are presented. A DAA with $d_{no} = 0.5\lambda$ and $1{,}000\lambda$ in length, consisting of 2,000 isotropic elements, is considered as an array demonstrator.

5.4.3.1 A 2,000-Element LSAA Demonstrator (Restricted Minimum d_n)

The application of ADT to DAA yielded a PSLL ~ −37.5 dB at 0° and −26.5 dB at 60° with $d_{avg} = 0.67\lambda$ for $\alpha_J = [1\ \ 1\ \ 1\ \ 2]$. The four regions, I, II, III, and IV, were assumed to be consisting of {285, 250, 250, 215} elements. The dilated length is $1{,}344.2\lambda$, extended by 34.4%. Table 5.11 summarizes the APMs and Figure 5.8(a) shows the simulated AF at 0° and 60° scan angles, and Figure 5.8(b) shows the variation in d_n across the four regions. It is observed that, due to the aperiodicity in an array, the grating lobes gets smeared over a wide angular extent. The PSLL observed on scanning at far angular region can be further minimized by smoothing the transition in d_n possibly by increasing the number of regions beyond 4 and applying different α_J. However, that may lead to a bit of increase in the number of combinations to be tried, but still it will be computationally faster than other techniques. The process can be made iterative by using the weighted least mean square error (WLMSE) criterion and comparing it with a desired pattern to achieve optimality.

5.4.3.2 A 2,000-Element LSAA with Restricted Aperture Length and Minimum d_n

The length is restricted to 999.5λ and the minimum $d_n = 0.5\lambda$ is chosen before applying ADT. Table 5.12 lists the LSAA's APMs for different combinations of α_j and number of elements in four regions obtained using ADT. The results are compared with the similar geometry designed by Dai et al. [36], using DE and Levy flight (DELF) technique, for LSAA with 1,540 elements having length of

Table 5.11
Oversized LSAA Performance (Fixed Spacing)

K	D_o (dBi)	HPBW (deg)	PSLL (dB)	First SLL (dB)	ASLL (dB)	d_{avg} (λ)	L (λ)
2,000	36.0	0.02	−37.5	−37.5	−62.9	0.67	9.744

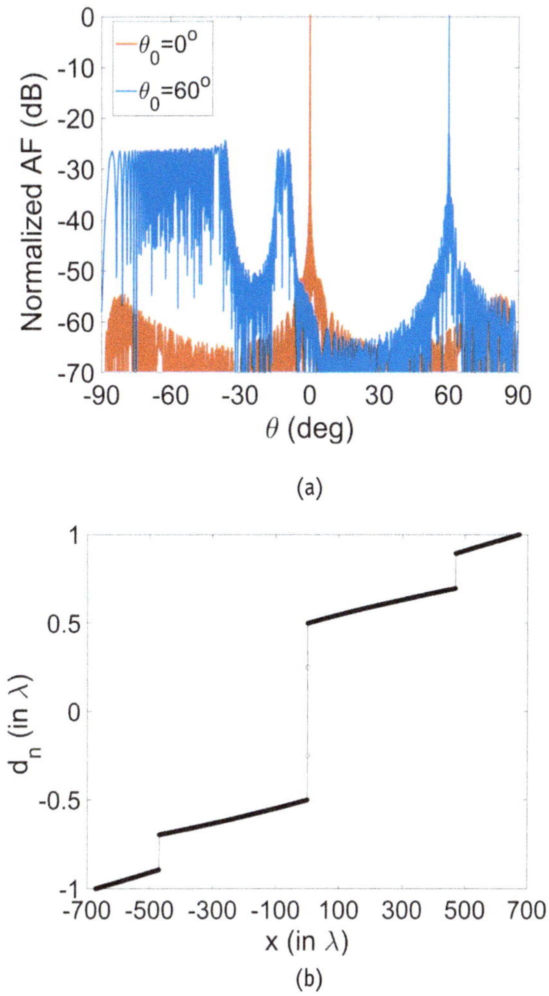

Figure 5.8
A 2,000-element LSAA ($d_{no} = 0.5\lambda$). (a) AF at 0° and 60° scan angles. (b) Plot of interelement spacing between array elements. (© 2020 IEEE. Reprinted with permission from [51].)

999.5λ and $d_{no} = 0.5\lambda$. It is observed that all of the combinations gives PSLL better than DELF with a higher η_{sp}. Only combination [1 1 1 4] yielded same PSLL but still it has shown a higher η_{sp}. Figures 5.9(a, b) shows the AF plot and variation of d_n with [1 1 1 2] combination yielding lowest PSLL. Figure 5.9(c) presents variation of D_o with the scan angle showing wide scan variation (3-dB variation up to 75°).

Table 5.12
Oversized LSAA Performance (Fixed Length and Spacing)

α_j	Regions [I II III IV]	D_o (dBi)	PSLL (dB)	ASLL (dB)	d_{avg} (λ)	η_{sp} (%)
[1 1 1 2]	[250 250 250 250]	35.1	−35.7	−60.6	0.62	18.8
[1 1 1 3]	[222 250 250 278]	35.0	−33.3	−53.2	0.63	20.8
[1 1 1 4]	[222 250 250 278]	34.9	−33.0	−54.4	0.64	21.8
[1 1 2 4]	[222 250 250 278]	34.8	−32.7	−55.2	0.66	24.8
[1 2 2 4]	[200 250 250 300]	34.6	−34.3	−56.0	0.68	27.3
DELF [36]	—	—	−33.0	—	—	23

5.4.4 Highly Sparse Linear Antenna Array ($d_{no} = 1\lambda$)

It was shown in Chapter 4 that an isophoric uniformly spaced LSAA with $d_{no} > 0.7\lambda$ demonstrates wide scan behavior due to reduction in the mutual coupling. However, large d_{no} leads to onset of GLs. ADT solves the problem by introducing the aperiodicity in d_n. A 37-element DAA with $d_{no} = 1\lambda$ (highly sparse) is considered as a demonstrator. The dotted black traces in Figures 5.10(a, b) plots AF at 0° and 60° scan showing the onset of GL.

The ADT simulations gives PSLL of –12.05 dB and –10 dB at 0° and 60° scan angles for $\alpha_J = [2\ \ 3\ \ 3\ \ 4]$ with no visibility of grating lobes in the visible region of space. In Figures 5.10(a, b), the solid trace displays AF demonstrating smearing of the grating lobes resulting into the suppression of SLL envelope, even after scanning. Figure 5.10(c) shows the variation of D_o with respect to the scan angle within 1.0 dB only, showing a wide scan behavior, which is completely free from the grating lobes. Table 5.13 summarizes APMs for DAA and LSAA at boresight. The $d_{avg} = 1.77\lambda$ is shown, which implies a high $\eta_{sp} = 77\%$. Thus, achieved PSLL with high sparsity as demonstrated is quite remarkable, referring to the predicted optimal levels of SLL [33]. Thus, the results obtained using ADT are nearly optimal.

Table 5.14 shows the variation of D_o and HPBW within ±0.64 dB from 0° to 60° angular extent with respect to the variation in the scan angle. The variation in D_o maximum at 30° and is 0.87 dB up to 75°. However, HPBW follows cosθ variation and shows scan capability up to 60°, where it doubles with respect to boresight and is four times at 75°, indicating a loss in the angular resolution.

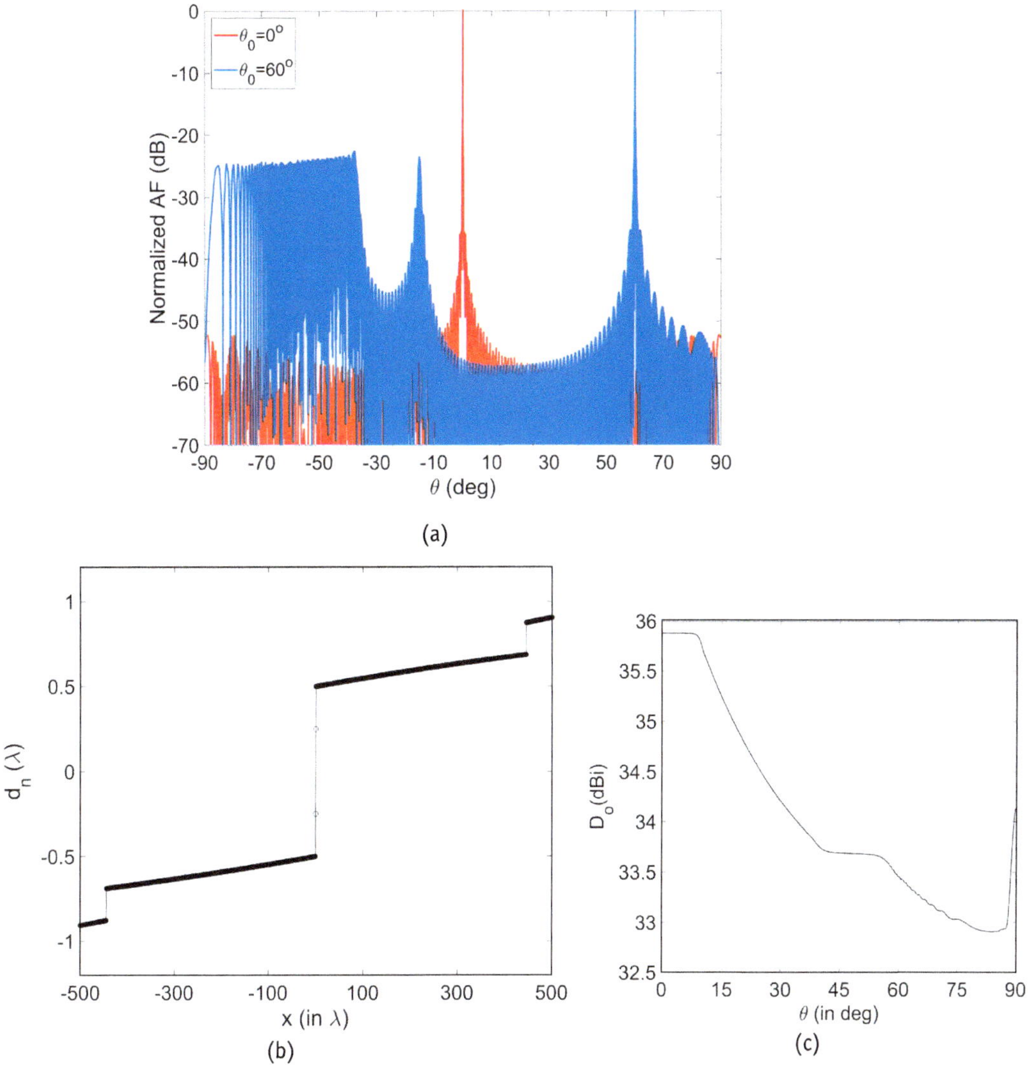

Figure 5.9
A 2,000-element A-LSAA with fixed length ($d_{no} = 0.5\lambda$) α_j = [1 1 1 2]. (a) AF at 0° and 60° scan angles. (b) Plot of interelement spacing between array elements. (c) Variation in D_o with respect to the scan angle. (© 2020 IEEE. Reprinted with permission from [51].)

5.5 Synthesis of Taylor-Like Distribution

It was observed in the previous section that ADT-led synthesis of nonuniform LSAAs leads to SLL lowering with APMs similar to dense LAAs. However, it was also observed that FSLL rises on scanning the array. This could be explained due to the subarrays'

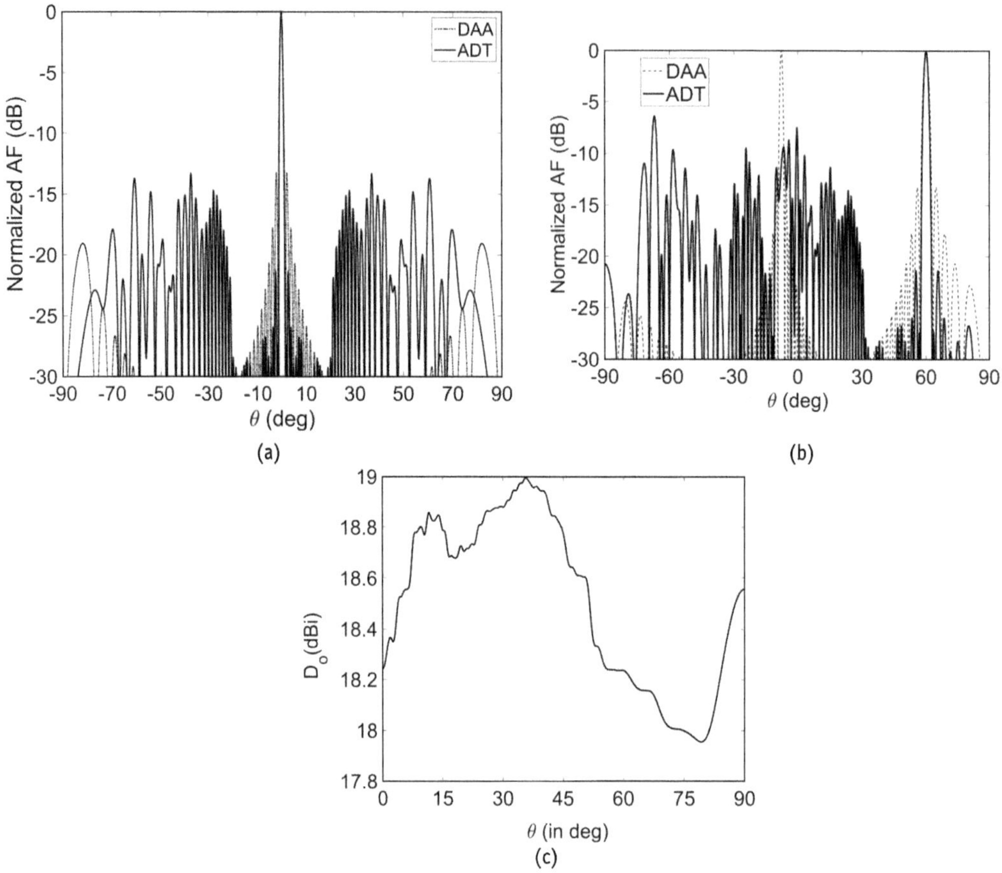

Figure 5.10
A 37-element LSAA ($d_{no} = 1\lambda$) radiative performance. (a) AF plot at boresight. (b) AF plot at 60° scan angle. (c) D_o versus scan angle. (© 2020 IEEE. Reprinted with permission from [51].)

Table 5.13
APMs of 37-Element Highly Sparse Array at Boresight

Case	D_o (dBi)	HPBW (deg)	PSLL (dB)	First SLL (dB)	ASLL (dB)	d_{avg} (λ)
DAA ($d_{no} = 1\lambda$)	18.7	1.2	GL at 90°	−13.25	−30.04	1
ADT	18.7	1.0	−13.29	−27.36	−23.39	1.77

Table 5.14
HPBW and D_o Variation with Respect to Scan Angle for 37-Element LSAA

θ_o (deg) →	0	15	30	45	60	75
D_o (dBi)	18.2	18.8	18.8	18.7	18.2	18.0
HPBW (deg)	1.0	1.1	1.2	1.5	2.2	4.2

formation in the array with larger regular intersubarray and interelement separations, which leads to quantized grating lobes (QGLs) [61]. It is known that the overall radiation pattern is a multiplicative product of element pattern (EP), subarray pattern (SP), and overall AF. Thus, on scanning since EP remains invariant, AF rises due to high SP. Figure 5.11 shows the effect of subarrays (SAs) on LSAA performance. The inset shows a 60-element LSAA (with N_a = 38), where due to nonuniformity in the spacings of elements, few SAs (11 in number) as shown with square boundaries are created with larger inter-subarray separations, say, 1.5λ, over here. Figure 5.11(a) shows the EP, SP, AF, and overall radiation pattern at boresight. Figure 5.11(b) shows the scanned pattern where FSLL rises compared to one at the boresight due to SP. Thus, it is required to alleviate the effect of the SAs in sparse antenna arrays to alleviate the FSLL. The regularity in spacing between active elements groups in the sparse region gives rise to a grating lobe at an angle governed

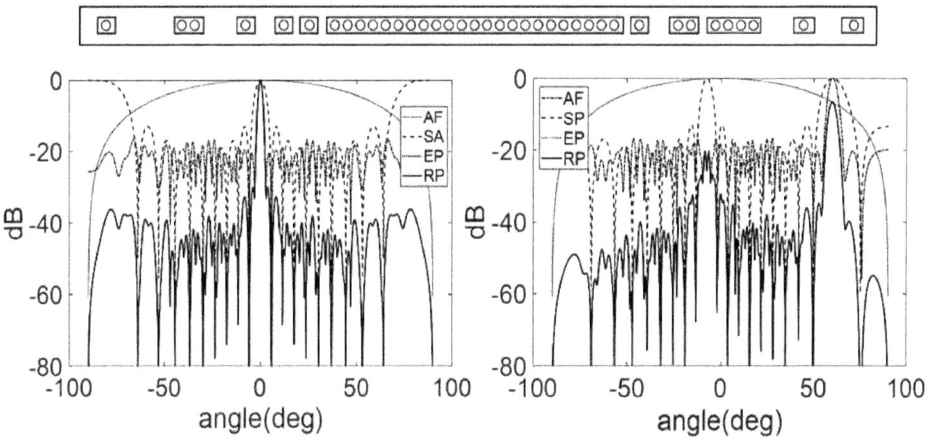

Figure 5.11
Effect of subarrays on LSAA performance.

by $d_{avg} \sim 1/(1 + \sin_{GL})$. Thus, these regularities need to be destroyed to alleviate GLs by aligning the peaks in subarrays' pattern, so as to contradict each other in the sense of the phase. This section presents a deterministic design procedure to tackle such problems and design highly sparse LSAAs with low SLL.

5.5.1 Design and Optimization Philosophy

The design strategy followed here is comparing the reference aperture current distribution with the simulated one obtained by the virtue of density tapering and minimizing the corresponding least mean square error (LMSE). Alongside the minimization of LMSE, the APMs of LSAA are made to closely agree with the denser counterpart. The array aperture efficiency, η_A, is defined as:

$$\eta_A = \frac{\left|\sum_{n=1}^{K} I_n\right|^2}{K \bullet \left|\sum_{n=1}^{K} I_n^2\right|} \tag{5.9}$$

In (5.9), $\eta_A = 100\%$ for isophoric DLAAs, with a PSLL< −13.2 dB. However, it is desired to have PSLL much smaller than −13 dB along with other APMs closer to dense LAA. The conventional amplitude tapering techniques [11] lowers PSLL but leads to a widening of HPBW and thus loss in the angular resolution. Further, it leads to a high RF loss due to involved feed networks required to apply the requisite weight settings to the antenna elements in an array [6]. A degree of comparison or array similarity factor (ASF) between the array efficiency of LSAA and DAA may be defined as:

$$ASF, \Delta \text{ (in \%)} = 100\left(1 - \frac{\eta_{A,\text{LSAA}}}{\eta_{A,\text{DAA}}}\right) \tag{5.10}$$

In (5.10), Δ helps to choose the minimum value of elements, N_{EL} to be positioned on LSAA's aperture yielding the desired $\eta_{A,\text{LSAA}}$, such that $\Delta \to 0\%$ (i.e., a minimal loss in the array efficiency). The matching of the continuous aperture current distribution having length $2a$ (in λ) with density tapered element locations is referred as the localization process. The localization process involves the matching of the cumulative amplitude current distribution, $I_C(= (1/K)\text{abs} \sum_{i=2}^{K} (I_{i-1} + I_i)$, (defined in general as per (5.11)) computed using a MATLAB function cumsum( ) with the desired reference cumulative current distribution. This serves as another criterion of optimality of designed LSAA.

$$I_C = \frac{1}{K}\int_{-a}^{x} \Gamma(\xi)\,d\xi \tag{5.11}$$

Another criterion of optimality of design is the beam parameter, comparing HPBWs defined as:

$$\Delta_{\text{BP}} = \frac{HPBW_{\text{LSAA}}}{HPBW_{\text{DAA}}} \tag{5.12}$$

HPBW strongly depends upon the spatial extent of antenna elements placed onto the aperture and are very weakly dependent on N_{EL}. However, there might still be a slight disagreement between LSAA and DAA in terms of HPBW, which is monitored by Δ_{BP}. Thus, the described synthesis technique uses parameters, η_A, ASF, Δ, and Δ_{BP} to check the optimality of the design as obtained for LSAA. The design guidelines as formulated are:

1. Statistical density tapering (SDT) is used to derive the reference model taper to attain high sparsity in LSAA with low SLL better than the optimal level achievable based on the sparsity [33]. SDT [28] depends upon the desired modal amplitude taper, I_n (Taylor's distribution with desired SLL, SLL_d) and a predefined selection probability factor, q. It demands allocating amplitude value, 1 or 0, based the product of normalized amplitude, a_n $(= I_n/\max\{I_n\})$ and q $(0 \le q \le 1)$. A random number, R_n, is generated (between 0 and 1), if $(a_n \bullet q) < R_n$, then elements is allocated 1 or else 0. In SDT, q decides η_{sp} and, as N_{EL} decreases, η_{sp} increases. However, this limits the lowering of PSLL limiting it to a value $\le 2/N_{\text{EL}}$.
2. The aperture current distribution obtained from SDT is chosen as a reference modal taper with high η_{sp}. The central part

of an array controls the NSLL region emulating $\sin x/x$ pattern giving PSLL ~ –13 dB in the NSLL region.

3. Local optimization: A_n^{sim} represents the cumulative current distribution for each iteration, which is compared with the modal current distribution, A_n^{mod}, obtained from the desired Taylor's distribution. The slope of the central region of A_n^{sim}- controls NSLL in proportion to the elements in the middle region of LSAA. Hence, to lower NSLL, more elements may be added in this region. The curves, A_n^{sim} and A_n^{mod}, are compared along the array length in discrete steps of $\lambda/2$ and the corresponding least mean square error (LMSE) and series error term, *err* terms are computed using (5.13) and (5.14).

$$LMSE = \min \frac{1}{N} \sum_{n=1}^{K} \left| A_n^{\text{mod}} - A_n^{\text{sim}} \right|^2 \tag{5.13}$$

$$err = \sum_{n=1}^{K} \left(A_n^{\text{mod}} - A_n^{\text{sim}} \right) < \pm \frac{1}{2M} \tag{5.14}$$

LSAA is characterized in terms of APMs, that is, D_o, HPBW, and SLL envelope, characterized in terms of NSLL, PSLL, ASLL, SLL at 1σ position (1σSLL), and SLL at 3σ position (3σSLL). The 1σSLL and 3σSLL are derived from $F(\theta)$ and ASLL defining the variance in results, given in (5.15) and (5.16).

$$1\sigma SLL = \sigma_1 = \frac{\sum_{l=1}^{L} \left| F(\theta_l) - ASLL \right|^2}{2L} + ASLL \tag{5.15}$$

$$3\sigma SLL = \sigma_3 = 3\frac{\sum_{l=1}^{L} \left| F(\theta_l) - ASLL \right|^2}{2L} + ASLL \tag{5.16}$$

These are to be evaluated at the angular locations, θ_l, *iff*$(F(\theta_l) - ASLL) > 0$. Here l denotes the sampled locations in θ at which AF is evaluated. The aim of the synthesis is to achieve ASLL < $1/K$ and 3σSLL < $3/K$ while maintaining $\Delta_{BP} \rightarrow 1$.

4. The lowering of PSLL over the complete angular region is adjusted by optimizing the density taper near the edges of LSAA guided by these steps:
   a. Adjusting the local similarity criterion such that $err < |1/2N_{EL}|$. Add active ($I_n = 1$) or zero ($I_n = 0$) elements at discrete locations in case $err > 0$ or if $err < 0$, respectively. The placement of 0 in middle of LAA shall be avoided. (N_{EL} is the number of elements in an array.)
   b. Active elements at edges with a length equal to DAA's maintains HPBW closer to DAA.
   c. Smoothing of simulated cumulative current distribution curve at discrete locations moving from the center towards edge.
5. Quadratic array dilation technique (QADT): A further fine-tuning may be achieved via QADT using (5.17). It can be applied to attain local similarity in the central part of LSAA as:

$$d_{max} = d_{min} + \beta\left(x - x_{middle}\right)^2 \tag{5.17}$$

$$\beta = \frac{d - d_{min}}{x_{max} - x_{min}} \tag{5.18}$$

   It is required to be mentioned here that QADT needs to be applied only in the central region surrounded by highly sparse region. QADT increases the d_{avg} by spreading the density distribution in the middle of the array region to alleviate the mutual coupling. Hence, it should not be applied to complete array. This may lead to an increase in the length of array and may cause increase in SLL $\propto 1/\sqrt{K}$ at angles away from main. This may appear as high SLL smeared over the far angular region away from the main beam.
6. Limited amplitude tapering at edges and/or in the center to few elements may further lower the PSLL.

Hence, the design technique proposed here is a sort of grid search-type placement method and controls the localization criterion in the sparse LAA in a fine and intuitive manner to obtain the desired radiation characteristics and the sparsity. The numerical studies further presented illustrate this methodology in an elaborative mannerism.

5.5.2 A 60-Element Linear Sparse Antenna Array (LSAA)

A 30λ-long, 60-element LAA with $d_n = 0.5\lambda$ is considered as the first case. The synthesis targets are, namely, $\eta_{sp} > 40\%$, $\Delta \to 0$, and with minimal possible SLL < $2/N_{EL}$. Dense LAA exhibits $D_o = 20.7$ dBi, HPBW = 1.6°, PSLL = –13.3 dB, and ASLL = –35.1 dB, and after applying Taylor's amplitude taper (SLL = –30 dB, $\bar{n} = 4$), it demonstrates HPBW of 2.0°.

Following the proposed design methodology, first, a suitable model taper is synthesized to serve as a reference for the optimization of density distribution of the elements using SDT subjected to optimization. Table 5.15 summarizes the simulated array parameters for isophoric LSAA obtained using SDT with reference to $PSLL_{\text{mod}}$ with q values of 1, 0.6, and 0.5.

The trends observed from Table 5.15 that form the initial SDT simulations are:

Table 5.15
Summary of SDT Simulations (60-Element LSAA)

q	Simulation No.	$PSLL_{\text{mod}}$ (dB)	D_o (dBi)	HPBW (deg)	PSLL (dB)	ASLL (dB)	η_{sp} (%)	d_{avg} (λ)	σ_1 (dB)	σ_3 (dB)	Δ (%)	Δ_{BP}
1	1	−20	19.5	2.0	−16.2	−27.2	25	0.65	−18.6	−13.2	14.8	1.25
	2	−25	19.2	2.0	16.2	−27.1	30	0.66	−18.3	−12.9	39.9	1.25
	3	−30	19.3	2.0	18.8	−26.3	28	0.68	−19.7	−13.5	44.4	1.25
	4	−35	18.6	2.4	−17.6	−26.5	40	0.64	−18.4	−12.7	33.9	1.5
	5	−40	19.0	2.2	−18.4	−25.6	45	0.58	−18.6	−12.9	41.4	1.37
	6	−45	18.1	2.8	−18.2	−26.3	48	0.64	−17.9	−12.3	34.9	1.75
	7	−50	18.2	2.8	−18.2	−25.7	45	0.64	−17.9	12.3	37.1	1.75
	8	−55	18.1	2.8	−18.3	−25.7	46	0.59	−18.4	12.3	33.3	1.75
	9	−60	18.1	2.8	−18.3	−25.1	46	0.59	−18.4	12.3	31.7	1.75
	10	−70	17.9	3.0	−17.5	−25.7	48	0.62	−17.8	12.0	36.2	1.88
	11	−80	17.8	3.0	−17.0	−24.7	50	0.68	−17.6	−11.9	36.9	1.88
	12	−90	17.8	3.0	−17.0	−23.9	50	0.68	−17.6	−11.9	35.5	1.88
	13	−100	17.9	3.0	−17.1	−23.9	48	0.6	−18.0	−12.1	36.4	1.88
	14	−110	−17.8	3.0	−17.0	−23.7	50	0.68	−17.6	−11.9	42.9	1.88
0.6	1	−35	−17.3	2.4	−13.1	−20.4	55	0.66	−15.0	−10.9	69.5	1.5
0.5	1	−35	−15.8	2.6	−6.5	−19.1	68	1.11	−11.6	−8.0	78.7	1.63

- In general, a decrease in D_o and an increase in parameters HPBW, Δ, Δ_{BP}, and ASLL are observed with the corresponding decrease in the $PSLL_{mod}$.
- PSLL and η_{sp} show an inverse relationship.
- Although SDT lowers the PSLL by thinning but it leads to a widening of HPBW, decreased D_o as the total energy stored in radiation pattern is equally distributed in the main beam and the sidelobe envelope in sync with the principle of the conservation of energy.
- It is observed that lowering of $PSLL_{mod}$ further than −50 dB, η_{sp} oscillate around 50% for $q = 1$. It is seen that for $PSLL_{mod} = -35$ dB, a PSLL < $2/N_E$ and $\eta_{sp} \sim 40\%$ are achieved. However, it does not satisfy the criterion on d_{avg} and Δ. Here, $d_{avg} = 0.64\lambda$ indicates strong MC showing failure in achieving sparsity. Hence, in order to reduce MC and increase sparsity, q is varied, which increased d_{avg} to 1.11λ for $q = 0.5$ (see Table 5.15), but accompanied with a high rise in PSLL (>$2/K$), reduction in D_o, high variance, and increased {Δ and Δ_{BP}}. Thus, SDT fails to achieve the design goals of high sparsity and low PSLL simultaneously.

Figures 5.12(a, b) show LSAA performance compared with DAA at 0° and 60° scan. It's observed from these and Table 5.15 that $q = 1$ gives the best approximation to the modal taper (PSLL = −17.6 dB, $d_{avg} = 0.64\lambda$, $\Delta_{BP} = 1.25$), but with a high MC. In order to reduce it, q is reduced to 0.6 and 0.5, but it increases PSLL (= −13.1 dB for q = 0.6 and −6.5 dB for $q = 0.5$, refer to Figure 5.12(a)) and increase in Δ_{BP} (=1.5 and 1.63 for $q = 0.6$ and 0.5, respectively). A similar trend of increase in PSLL with q is shown for 60° scan angle (Figure 5.12(b)). It is also seen that, for $q = 0.5$ (high η_{sp} and d_{avg}), there is a rise in GL due to regularity in spacing between subarrays of filled locations in the array aperture (see top inset in Figure 5.12(d)).

The comparison of A_n^{sim} and A_n^{mod} is shown in Figure 5.12(c) where A_n^{sim} curves are plotted for $q = 1$, 0.6, and 0.5. Figure 5.12(d) plots *err* with respect to element locations for $q = 1$, 0.6, and 0.5 providing a check on the localization matching of the element locations to satisfy $err < |1/2N_{EL}|$ criterion. It is apparent that A_n^{sim} ($q = 1$) closely follows A_n^{mod} except for the disparities near the LSAA's edges. This is due to the underlying phenomenology of the SDT that it tries to position the maximum number of elements in the center of array with empty locations near the edges that lead to this disagreement. This phenomenon is also supported by the *err* plot in Figure 5.12(d). Thus, it is evident that SDT cannot achieve high sparsity (large $d_{avg} \geq 1\lambda$) and low PSLL simultaneously.

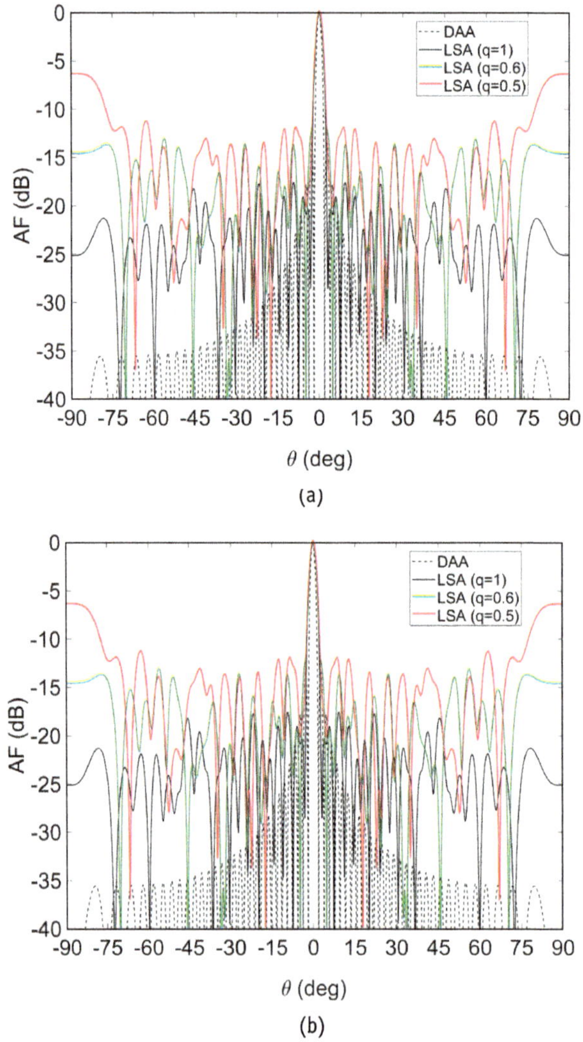

Figure 5.12

Performance characteristics of a 60-element LAA using SDT. (a) AF at 0° scan angle. (b) AF at 60° scan angle. (c) Cumulative amplitude distribution versus element location. (d) Least mean square error versus element location.

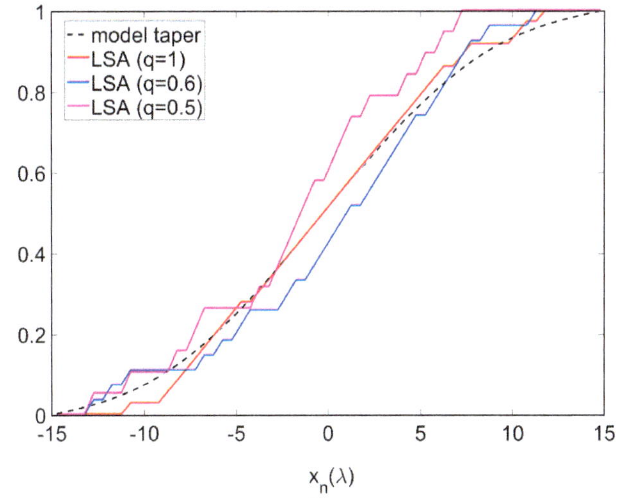

(c)

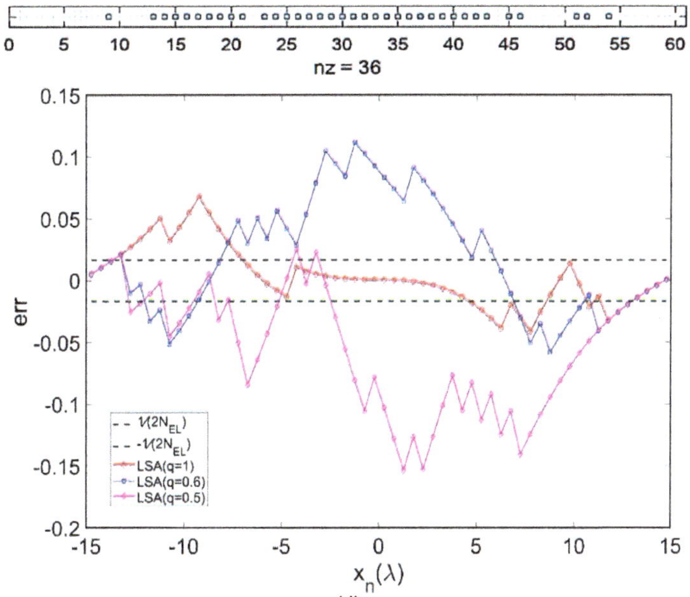

(d)

Figure 5.12
(Continued)

5.5.3 Optimization of a 60-Element LSAA

The designed LSAA using SDT in the previous section is chosen as the reference taper for the further optimization based on the following steps:

1. The lengths of LSAA and DAA are kept the same (fixed length) to keep $\Delta_{BP} \rightarrow 1$.
2. The minimization of LMSE and *err* is done by localized optimization at the discrete locations in LSAA (~$0.5n\lambda$, $n = -N_{-n}$, ..., N_{-n}), ensuring that $err < |1/2M|$.
3. Limited QADT is applied to the central 20 elements to increase d_{avg}.
4. The amplitude taper is applied to a few elements in an array at the edges only.

Table 5.16 summarizes the performance of the LSAA after optimization and compares the performance of DLAA and LSAA. SDT-D(O) represents isophoric LSAA synthesized using the deterministic proposal, keeping the length the same as DLAA. SDT-D(O-A) represents the effect of amplitude tapering on this array with taper applied to the last element on either side of the LSAA. SDT represents isophoric LSAA designed using only SDT and SDT-L represents results by keeping the length same as DLAA. DAA(A) represents amplitude tapered (Taylor's –30 dB, $\bar{n}$ = 4) LAA. It is clearly shown in Table 5.16 that the lowest PSLL of –17.6 dB is achieved for SDT (q = 1) without restriction on the LSAA's length with d_{avg} ~ 0.64λ (i.e., high mutual coupling). However, with a restricted length, PSLL ~ –12.5 dB is achieved with d_{avg} ~ 0.86λ, that is, medium sparsity (see Chapter 4). Thus, PSLL and sparsity are contradicting each other. Further, the results for q = 0.6 and 0.5 show increase in sparsity (d_{avg} >> 0.5λ) but with higher respective PSLL of –13.1 and –6.5 dB. Further, with restricted length, a PSLL of –10.5 both for q = 0.6 and 0.5 is achieved.

However, SDT-D(O) shows a much lower PSLL ~ –15.6 dB (better than 5.1 dB with respect to SDT-L (q = 0.6 and 0.5)) with a high sparsity, that is, d_{avg} ~ 1λ [12], indicating low mutual coupling. Further, SDT-L(O-A) shows further lowering of PSLL by 1.3 dB by just marginally tapering only one element on opposite edges of LSAA. The other design targets $\Delta \rightarrow 0$ and $\Delta_{BP} < 1.5$ are also achieved as well. The value of Δ is substantially lower than the other cases and

Table 5.16
Optimized 60-Element LSAA Simulations

Simulation No.	Case	D_o (dBi)	HPBW (deg)	PSLL (dB)	ASLL (dB)	η_{sp} (%)	d_{avg} (λ)	σ_1 (dB)	σ_3 (dB)	Δ (%)	Δ_{BP}
1	DAA	20.8	1.6	−13.3	−35.2	0	0.5	−23.4	−15.6	—	—
2	DAA (*A*)	20.1	2.0	−30.0	−45.1	0	0.5	−24.3	−15.3	23.2	1.25
3	SDT (q = 1)	18.6	2.4	−17.6	−26.5	40	0.64	−18.4	−12.7	33.9	1.50
4	SDT-L (q = 1)	19.6	2.2	−12.5	−21.5	41	0.86	−17.3	−12.3	28.0	1.37
5	SDT (q = 0.6)	17.3	2.4	−13.1	−20.4	55	0.66	−15.0	−10.9	69.5	1.50
6	SDT-L (q = 0.6)	16.9	2.0	−10.5	−18.1	58	1.22	−15.0	−10.9	48.5	1.25
7	SDT (q = 0.5)	15.8	2.6	−6.5	−19.1	68	1.11	−11.6	−8.0	78.7	1.63
8	SDT-L (q = 0.5)	16.4	2.2	−10.5	−17.8	63	1.43	−14.7	−10.6	54.7	1.37
9	SDT-D (*O*)	18.7	2.4	−15.6	−20.7	50	1.0	−16.5	−11.5	8.6	1.50
10	SDT-D (*O-A*)	18.7	2.8	−16.9	−21.5	50	1.0	−16.8	−11.5	8.6	1.75
11	SDT-D (*O*)	20.6	1.6	−20.6	−29.1	17	0.59	−19.6	−13.8	2.9	1.00

A: amplitude tapering; SDT-D: SDT complemented with deterministic proposal; O: optimized.

closer to zero. Another case shown in Simulation No. 11 exhibits D_o = −20.6 dB, HPBW = 1.6°, $\Delta \sim 2.9$, and $\Delta_{BP} = 1$, which shows the performance of the LSAA almost identical as DAA with much lower PSLL < −20.6 dB. However, it shows $\eta_{sp} \sim 17\%$ and $d_{avg} \sim 0.6\lambda$. This is an excellent result for a medium sparse array.

Figure 5.13 shows the performance of the optimized LSAA. Figure 5.13(a) shows the comparison of cumulative current distributions demonstrating a very good match between A_n^{sim} and A_n^{mod}. This is also confirmed by the *err* plot in Figure 5.13(b). Figure 5.13(c) shows the AF at 0° and 60° angles illustrating no high SLL peaks and onset of grating lobes. Thus, LSAA demonstrates a wide-angle scanning capability with low PSLL. The design is validated with experimental modeling using Ansys HFSS, utilizing CB-MPA.

Figure 5.13(d) shows validation of the design performance with the electromagnetic simulations using based on experimental modeling done using Ansys HFSS. Two models are generated using two different antenna elements, CBMPA and GTSA. These elements' performance was discussed earlier. It is observed that the array radiation pattern obtained is having a good agreement with the MATLAB simulation based on isotropic elements. The only variation is

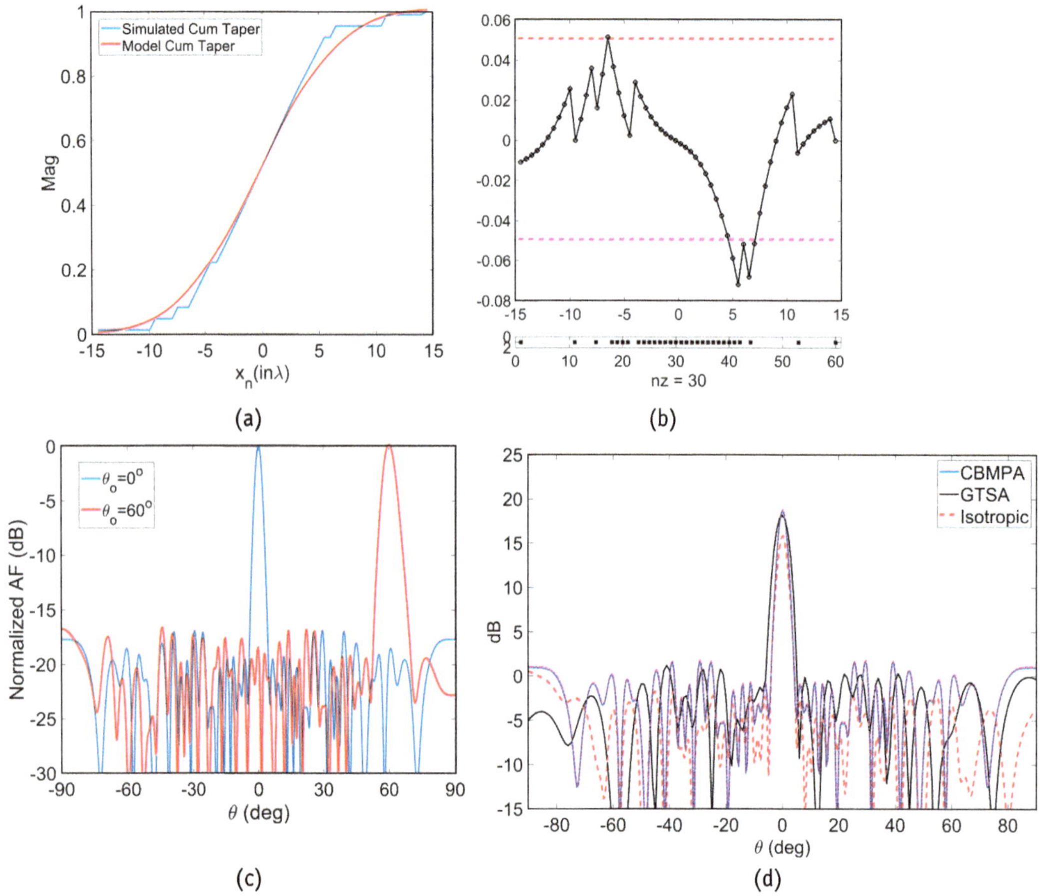

Figure 5.13
AF of the final optimized 60-element LSAA with limited amplitude tapering. (a) Cumulative curves comparison. (b) The *err* curve. (c) AF plot. (d) Validation with em simulations.

observed in the directivity because of the difference in the individual gain of the antenna elements (CBMPA and GTSA have a gain of 4–6 dBi, which is higher than isotropic elements). Hence, an excellent agreement of results is shown proving the accuracy of the proposed synthesis technique. The SLL is also in agreement within 1 dB, which is excellent and demonstrates the high accuracy of the design procedure proposed here. Thus, it is seen that, due to large d_{avg}, the effect of the MC is not evident from the em simulations as well.

Table 5.17
SDT Simulations (200-Element LSAA) Versus the Modal Taper Chosen (q = 1)

Simulation No.	$PSLL_{mod}$ (dB)	$\eta_{A,SAA}$(%)	$\eta_{A,TAY}$(%)	Δ (%)	$PSLL_{sim}$ (dB)	η_{sp} (%)
1	−20	73.5	95.3	22.8	−19.3	26.5
2	−30	65.5	85.7	23.5	−19.0	34.5
3	−40	56.5	79.4	28.8	−18.6	43.5
4	−50	52.0	75.3	30.9	−17.8	48.0
5	−60	50.0	72.6	31.1	−18.5	50.0
6	−70	46.5	70.8	34.3	−18.9	53.5
7	−80	45.0	69.5	35.2	−18.5	55.0
8	−90	44.5	68.5	35.0	−17.6	55.5

5.5.4 A 200-Element Linear Sparse Array Antenna (LSAA)

A 100λ-long LSAA, consisting of 200 elements, distributed over a 0.5λ array grid is considered to show the large array simulation capabilities of the present proposal. Table 5.17 shows the probable choices of modal taper and corresponding APMs based on SDT. Table 5.18 summarizes the performance of LSAA compared with uniformly fed and uniformly spaced LAA (UF-USLAA). A comparison is also done with the results of a similar 200-element array obtained using probability tapering IFT (PTIFT) approach [34]. PTIFT shows PSLL of –23.3 dB with 34% thinning, whereas the present technique demonstrates a comparable PSLL of –23.1 dB with much higher η_{sp} = 50% and d_{avg} ~ 1.0λ, indicating high sparsity.

Table 5.18
Array Performance of 200-Element LSAA (Fixed Length)

Case	D_o (dBi)	HPBW (deg)	PSLL (dB)	ASLL (dB)	η_{sp} (%)	d_{avg} (λ)	σ_1 (dB)	σ_3 (dB)	Δ (%)	Δ_{BP}
DAA	26.0	0.4	−13.3	−42.8	0	0.5	−29.1	−20.8	—	—
I	24.9	0.4	−22.2	−30.2	30	0.77	−23.8	−18.2	26	1.0
II	23.1	0.4	−23.1	−25.2	50	1.0	−21.2	−16.2	30.5	1.0
PTIFT	—		−23.3		34	—	—	—	—	—

The angular resolution is the same as an isophoric DAA with a much lower PSLL. In case I, PSLL = –22.2 dB and η_{sp} = 30% similar to PTIFT is shown along with d_{avg} ~ 0.77λ, indicating medium sparsity [12]. Figures 5.14(a, b) show the AF plots and cumulative amplitude plots for the optimized LSAA.

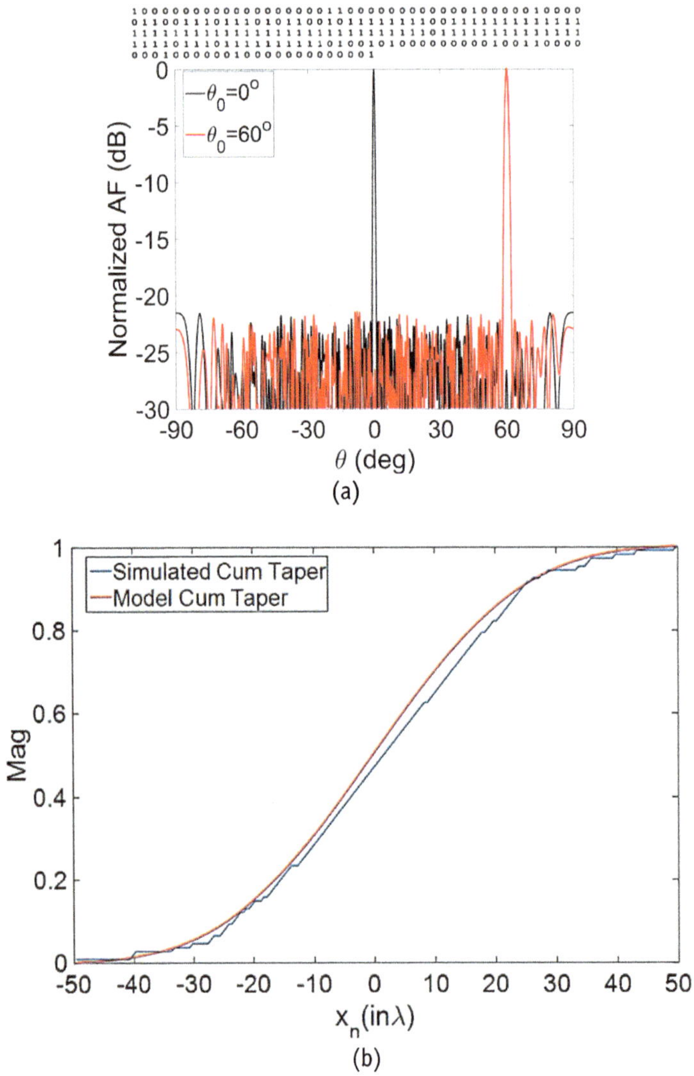

Figure 5.14
AF of a final optimized 200-element LSAA with 46.5% sparsity. (a) AF at 0° and 60° scan angles and top shows the 1s and 0s distribution. (b) Comparative cumulative amplitude distribution.

5.5.5 A 3,600 (60 × 60) Element Planar Sparse Array Antenna (PSAA)

The previous sections established the appropriateness of the proposal in synthesis of LSAAs. This section shows the extension of the proposal towards a synthesis of a planar sparse array antenna (PSA) as well. A 60-element LSAA from previous section is considered along the x- and y-axis and used for generating a PSAA.

Figure 5.15(a) shows the layout of the PSAA as obtained and Figure 5.15(b) shows the AF at boresight as plotted in an uv-coordinate

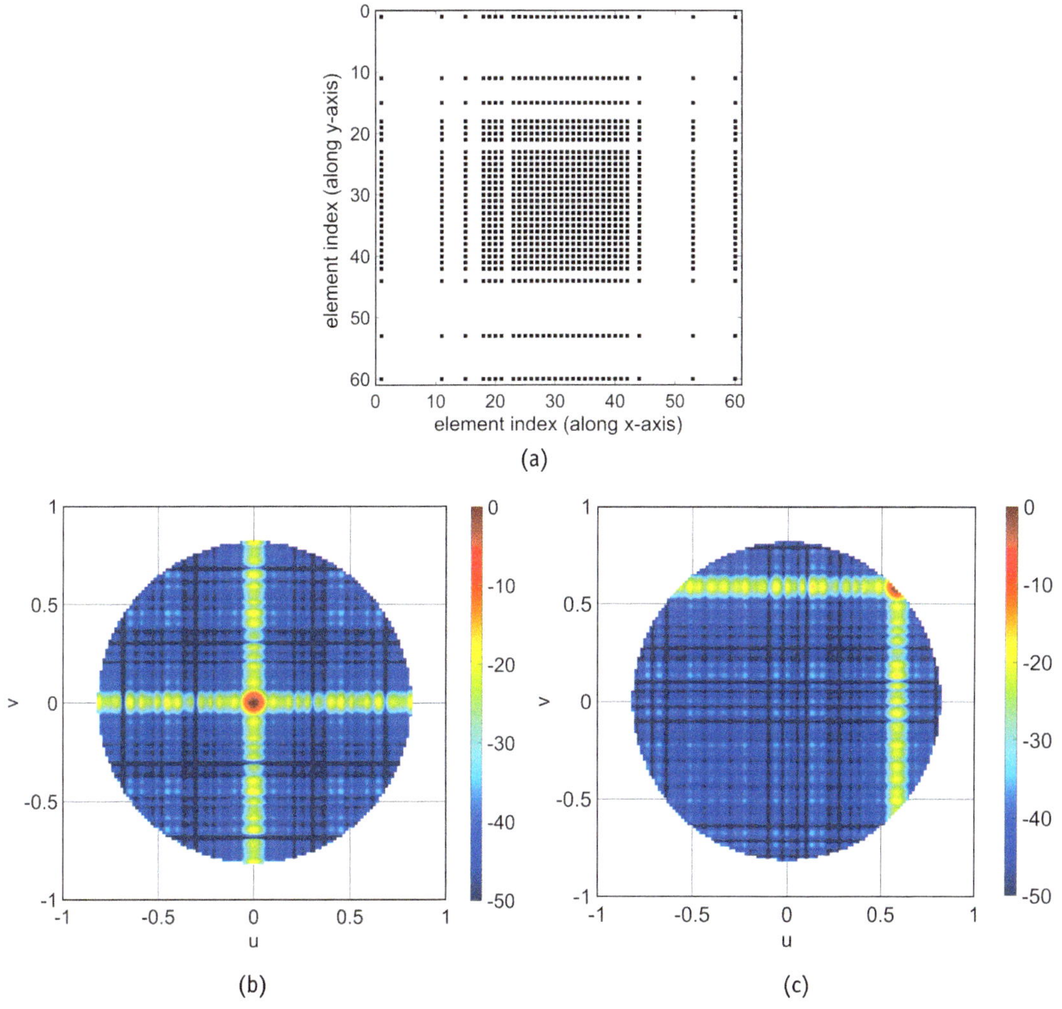

Figure 5.15
A 3,600-element sparse array (color bar is in decibels). (a) Planar sparse array antenna layout. (b) AF at boresight. (c) AF at 60° scan angle.

system ($u = \sin\theta\cos\varphi$; $v = \sin\theta\sin\varphi$. The NSLL < –20 dB with the worst PSLL of –18 dB beyond a 50° angle away from the main lobe at boresight (Figure 5.15(b)) is achieved. A highly sparse array with 900 active elements, showing a very high sparsity level of $\eta_{sp} = 75\%$, and $d_{avg} > 1\lambda$ is realized in comparison to DAA consisting of 3,600 elements. Figure 5.15(c) shows the AF at a (60°, 45°) scan angle. The worst PSLL < –17.2 dB is shown even on scanning along the intercardinal planes and no onset of GLs is observed. Thus, wide scanning capabilities of synthesized planar sparse arrays are also demonstrated. The effectiveness of the present deterministic proposal in a synthesizing linear as well as planar highly sparse arrays with low SLL is demonstrated.

5.6 Conclusions

This chapter discussed the various design techniques for linear sparse array antennas supported with case studies. I have chosen the selective techniques from the literature and my own work for demonstration for portraying the characteristics of linear sparse array antennas. The advantages of the antennas over their denser counterparts are being brought out with the help of numerical test case studies.

References

[1] Oweis, A. I., et al., "Patch and Monopole Antennas in Linear Coprime Arrays for Direction of Arrival Estimation Using Compressed Sensing," *IET Microwaves, Antennas & Propagation*, Vol. 14, No. 2, 2019, pp. 209–214.

[2] Oweis, A. I., et al., "A CoPrime Array of Patch Antennas for DOA Estimation in Mobile Handheld Devices," *2018 IEEE International Symposium on Antennas and Propagation & USNC/URSI National Radio Science Meeting*, Boston, MA, 2018, pp. 1305–1306.

[3] Hoydis, J., S. ten Brink, and M. Debbah, "Massive MIMO: How Many Antennas Do We Need?" *2011 49th Annual Allerton Conference on Communication, Control, and Computing (Allerton)*, Monticello, IL, 2011, pp. 545–550.

[4] Pinchera, D., et al., "Antenna Arrays for Line-of-Sight Massive MIMO: Half Wavelength Is Not Enough," *Electronics*, Vol. 6, No. 3, 2017.

[5] https://en.wikipedia.org/wiki/Instrument_landing_system_glide_path.

[6] Fourikis, N., *Phased Array-Based Systems and Applications*, New York: John Wiley & Sons, 1997.

[7] Weber, R. H., and R. Weber, *Internet of Things*, Vol. 12, New York: Springer, 2010.

[8] Booysen, A. J., "Circular Aperture Pattern Synthesis from Collapsed Equivalent Line-Source Distributions," *IEEE Transactions on Antennas and Propagation*, Vol. 52, No. 11, 2004.

[9] Balanis, C. A., *Antenna Theory: Analysis and Design*, Second Edition, New York: John Wiley & Sons, 1997.

[10] Stutzman, W., "Bibliography for Antennas: A List of Every English-Language Antenna Book Ever Written," *IEEE Antennas and Propagation Magazine*, Vol. 50, No. 4, August 2008.

[11] Elliott, R. S., *Antenna Theory and Design*, New York: Wiley-IEEE Press, 2003.

[12] Kumar, P., A. Kedar, and A. K. Singh, "Design and Development of Low Cost Low Sidelobe Level Slotted Waveguide Antenna Array in X-Band," *IEEE Transactions on Antennas and Propagation*, Vol. 63, No. 11, 2015, pp. 4723–4731.

[13] Kedar, A., et al., "Wide Band Low Profile Linear Microstrip Antenna Array with Cosecant Squared Shaped Beam Pattern," *Microwave and Optical Technology Letters*, Vol. 49, No. 4, 2007, pp. 963–965.

[14] Kumar, P., A. Kedar, and N. K. Vishwakarma, "Genetic Algorithm Aided Synthesis of Low SLL Antenna Array Utilizing Position Optimization Based on Zoning Technique," *INCAP 2018*, Hyderabad, India, December 16–19, 2018.

[15] Kedar, A., and A. K. Singh, "CAD of Wideband Printed Transmission Line Transformers Using Particle Swarm Optimisation," *International Journal of Swarm Intelligence Research*, Vol. 1, No. 2, 2014, pp. 193–208.

[16] Haupt, R. L., "Thinned Arrays Using Genetic Algorithms," *IEEE Transactions on Antennas and Propagation*, Vol. 42, No. 7, 1993, pp. 993–999.

[17] Khodier, M. M., and C. G. Christodoulou, "Linear Array Geometry Synthesis with Minimum Sidelobe Level and Null Control Using Particle Swarm Optimization," *IEEE Transactions on Antennas and Propagation*, Vol. 53, No. 8, 2005, pp. 2674–2679.

[18] Pozar, D., *Microwave Engineering*, Fourth Edition, New York: John Wiley & Sons, 2011.

[19] Unz, H., "Linear Arrays with Arbitrarily Distributed Elements," *IRE Transactions on Antennas and Propagation*, Vol. 8, No. 2, 1960, pp. 222–223.

[20] King, D. D., R. F. Packard, and R. K. Thomas, "Unequally Spaced Broad-Band Antenna Arrays," *IRE Transactions on Antennas and Propagation*, Vol. 8, No. 4, 1960, pp. 38–81.

[21] Andreason, M. G., "Linear Arrays with Variable Interelement Spacings," *IRE Transactions on Antennas and Propagation*, Vol. 10, No. 2, March 1962, pp. 137–143.

[22] Harrington, R. F., "Sidelobe Reduction by Nonuniform Element Spacing," *IRE Transactions on Antennas and Propagation*, Vol. 9, No. 2, 1961.

[23] Ishimaru, A., "Theory of Unequally-Spaced Arrays," *IRE Transactions on Antennas and Propagation*, Vol. 10, No. 6, 1962, pp. 691–702.

[24] Leeper, D. G., "Isophoric Arrays—Massively Thinned Phased Arrays with Well-Controlled Sidelobes," *IEEE Transactions on Antennas and Propagation*, Vol. 47, No. 12, 1999, pp. 1825–1835.

[25] Kumar, B. P., and G. R. Branner, "Design of Unequally Spaced Arrays for Performance Improvement," *IEEE Transactions on Antennas and Propagation*, Vol. 47, No. 3, 1999, pp. 511–523.

[26] Dickhoff, J. H., et al., "Pattern Synthesis of Linear, Sparse Array Antennas," *2006 European Radar Conference*, Manchester, U.K., September 13–15, 2006, pp. 84–87.

[27] Toso, G., M.C. Viganó, and P. Angeletti, "Null-Matching for the Design of Linear Aperiodic Arrays," *2007 IEEE Antennas and Propagation Society International Symposium*, Honolulu, HI, June 10–15, 2007.

[28] Skolnik, M.I., "Non-Uniform Arrays," in *Antenna Theory*, R. E. Collin and F. Zucker (eds.), New York: McGraw-Hill, 1969.

[29] Hoorfar, A., and C. Israel, "Nature-Inspired Optimization of Aperiodic Metasurfaces for Antenna Beam-Shaping," *2019 IEEE International Symposium on Antennas and Propagation and USNC-URSI Radio Science Meeting*, Atlanta, GA, July 7–12, 2019, pp. 1035–1036.

[30] Chen, K. S., Z. S. He, and C. L. Han, "A Modified Real GA for the Sparse Linear Array Synthesis with Multiple Constraints," *IEEE Transactions on Antennas and Propagation*, Vol. 54, No. 7, 2006, pp. 2169–2173.

[31] Goudos, S. K., et al., "Sparse Linear Array Synthesis with Multiple Constraints Using Differential Evolution with Strategy Adaptation," *IEEE Antennas and Wireless Propagation Letters*, Vol. 10, No. 7, 2011, pp. 670–673.

[32] Cen, L., et al., "Linear Aperiodic Array Synthesis Using an Improved Genetic Algorithm," *IEEE Transactions on Antennas and Propagation*, Vol. 60, No. 2, 2012, pp. 895–902.

[33] Hooker, J. W., and R. K. Arora, "Optimal Thinning Levels in Linear Arrays," *IEEE Antennas and Wireless Propagation Letters*, Vol. 9, 2010, pp. 771–774.

[34] Gu, L., "Linear Array Thinning Using Probability Density Tapering Approach," *IEEE Antennas and Wireless Propagation Letters*, Vol. 18, No. 9, 2019, pp. 1936–1940.

[35] Jiang, Y., S. Zhang, and Q. Guo, "An Effective Two-Step Approach to the Synthesis of Uniform Amplitude Linear Arrays," *IEEE Antennas and Wireless Propagation Letters*, Vol. 16, 2017, pp. 437–440.

[36] Dai, D., et al., "An Effective Approach for the Synthesis of Uniformly Excited Large Linear Sparse Array," *IEEE Antennas and Wireless Propagation Letters*, Vol. 17, No. 3, 2018, pp. 377–380.

[37] Pincherra, D., "On the Trade-Off Between the Main Parameters of Planar Antenna Arrays," *Electronics*, 2020, pp. 1–17.

[38] Caorsi, S., et al., "Peak Sidelobe Level Reduction with a Hybrid Approach Based on GAs and Difference Sets," *IEEE Transactions on Antennas and Propagation*, Vol. 52, No. 2, 2004, pp. 1116–1121.

[39] D'Urso, M., and T. Isernia, "Solving Some Array Synthesis Problems by Means of an Effective Hybrid Approach," *IEEE Transactions on Antennas and Propagation*, Vol. 55, No. 3, 2007, pp. 750–759.

[40] Eberhart, R. C., and J. Kennedy, "A New Optimizer Using Particle Swarm Theory," *Proceedings of the Sixth International Symposium on Micro Machine and Human Science*, Nagoya, Japan, October 4–6, 1995, pp. 39–43.

[41] Kennedy, Y. J., R. Eberhart, and Y. Shi, *Swarm Intelligence*, Burlington, MA: Morgan Kaufmann Publishers, 2001.

[42] Boeringer, D. W., and D. H. Werner, "Particle Swarm Optimization Versus Genetic Algorithms for Phased Array Synthesis," *IEEE Transactions on Antennas and Propagation*, Vol. 52, No. 3, 2004, pp. 771–779.

[43] Kedar, A., and A. K. Singh, "CAD of Wideband Printed Transmission Line Transformers Using Particle Swarm Optimisation," *International Journal of Swarm Intelligence*, Vol. 1, No. 2, 2014, pp. 193–208.

[44] Vishwakarma, N., et al., "Design Considerations for a Wide Scan Cavity Backed Patch Antenna for Active Phased Array Radar," *2011 Indian Antenna Week (IAW)*, Kolkata, India, December 18–22, 2011, pp. 1–4.

[45] www.altair.com/feko.

[46] www.ansys.com/hfss.

[47] Keizer, W. P. M. N., "Linear Array Thinning Using Iterative FFT Techniques," *IEEE Transactions on Antennas and Propagation*, Vol. 56, No. 8, 2008, pp. 2757–2760.

[48] Keizer, W. P. M. N., "Element Failure Correction for a Large Monopulse Phased Array Antenna with Active Amplitude Weighting," *IEEE Transactions on Antennas and Propagation*, Vol. 55, No. 8, 2007, pp. 2211–2218.

[49] Kedar, A., "A Wide Band Wide Scan Low Cross Polarized Grooved Tapered Slot Antenna Array," *IET Microwave Antennas and Propagation*, Vol. 13, No. 2, 2019, pp. 178–184.

[50] Von Aulock, W. H., "Properties of Phased Arrays," *Proceedings of the IRE*, Vol. 48, No. 10, 1960, pp. 1715–1727.

[51] Kedar, A., "Deterministic Synthesis of Wide Scanning Linear Sparse Antenna Arrays," *IEEE Transactions on Antennas and Propagation*, Vol. 67, No. 2, 2020, pp. 905–914.

[52] Toso, G., and R. Mailloux, "Special Issue on Innovative Phased Array Antennas Based on Non-Regular Lattices and Overlapped Subarrays," *IEEE Transactions on Antennas and Propagation*, Vol. 62, No. 4, 2014, pp. 1546–1795.

[53] Angeletti, P., G. Toso, and G. Ruggerini, "Sparse Array Antennas with Optimized Elements Positions and Dimensions," *The 8th European Conference on Antennas and Propagation*, The Hague, Netherlands, April 6–11, 2014.

[54] Bucci, O. M., et al., "Isophoric Sparse Arrays Ensuring Global Coverage in Satellite Communications," *IEEE Transactions on Antennas and Propagation*, 2014, Vol. 62, No. 4, pp. 1607–1618.

[55] Bucci, O. M., A. Perna, and D. Pinchera, "Synthesis of Isophoric Sparse Arrays Allowing Zoomable Beams and Arbitrary Coverage in Satellite Communications," *IEEE Transactions on Antennas and Propagation*, Vol. 63, No. 4, 2015, pp. 1445–1457.

[56] Bucci, O. M., et al., "Isophoric Sparse Arrays Ensuring Global Coverage in Satellite Communications," *IEEE Transactions on Antennas and Propagation*, Vol. 62, No. 4, 2014, pp. 1607–1618.

[57] Kedar, A., and L. P. Ligthart, "Wide Scanning Characteristics of Sparse Phased Array Antennas Using an Analytical Expression for Directivity," *IEEE Transactions on Antennas and Propagation*, Vol. 67, No. 2, 2019, pp. 905–914.

[58] Cen, L., "Linear Aperiodic Array Synthesis Using an Improved Genetic Algorithm," *IEEE Transactions on Antennas and Propagation*, Vol. 60, No. 2, 2012, pp. 895–902.

[59] Sadiku, M. N. O., *Numerical Techniques in Electromagnetics with MATLAB*, Third Edition, Boca Raton, FL: CRC Press, 2009.

[60] Karmakar, N. C., "Investigation into a Cavity-Backed Circular-Patch Antenna," *IEEE Transactions on Antennas and Propagation*, Vol. 50, No. 12, 2002, pp. 1706–1715.

[61] Mailloux, R. J., *Phased Array Antennas*, Second Edition, Norwood, MA: Artech House, 2005.

6

Concentric Ring Sparse Array Antennas

6.1 Introduction

Concentric ring antenna arrays (CRAs) are designed by arranging the coplanar co-centric rings together in an aperture. An isophoric (equally fed elements) CRA demonstrates azimuthally symmetric pattern with low SLL, that is, a PSLL < –17.6 dB compared to isophoric DAA (PSLL < –13.2 dB) [1, 2] of a rectangular or square shape. The applications of CRAs are numerous such as direct radiating arrays (DRAs) [3] or 2-D steering (phased) arrays [4] in radars and satcom. Isophoric CRAs do not employ any amplitude tapering, thus removing the necessity of unequal power division, which reduced the RF losses. In the case of active phased arrays, it makes all the amplifiers operate in similar dynamic conditions, thereby increasing the PAE. However, they possess high PSLL, which led to various efforts for SLL reduction in the past [5–13]. These efforts are broadly categorized as: thinned [10, 11] (few elements turned off) and space or density tapered (or sparse) [3–9, 13, 14] (nonuniform placement of array elements) CRAs. The former gives a nonsymmetric radiation pattern due to the randomly absent elements, whereas the latter introduce aperiodicity by nonuniform separation between the constituent rings while having symmetrically located elements and thus exhibiting an azimuthal symmetry. A special issue of *IEEE Transactions* [13] is dedicated to the sparse antenna arrays (SAAs) presenting many related papers. This chapter discusses the design and development of sparse concentric ring antenna arrays and studies their characteristics.

6.2 Review of the Literature

In general, the synthesis of SAAs is a highly complex nonlinear problem due to the nonlinear relationship between the locations of the elements and the array factor (AF). The existing techniques for the synthesis of such arrays basically can be categorized into:

- Synthesis of array geometry based on the designers' prior experience;
- Utilization of analytical methods such as Dolph and Taylor [1];
- Utilizing density or space-tapering techniques [8, 9, 14];
- Computer-aided trial-and-error methods such as dynamic programming [13];
- Stochastic or heuristic techniques such as genetic algorithm [10, 11], particle swarm optimizer, simulated annealing [4, 6, 13].

All of these techniques are computationally exhaustive and the exhaustion time grows up exponentially with the increase in the size (number of elements) of the antenna array. So the deterministic techniques [15–19] are quite attractive and helpful in reducing the exhaustion rate of the synthesis. Nowadays, the best of both the techniques, providing an optimal solution, called the hybrid techniques [20–23], is used. However, there is no universal search algorithm [24] that serves all of the design criteria. Thus, an algorithm needs to be chosen cognitively to synthesize an array fulfilling the set design criteria. This chapter focuses more upon deterministic techniques for the synthesis of sparse CRAs (SCRAs) because they provide beautiful insight into the physics of the antenna arrays and how the dimensional variables control its performance. Also, the wideband and wide scan characteristics of the SAAs with the aid of few numerical case studies are presented and discussed in detail.

6.3 Types of Circular Sparse Arrays

A circular sparse array antenna (CSAA) can be designed with the radiating elements arranged either in the concentric rings or in the form of radial arms. The various design requirements of these structures considered during the analysis and synthesis are summarized here:

- Upper-hemispherical scanning is required (i.e., $0 < \varphi_{\text{azimuth}} < 360°$).
- HPBW in the plane of scanning should vary within the strict limits. Wideband arrays capable of wide angular scanning necessitate a wide-angular/symmetrical element pattern (isotropic over the upper hemisphere) achievable over a wide frequency range. This design requirement is important for the realization of wide scanning antenna arrays.
- Array aperture should be circular and the elements should be located uniformly in each of the rings.
- Multiring planar arrays are considered (a special case is a radial array but with worse characteristics than concentric ring array versions).

Figures 6.1(a, b) represents graphically two versions of circular sparse arrays mentioned earlier (i.e., sparse arrays with ring and

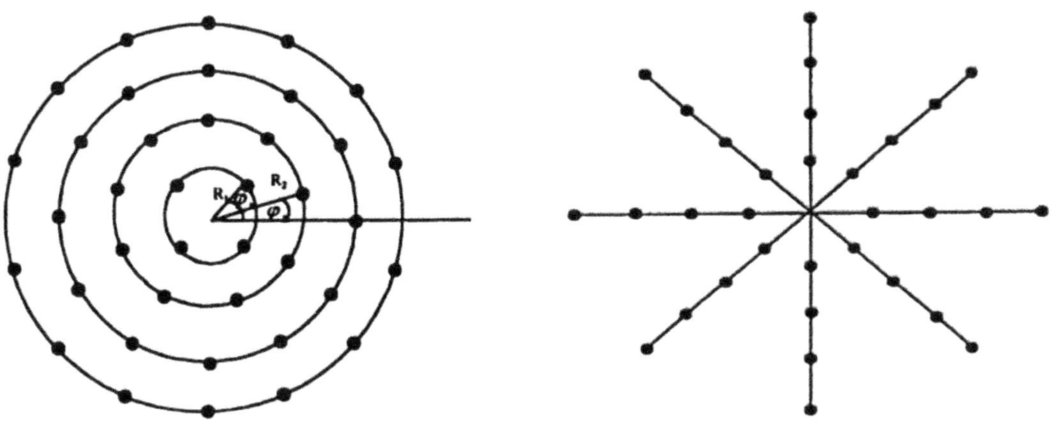

Figure 6.1
Circular antenna arrays: (a) concentric ring version and (b) radial arms version.

radial structures, respectively). The first one has uniformly distributed elements per ring and the second one has an equal number of elements per ring, respectively. The first one allows for a more uniform distribution of elements over the aperture, and the second one has a nonuniform distribution over the aperture.

The synthesis of these arrays can be basically summarized into two main steps to be followed that form the basis for all of the existing or forthcoming methods for the purpose. These are:

- Emulating the surface current (continuous) distribution on the circular aperture by deciding the number of rings or radial arms. The continuous current distribution to achieve low SLL may be derived from any analytical techniques such as Dolph or Taylor.
- Spatial distribution of the radiating elements on these rings (or arms) to discretize the continuous current rings while addressing the SLL requirements.

The array factor of continuous rings CRA is given as:

$$F_k = \sum_{k=1}^{M} A_k J_o\left(2\pi R_k \sin\theta\right) \tag{6.1}$$

In (6.1), R_k is the radii of the ring, M is the total number of rings, J_o is the Bessel function of the first type with zero order, A_k is the excitation weight applied to a ring, and k and θ are the angle of observation.

Figure 6.2 shows the variation of PSLL (dB) with the variation in the number of concentric rings in CSAA for different targeted PSLL (i.e., $PSLL_d = -13$ dB to -40 dB). It is observed that a minimum number of rings are required to achive low PSLL closer to $PSLL_d$, preferably odd in number. This will be further proven by analysis presented herein the chapter. It is also seen that by employing too many rings does not lower the PSLL and a minimum number of rings anywhere between 1 and 20 is sufficient to emulate the continuous current distribution. For instance, for $PSLL_d = -13$, -20, -30, and -40 dB only 3, 7, and 17 numbers of rings are sufficient. Although the simulated PSLL is not very close to $PSLL_d$, the further lowering of PSLL is carried out by discretization of these continuous

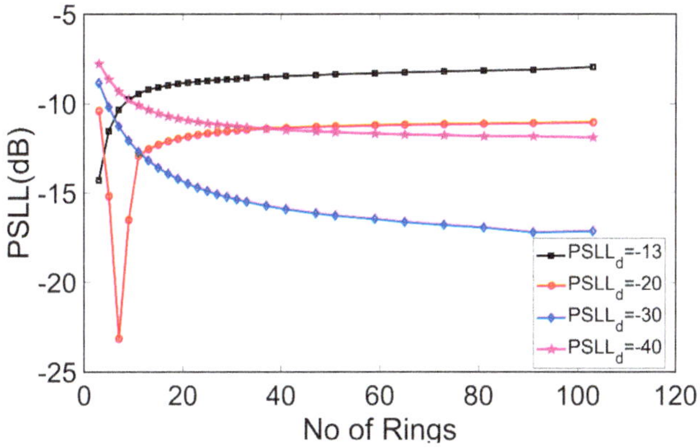

Figure 6.2
Relationshp between PSLL and the number of rings.

rings by employing spatial or density distribution of elements on these rings, as shown later.

The different control variables available to optimize CRAs' perfromance are:

- Number of rings, M;
- Number of elements, N_{TOT}, distributed on the circular aperture, where $N_{TOT} = N_1 + N_2 + \ldots + N_k$ and N_k is the number of elements in the kth ring;
- Interelement spacing in each of the rings;
- Interring separation;
- Excitation amplitude, A_k, assigned to each of the rings. In isophoric sparse CRAs, these are chosen as uniform equal to unity.

In the prior literature, it is well-established that the array factor of CRAs is expressible in the form of Bessel function [15, 17, 19] of the first kind restricted to the order zero, stating a negligible contribution from higher-order Bessel functions. The results with these approaches are shown with no or limited scan performance over the bandwidth. Later [25], I demonstrated that the incorporation of higher-order Bessel functions in AF is important for synthesizing the wide-scanning wideband CRAs, especially for phased array antenna applications. A generalized representation of AF of CRA in terms of the Bessel function was derived and presented [25]. This chapter deliberates an explanation of the properties of CRA and their synthesis using this approach as shown in the later sections.

6.3.1 The Array Factor of CRAs

Figure 6.3 shows the layout and associated coordinate system of an M-ring CRA, consisting of identical radiating elements (REs). The radii of the rings are represented as a_k (k = 1, 2, M). Each of these rings consists of N_k elements, where each of the elements has a unique location given by (a_{kn}, φ_{kn}), where $\varphi_{kn}(n = 2,\ldots,N_k)$ gives the angular location given by (6.2).

$$\varphi_{kn} = \varphi_{k1} + \frac{2\pi n}{N_k} \tag{6.2}$$

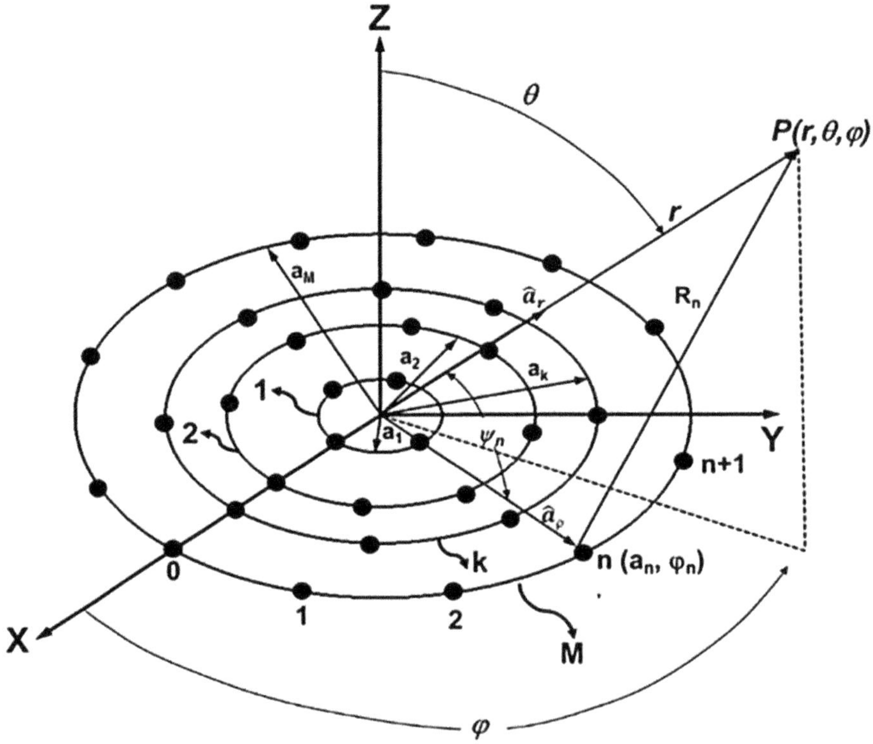

Figure 6.3

Concentric ring antenna array and the associated coordinate system. (© 2019 IEEE. Reprinted with permission from [25].)

In (6.2), φ_{k1} gives the angular location of the first element in the kth ring and is chosen as an optimization variable for the SLL minimization. It represents the rotation of the rings with respect to the other concentric rings.

$$\varphi_{k1} \begin{array}{l} = 0, \text{ if } \varphi = 0 \ (x\text{-axis}) \ (\text{i.e, the first element is on } x\text{-axis}) \\ \neq 0 \text{ if shifted with respect to the } x\text{-axis} \end{array} \tag{6.3}$$

AF for a circular array (CRA with one-ring only) is defined as:

$$F_{1A}(\theta,\varphi) = \sum_{n=1}^{N_1} I_{1n} e^{j\left[\beta a_1 \sin\theta \cos(\varphi - \varphi_{1n}) + \alpha_{1n}\right]} \tag{6.4}$$

In (6.4), the subscript 1 represents a single-ring CRA, F_{1A} denotes its AF, I_{1n} is the current excitation of the nth element in the ring located at φ_{1n}, and α_{1n} is the associated beam-steering coefficient applied to n^{th} element (with respect to the array center). Here, $\beta = 2\pi/\lambda$, and $\lambda = c/f$, $c = 3.10^8$ m/s is the free-space velocity of light and f is the operational frequency (in hertz) and α_{kn} ($k = 1$ for single-ring CRA) is the phase factor, which is in general defined as:

$$\alpha_{kn} = -\beta a_k \sin\theta_o \cos(\varphi_o - \varphi_{kn}) \tag{6.5}$$

In (6.5), (θ_o, φ_o) defines the scan angle. For a multiring CRA, (6.2) is written in a generalized form as:

$$F_A(\theta,\varphi) = \sum_{k=1}^{M} \sum_{n=1}^{N_k} I_{kn} e^{j\left[\beta a_k \sin\theta \cos(\varphi - \varphi_{kn}) + \alpha_{kn}\right]} \tag{6.6}$$

A new set of variables is defined as given in (6.7):

$$\rho_k = a_k \sqrt{(u - u_0)^2 + (v - v_0)^2} \tag{6.7}$$

$$\xi = \tan^{-1} \frac{(v - v_o)}{(u - u_o)} \tag{6.8}$$

$$u = \sin\theta \cos\varphi; \ v = \sin\theta \sin\varphi \tag{6.9}$$

$$u_0 = \sin\theta_0 \cos\varphi_0;\ v_0 = \sin\theta_0 \sin\varphi_0 \tag{6.10}$$

Therefore, the utilization of (6.7) through (6.10) leads to the compact form of (6.5) as:

$$F_A(\theta,\varphi) = \sum_{k=1}^{M}\sum_{n=1}^{N_k} I_{kn} e^{j\beta\rho_k \cos(\varphi_{kn}-\xi)} \tag{6.11}$$

For a multiring isophoric CRA (i.e., $I_{kn} = I$), utilizing (6.12) [26], (6.11) is reformulated and given by the expression in (6.13):

$$e^{jx\cos(\vartheta)} = \sum_{m=-\infty}^{\infty} j^m \mathcal{I}_m(x) e^{jm\vartheta} \tag{6.12}$$

$$F_A(\theta,\varphi) = I\sum_{k=1}^{M} N_k \sum_{m=-\infty}^{\infty} \mathcal{I}_{mN_k}(\beta\rho_k) e^{\left[jmN_k\left(\frac{\pi}{2}-\xi+\varphi_{k1}\right)\right]} \tag{6.13}$$

Thus, (6.13) is a generalized expression of AF, completely expressible in terms of Bessel function ($\mathcal{I}_m$) of the first kind with order m. Further, (6.13) may be split into two major terms as shown highlighted inside the square brackets:

$$F_A(\theta,\varphi) = I\sum_{k=1}^{M} N_k g_k(\theta,\varphi)\left[\mathcal{I}_o(\beta\rho_k) + \sum_{\substack{m=-\infty \\ m\neq 0}}^{\infty} \mathcal{I}_{mN_k}(\beta\rho_k) e^{\left[jmN_k\left(\frac{\pi}{2}-\xi+\varphi_{k1}\right)\right]}\right] \tag{6.14}$$

- In (6.14), g_k is the element pattern chosen identical for all REs; thus, $g_k = g$ taken as $\cos^\kappa$, where $\kappa = 1$ represents isotropic REs.
- The scaling factor, $\beta\rho_k$, implicitly considers the frequency dependence and scales up or down the AF proportionately to increase or decrease in f leading to a frequency-independent synthesis of CRA.
- Thus, frequency bandwidth (FBW ~ $2(f_H - f_L)/f_H + f_L$) of CRA is solely based upon the selected REs. This implies that the actual realizable bandwidth will depend upon the practical

radiator's bandwidth. Hence, a choice of a wideband radiator is essential to ensure the CRA's wide bandwidth.

Thus, (6.14) expresses the array factor of CRA in terms of Bessel function considering the array topology, the complex weight excitations, and the element pattern.

- In (6.14), the first term has only radial (a_k) dependence, whereas the second term has radial as well as rotational (φ-) dependence.
- At boresight, if, $\varphi_{k1} = 0$, in the second term, exp(•) ≈ 1, thus giving only radial dependence. The rotational dependence becomes effective only if $\varphi_{k1} \neq 0$ (i.e., an offset in the angular location of the first element of the kth ring with respect to the x-axis). The effect of offset is pronounced at far angles away from the main beam, especially while scanning, ignored earlier [15, 19], due to a focus on the limited scan applications. However, in the phased array radars, a wide-angular scanning with a low SLL is highly desirable and a functional requirement. Thus, an offset optimization is employed herein the proposed technique presented later to optimize the far SLLs as well.

As per the standard synthesis practice for arrays [1, 27], the synthesis is done at boresight, that is, $(\theta_o, \varphi_o) = (0°, 0°)$, leading to $\rho_k = a_k \sin\theta$; $\xi = \varphi$ (using (6.6)).

6.4 Design Guidelines Framework

The previous section presented generalized expression as derived for the array factor of CRA based on the Bessel functions of the first kind of order m. This expression may be used for deriving the design guidelines framework to design concentric ring antenna arrays. The design presents a controlled SLL minimization over the complete angular sector of a far-field radiation pattern of an array by dividing it into three regions, namely, near SLL (NSLL) from 0° to 20°, non-near SLL (NNSLL) from 20° to 60°, and far SLL (FSLL), beyond 60°. The design guidelines are presented in Sections 6.4.1 through 6.4.5.

6.4.1 First Term in (6.14)

Figure 6.4 shows that $\mathcal{I}_o(\beta\rho_k)$ has an oscillatory behavior [26], which has a maximum (equal to 60 for $N_{tot} = 60$), when its argument equals its order, that is,

$$\begin{aligned} \beta\rho_k &= \beta a_k \sin\theta = 0 \\ &\Rightarrow \theta = 0^\circ \text{ (i.e., at boresight angle)} \end{aligned} \tag{6.15}$$

$\mathcal{I}_o(\beta\rho_k)$ decays as $\beta\rho_k$ differs from 0 and becomes negligible at far angles. In Figure 6.4, the leftmost figure on the top plots the argument, $\beta\rho_k$, showing an increase with the increase in the angle. The right-side figure (in Figure 6.4) on the top shows the magnitude plot of $\mathcal{I}_o(\beta\rho_k)$ and the two plots in bottom of Figure 6.4 shows the corresponding normalized plots in decibels and magnitude forms with respect to the argument. It is seen that at far angles, the level falls more than 20 dB down. At boresight:

$$\mathcal{I}_o(\beta\rho_k) \rightarrow \mathcal{I}_o(\beta a_k \sin\theta) \tag{6.16}$$

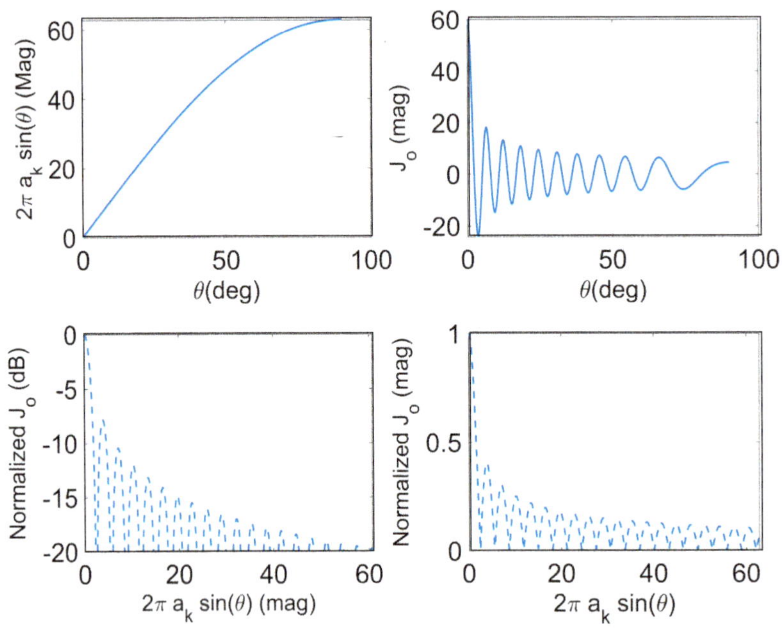

Figure 6.4
Oscillatory behavior of $\mathcal{I}_o(\beta\rho_k)$ ($k = 60$ and $a_k = 10\lambda$).

Equation (6.16) gives a rotationally symmetric pattern (no φ depenedence) with a high PSLL in NSLL region and a lowered PSLL in NNSLL and FSLL regions. Thus, a_k shall be chosen deterministically such that the sum of the maximas of $\mathcal{I}_o(\beta\rho_k)$ due to the individual rings leads to a minimal value of $\sum_{k=1}^{M}\left[\mathcal{I}_o\left(\beta\rho_k\right)\right]$, the first design guideline. The process will be explained later with numerical studies.

6.4.2 Maximum of AF

Thus, the maximum of the array factor, $[AF]_{\max}$, assuming there is no mutual coupling (MC), is given as:

$$\left|F_A(\theta,\varphi)\right|_{\max} = \sum_{k=1}^{M} N_k I\left[\mathcal{I}_o\left(\beta\rho_k\right)\right]_{\max} \tag{6.17}$$

Thus, the first term defines the directivity of the CRA.

6.4.3 Second-Term in (6.14)

The second term in (6.14) is given in (6.18).

$$\sum_{k=1}^{M} N_k \left[\sum_{\substack{m=-\infty \\ m\neq 0}}^{\infty} \mathcal{I}_{mN_k}\left(\beta\rho_k\right) e^{\left[jmN_k\left(\frac{\pi}{2}-\xi+\varphi_{k1}\right)\right]}\right] \tag{6.18}$$

This term consists of two terms, $\mathcal{I}_{mN_k}(\beta\rho_k)$ (a_k dependence) and exponential multiplicative term (with φ dependence).

- The magnitude of $\mathcal{I}_{mN_k}(\beta\rho_k)$ is small if $\beta\rho_k << mN_k$ and approaches its maximum as $\beta\rho_k \rightarrow mN_k$ and decays afterwards gradually (see Figure 6.5). Figure 6.5 shows a typical case representing the behavior of the term, which may vary depending upon the argument variables in (6.18).
- Thus, when $\beta\rho_k = mN_k$, combined SLL due to the contribution of first and the second terms will be higher (see Figure 6.6).
- In the case of a large array aperture (for example, $R_{ap} > 5\lambda$), this location will be in NNSLL region, whereas for a smaller aperture (for example, $R_{ap} < 5\lambda$), it will be in FSLL region. This is an important key to the SLL minimization procedure.

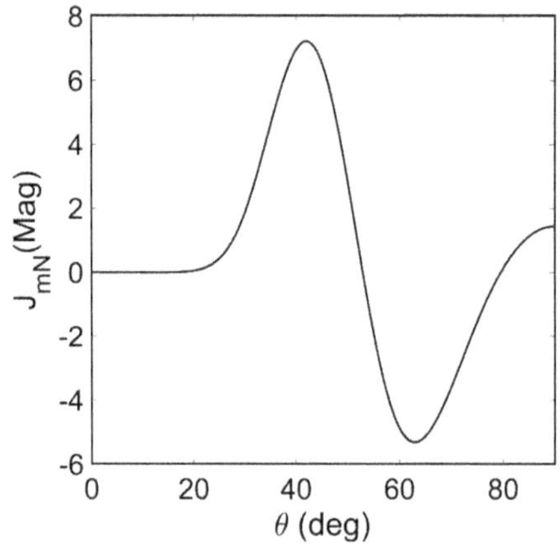

Figure 6.5
Plot of the second term, (6.18).

- The SLL differs with the variation in φ leading to a nonsymmetry in NNSLL/FSLL regions (see Figures 6.5 and 6.6). The prior literature [17, 19, 22] does not consider this fact and demonstrates perfect rotational symmetry, which does not hold universally, especially in the large arrays. Thus, the higher-order Bessel function terms leads to nonlinearity in azimuthal pattern and high PSLL in NNSLL and FSLL regions [25]. Hence, a CRA design should consider the higher-order

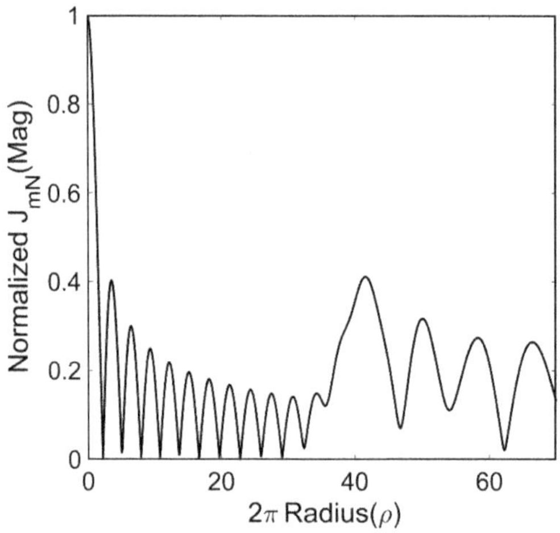

Figure 6.6
Combined plot of the two terms.

Bessel functions for synthesizing the complete angular region of the array factor with low PSLL, which is highly desirable for wide-scan CRAs (with scan volume better than 60° away from boresight).

- The SLL minimization can be achieved by adjusting a_k, N_k, and φ_{k1} with the underlying philosophy to minimize the net sum of contributions to the second term from the individual rings in a CRA.

6.4.4 Simplified Version of (6.14)

The contribution from the second term for the values of m, $|m|$ > 1 is effective only in the FSLL region, where SLL is already very low with respect to the main beam and thus may be ignored in the analysis. Hence, overlooking the $|m|$ > 1 terms, the second term in (6.15) may be further simplified as (assuming $\psi_p = \pi/2 - \xi + \varphi_{ok}$) depending upon whether the number of the elements in the rings, N_k, in a CRA is even or odd.

For odd values of N_k,

$$\mathcal{I}_{N_k}(\beta\rho_k)\left(e^{jN_k\psi_p} - e^{-jN_k\psi_p}\right) = 2j\mathcal{I}_{N_k}(\beta\rho_k)\sin\left(N_k\psi_p\right) \tag{6.19}$$

For even values of N_k,

$$\mathcal{I}_{N_k}(\beta\rho_k)\left(e^{jN_k\psi_p} + e^{-jN_k\psi_p}\right) = 2\mathcal{I}_{N_k}(\beta\rho_k)\cos\left(N_k\psi_p\right) \tag{6.20}$$

Thus, it is observed that the second term is imaginary and real for odd and even values of N_k, respectively. Hence, it is in phase quadrature with the first term, for odd N_k. Thus, the odd contribution will lead to the minimization of $F_A(\theta, \varphi)$ and thus SLL.

Henceforth, for odd values of N_k, (6.14) is rewritten as:

$$F_A(\theta,\varphi) = \sum_{k=1}^{M} N_k I\left[\mathcal{I}_o(\beta\rho_k) + 2j\mathcal{I}_{N_k}(\beta\rho_k)\cos\left\{N_k\left(\frac{\pi}{2} - \xi + \varphi_{k1}\right)\right\}\right] \tag{6.21}$$

This is an approximate expression for the array factor for a sparse CRAs with odd N_k, used for the initial design purpose. Thus, it is an important deterministic guideline formulated here that the number of elements chosen shall be odd in the value for SLL reduction.

6.4.5 Effect of Rotation

In (6.21), φ_{k1} variable controls the mutual rotation of the rings with respect to each other, which leads to a nonuniform angular spread of the radiating elements on the CRA's aperture. This reduces the density of elements from the center towards the edge in the projections of CRA for linear φ-cuts [Ch. 6, pp. 243–249, 27], varying with respect to φ-cut resulting in virtual amplitude tapered linear arrays with low SLL. The concept will be elaborated in Chapter 7 on planar sparse array antennas. It also reduces MC between the elements as it increases the d_{avg}, thereby mitigating the scan loss and facilitating design of wide-scan low SLL CRAs.

Thus, the guidelines in Sections 6.4.1 through 6.4.5 framed above can be used for the design of wideband, wide scan concentric ring sparse array antennas with low sidelobe levels. The next few sections present the design hierarchy proposed and few numerical case studies.

6.5 Design Process

As has been elaborated in the previous section that the complete synthesis problem can be split into two regions for the SLL minimization over the complete angular spectrum of the array factor. These are:

1. *NSLL region:* In this region, the design and optimization are done using the first term of (6.21), that is, a minimization of Real($F_A(\theta, \varphi)$).
2. *NNSLL region:* In this region, the design and optimization are done using the complete AF expression in (6.21), that is, a minimization of $F_A(\theta, \varphi)$.

The design goals set for the optimization are:

- Directivity, D_o;
- Half-power beamwidth (HPBW);
- SLL (PSLL and ASLL).

The design process assumes that CRA consists of isotropic radiators, radiating only in the upper hemisphere (assuming a metallic back plane). The design process is summarized as follows:

1. The total number of elements, N_{TOT}, in array is obtained as:

$$N_{\text{TOT}} = 10^{(D_o - D_e)/10} \tag{6.22}$$

   D_e is the element directivity (=2(3 dBi) for isotropic radiators with metal back plane). Here MC may be ignored as the concerned CRA is highly sparse (i.e., $\eta_{\text{sp}} > 70\%$ and/or $d_{\text{avg}} > 2\lambda$) [14]. Thus, depending upon the desired directivity, the total number of elements may be derived distributed over the CRA aperture.

2. The radii of complete array aperture, R_{ap}, is derived from the Hansen's tables for Taylor-weighted circular aperture [28]:

$$R_{ap} = \frac{\sigma\lambda}{2} \bullet \frac{HPBW_o}{HPBW} \tag{6.23}$$

   Herein, σ is the dilation factor, $HPBW_o$ is the ideal HPBW (for desired PSLL) and HPBW is the desired goal. Thus, R_{ap} sets the upper limit for the radii a_k while synthesizing CRAs.

3. The number of rings, M, is predefined by the array designer based on criteria defined in Section 6.3. This will have impact upon the d_{avg}. In the case of phased array antennas, this will control the complexity and the cost of the active circuitry as well.

4. The radii of rings are obtained as per the relation given below:

$$a_1 = R_{ap} \frac{1 - GR}{1 - GR^{M-1}} \tag{6.24}$$

$$a_k = a_1 \bullet GR^{k-1} \tag{6.25}$$

   $GR = 1.618$ is the golden ratio, a_1 is the innermost ring's radii and $a_k \le R_{ap}$. Thus, deterministically computed radii using (6.24) and (6.25) ensure that the rings emulate the continuous aperture current distribution optimally by spreading across the aperture.

5. Number of elements, N_k: The choice of N_k is made while adhering to two important design constraints:
   a. $\sum_{k=1}^{K} N_k = N_{\mathrm{TOT}}$. The total sum of the elements in the rings shall not exceed N_{TOT}.
   b. The interelement distance, $d_{ke} = 2\pi a_k/N_k$ shall adhere to the range, $d_{e1} < d_{ke} < d_{e2}$, (d_{e1} and d_{e2} not less than 1λ) to ensure highly sparse array. The values of $d_{e1/e2}$ may be varied depending upon the desired sparsity in array or may be configured to realize a DAA as well.

Therefore, $[N_k]$ consists of possible choices for N_k (only odd values) in the kth ring satisfying condition 5b. The possible combinations of elements from each of the rows of the matrix are studied, which satisfies condition 5a. Out of these, one combination that yields a minimum PSLL utilizing the design guidelines framework is chosen.

A generalized expression for directivity of phased array antennas earlier proposed in Chapter 4 is used for computation of D_o. It considers the effect of array geometry, mutual coupling, the scan angle, and complex excitations on the CRA's performance. Thus, the ignorance of mutual coupling while deriving the AF of CRA based on the Bessel function is taken care of through this expression while estimating the directivity and its variation with the scan angle. The design may be further optimized by varying a_K and φ_{k1} utilizing the design guidelines in Section 6.4.

6.6 Numerical Case Studies

This section discusses a few case studies illustrating the described procedure in the previous sections and display sparse CRA characteristics deterministically.

6.6.1 A Three-Ring CRA

A three-ring CRA ($M = 3$) is designed using design procedure stated before. The design goals are set as:

- $D_o \geq 20$ dBi;
- HPBW $\leq 3°$;

- PSLL ≤ –20 dB;
- $d_{avg} \geq 2\lambda$ and $\eta_{sp} \geq 70\%$.

The computation of design parameters using Section 6.5 is presented next:

$$N_{TOT} = 10^{(20-2)/10} \sim 63 \text{ (odd using (6.18))} \tag{6.26}$$

$D_e = 2$ dBi is chosen considering the average gain of isotropic RE with ±0.5-dB variation across the hemisphere. Next, for a Taylor's distribution (–20 dB) [28],

$$R_{ap} = \frac{\sigma\lambda}{2} \bullet \frac{HPBW_o}{HPBW} = \frac{1.1398}{2}\left(\frac{51.17}{3}\right) \sim 10\lambda \text{ (using (6.19))} \tag{6.27}$$

A dense 3-ring CRA having aperture radii of 10λ, with $d_{ke} = 0.5\lambda$, will consist of 253 elements. In reference to it, a sparsity, $\eta_{sp} = (1 - 63/253)100 \sim 75\%$ (>70%), is achieved. The radii, a_k ($k = 1, 2$, and 3), are calculated using (6.25):

$$a_1 = 10\lambda \frac{1 - 1.618}{1 - 1.618^{3-1}} = 3.82\lambda \tag{6.28}$$

$$a_2 = a_1 \bullet 1.618^{2-1} = 6.18\lambda \tag{6.29}$$

$$a_3 = a_1 \bullet 1.618^{3-1} = 9.99\lambda \tag{6.30}$$

Hence, the radii are chosen as $a_k = \{3.82, 6.18, 9.99\}$ (in λ). Figures 6.7 and 6.8 show the variation of the first term of (6.21) with respect to the variation in the number of elements, N_k, and the radii of rings, a_k. It is observed that, with the variation in N_k, the magnitude of the first term varies and this variation can be used to decide the overall gain of CRA; for example, here it is chosen as to be 63 (the sum of the three rings' contribution).

Figure 6.8 shows that the main lobe width is controlled by the radii of the ring and the different plots show that the different curves corresponding to the different radii are not cophasal and this feature may be used to minimize the SLL by aligning maxima

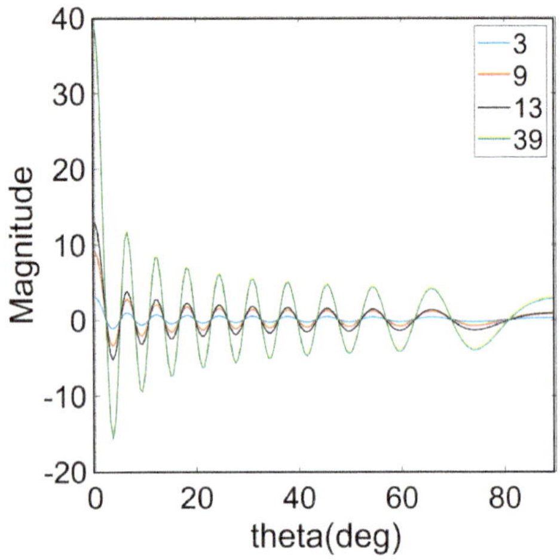

Figure 6.7
Variation of the first term with respect to the number of elements in a ring.

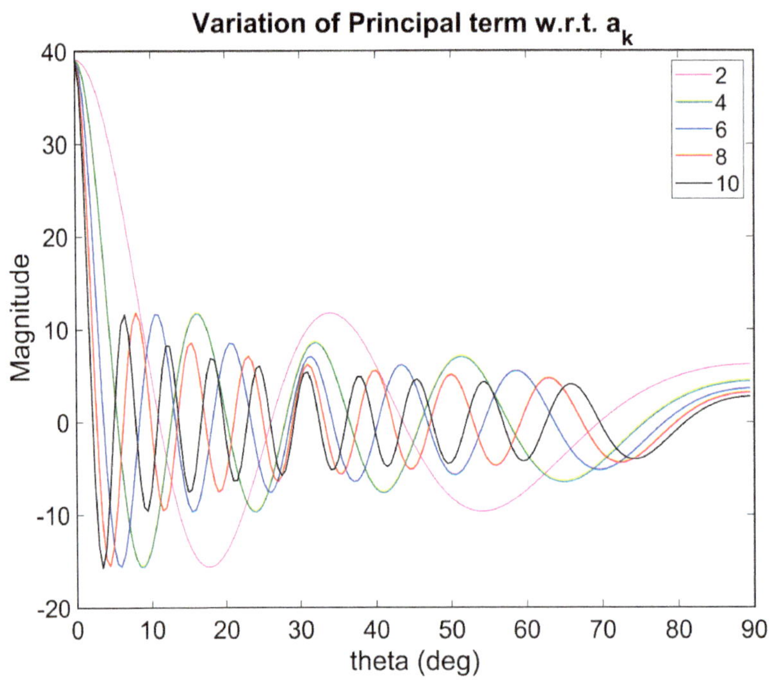

Figure 6.8
Variation of the first term with respect to the radii of the ring.

and minima of AF due to different rings to reduce the overall magnitude. The matrix $[N_k]$ is calculated using step 5 in Section 6.5, considering $\{d_{e1}, d_{e2}\} = \{1.2\lambda, 3\lambda\}$ (to achieve larger value of d_{avg}) as:

$$N_k = \begin{bmatrix} 0 & 11 & 13 & 17 & 19 & 0 \\ 13 & 17 & 19 & 23 & 29 & 31 \\ 23 & 29 & 31 & 37 & 41 & 43 \end{bmatrix} \tag{6.31}$$

In (6.31), $[N_k]$ satisfies condition 5b in Section 6.5 giving possible values for N_k in the three rings. Out of these combinations, the ones with $N_{TOT} \neq 63$ are simply rejected. A bookkeeping is necessary for rest of the combination to record PSLL using (6.11) or (6.21). The final combination selected is {13, 13, 37} as it gave a minimum PSLL. Table 6.1 summarizes the simulated design parameters.

Figure 6.9 shows the individual contributions from each of the three rings to the real part, the imaginary part, and the overall AF in (6.21). Figure 6.9(a) shows the plot of real(F_A) or Re(F_A) with a maximum at 0° ($\beta\rho_k = 0$) and decaying sidelobes. The maximas of rings 2 and 3 are in the opposite sense of phase with those of ring 3 leading to a minimal sum, that is, low SLL in the NSLL region (see the inset in Figure 6.9(a)). Figure 6.9(b) shows the plots of imag(F_A) or Im(F_A) with peaks corresponding to three rings appearing after 20° = asin $(N_2/(\beta a_2))$ (i.e., in the NNSLL region). It shows the effect of interference from higher-order indices, $m = \pm 1$ due to the second term in (6.17). The respective peaks add up in the same sense of phase leading to high PSLL; see the red curve in Figure 6.10(a). The combined plot shows high PSLL in NNSLL region due to the addition of Im(F_A) with Re(F_A). Also, these interferences lead to a nonsymmetric pattern with respect to φ-variation in NNSLL region,

Table 6.1
Design Parameters of 3-Ring CRA

Ring Index k	Number of Elements N_k	Ring Radii, a_k (in λ)	d_{ke} (λ)
1	13	3.82	1.846
2	13	6.18	2.987
3	37	9.99	1.698

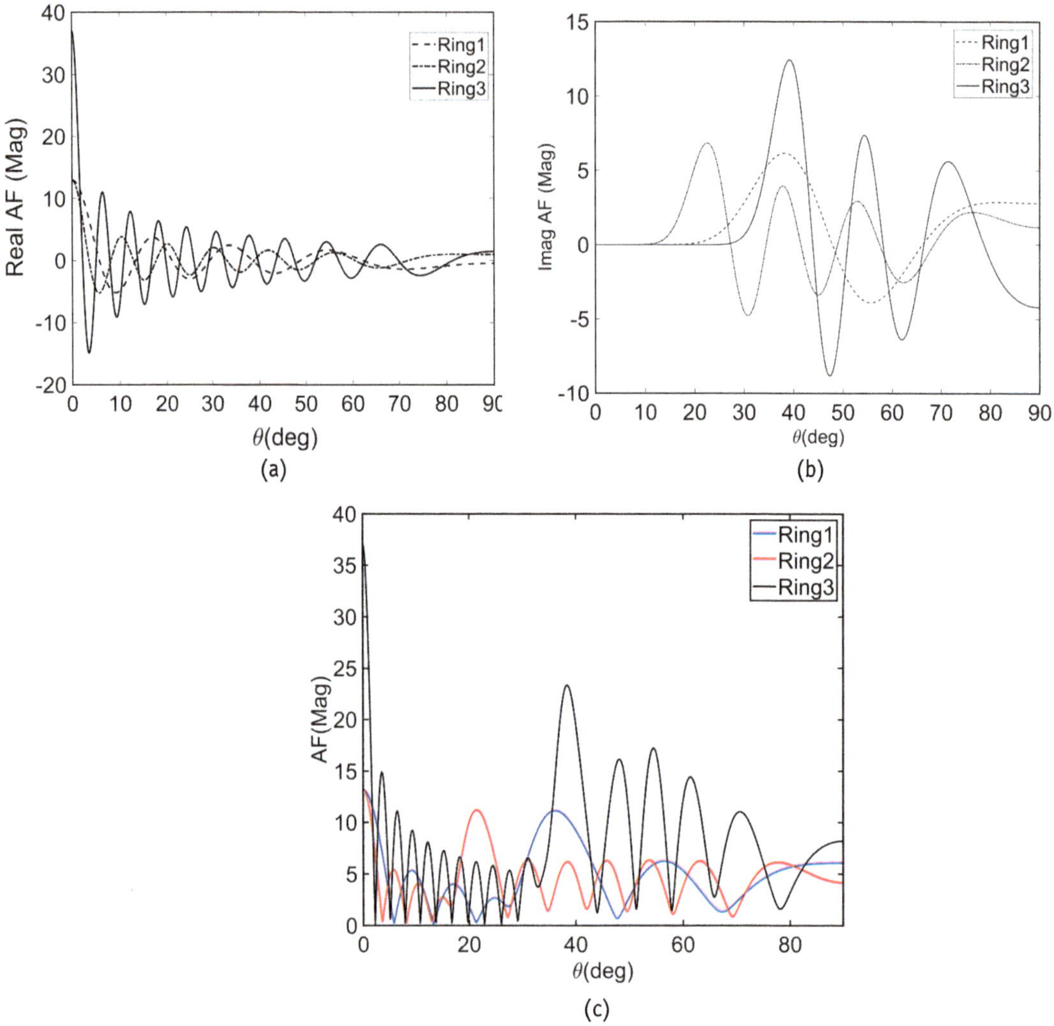

Figure 6.9

Array factor plot of the individual rings of the 3-ring CRA. (a) Real contribution of the three individual rings from the first term. (b) Imaginary contribution of the three individual rings from second term. (c) AF of the individual rings combining the two terms from (6.21). (© 2019 IEEE. Reprinted with permission from [25].)

shown in Figures 6.10(a–c). The average of d_{ke} gives $d_{avg} \sim 2.17\lambda$. Thus, $\eta_{sp} = 75\%$ and $d_{avg} > 2\lambda$ shows that designed CRA is highly sparse.

Figure 6.10(e) shows AF (in dB) for φ-cuts varying from 0° to 90° in steps of 15°. Figure 6.10(f) shows its 3-D plot with an increment of 0.1° in θ and φ (in the uv plane) showing an ASLL ~ –17.42

dB. Figure 6.10(e) shows PSLL < −14.74 dB and −11.46 dB in NSLL and NNSLL regions, respectively. However, Figure 6.10(f) shows that a PSLL of −9.06 dB is observed on taking finer samples in φ in the NNSLL region due to sparse topology. D_o = 22.4 dBi is computed using the generalized expression of directivity at boresight

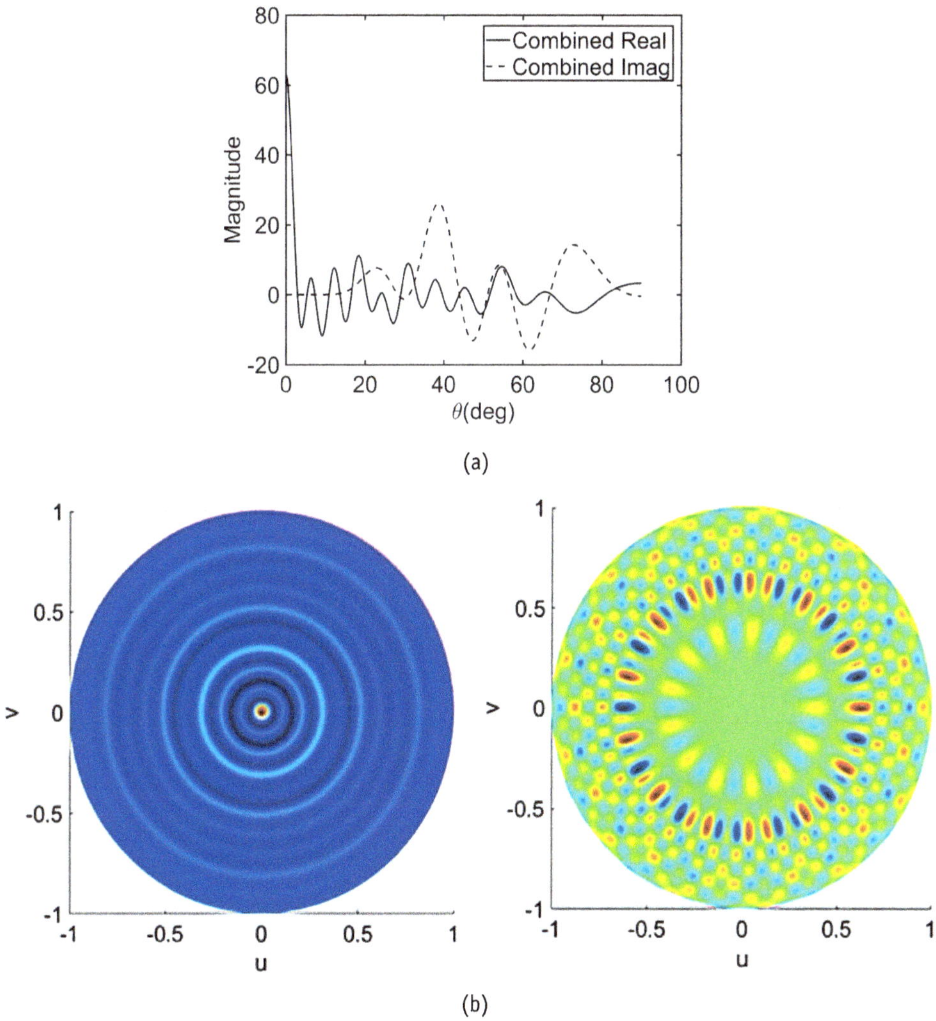

Figure 6.10
A 63-element, 3-ring CRA performance. (a) Combined magnitude of real and imaginary parts of AF from (6.17). (b) Combined magnitude of real (left plot) and imaginary (right plot) parts of AF in 3-D format. (c) Plot of magnitude of AF and AF in dB for the 63-element CRA. (d) Geometry layout of 63-element CRA. (e) Plot of AF across different φ-cuts. (f) The 3-D plot of the array factor. (© 2019 IEEE. Reprinted with permission from [25].)

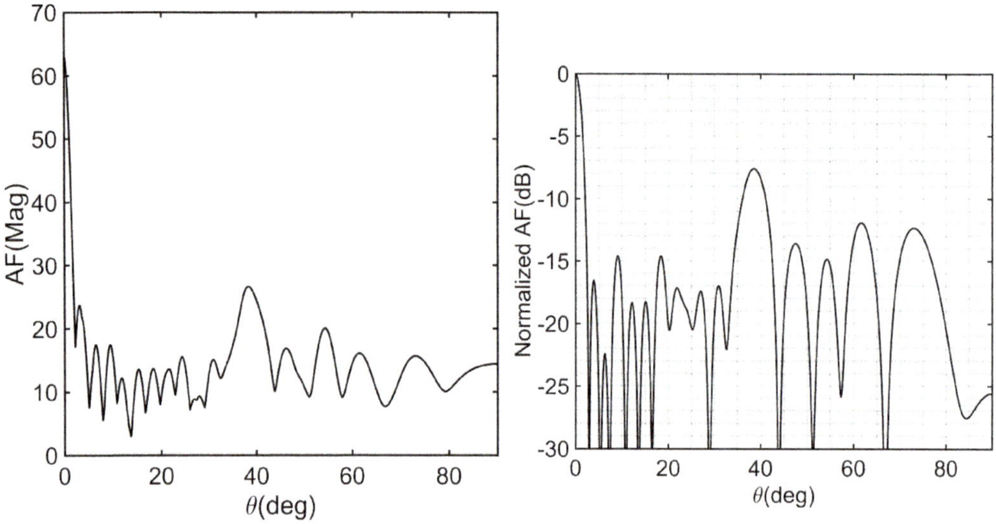

(c)

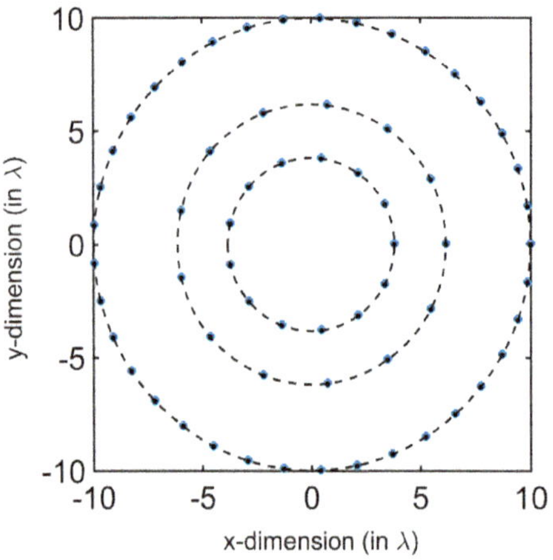

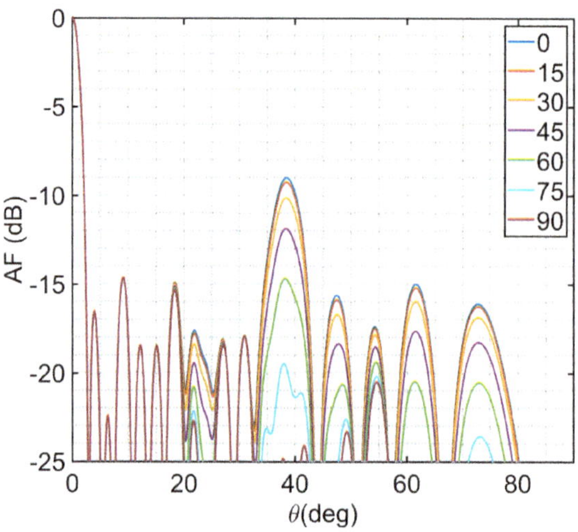

(d)

(e)

Figure 6.10
(Continued)

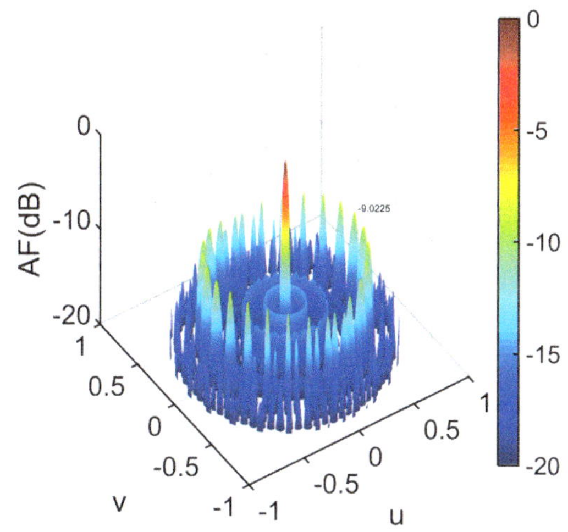

(f)

Figure 6.10
(Continued)

including the effect of MC. In a sparse array, statistical estimates of PSLL and ASLL are (3 to 6)/N_{TOT} (~−13.22 to −10.2 dB) and 1/N_{TOT} (~−17.99 dB), respectively [14, 15]. Thus, the designed CRA demonstrates results close to the optimal. The high PSLL in the NNSLL region is further optimized by varying, a_k and φ_{k1}, as will be shown next.

6.6.1.1 Deterministic Optimization in the NNSLL Region

The optimization of PSLL in the NNSLL region is carried out by tuning φ_{k1} ($\varphi_{k1} \neq 0$) and a_k based on the design guidelines in Section 6.4. The values of a_k are adjusted so that the contributions to the first term in (6.11) or (6.24) from individual rings add up to a minimal value. Further, φ_{k1} is varied in finer steps leading to the different projections along φ-cuts. It results into the low PSLL linear arrays along these φ-cuts. Thus, a low SLL envelope across the hemisphere in the NNSLL region is achieved by organizing respective maximas of Im(F_A) corresponding to the different rings arranged in an opposite sense of phase.

The variation of a_k and φ_{k1} concerning the inner two rings is varied in finer steps of $\lambda/10$ and $2\pi/(N_2 + N_3)$, respectively. The corresponding values of PSLL across a 360° variation in φ with incremental steps of 1° is noted. This could be accomplished by any iterative technique and sorting the results in the ascending order of PSLL. The combination yielding a minimum PSLL is chosen out of the various iterations. Here, initially, a_1 is varied followed by φ_{21} variation. This is followed by the variation in a_2 and φ_{31} in succession. The sequence shown is case-specific and may be varied for different cases. At each step of the optimization, the PSLL is noted. Table 6.2 summarizes the details of the steps followed during the optimization process and the final design parameters as obtained, where s_2 and s_4 vary from 1 to 26 (360° variation), and s_1 and s_3 vary from 1 to 10 (λ variation).

Figure 6.11 shows the optimized CRA's performance. Figure 6.11(b) shows the 2-D AF plot across different φ-cuts in the steps of 15°. Figure 6.11(b) shows appreciable reduction in the imaginary contribution in the NNSLL region (see Figure 6.11(c)). It shows that with variation in φ_{k1} there is no variation $\text{Re}(F_A)$ while $\text{Im}(F_A)$ varies, confirming the effect of the higher-order Bessel function. Figure 6.11(a) shows the optimized layout of 3-ring CRA. Figure 6.11(d) shows a 3-D plot of AF showing PSLL < –12.41 dB(<$3/N_{\text{TOT}}$) in the NNSLL region, a significant improvement by 3.98 dB in PSLL, along with ASLL < –19.8 dB (<$1/N_{\text{TOT}}$). The use of (4.15) from Chapter 4

Table 6.2
Optimization Process and Design Results

Steps	Process	Radii (λ) and Rotation Angle, φ_{k1} (rad)	d_{ke} (λ)
I	$a_1 \rightarrow a_1 \pm s_1 \bullet \frac{\lambda}{10}$ $\varphi_{21} \rightarrow \varphi_{21} \pm s_2 \bullet \frac{2\pi}{26}$	$a_1 = 3.5\lambda$; $a_2 = 5.88\lambda$; $a_3 = 9.936\lambda$; $\varphi_{01} = 0$; $\varphi_{02} = 0.242$; $\varphi_{03} = 0$	$d_{1e} = 1.69$; $d_{2e} = 2.84$; $d_{3e} = 1.68$ $d_{avg} = 2.16$
II	$a_2 \rightarrow a_2 \pm s_3 \bullet \frac{\lambda}{10}$ $\varphi_{31} \rightarrow \varphi_{31} \pm s_4 \bullet \frac{2\pi}{26}$	$a_1 = 4.056\lambda$; $a_2 = 5.88\lambda$; $a_3 = 9.936\lambda$; $\varphi_{01} = 0$; $\varphi_{02} = 0.242$; $\varphi_{03} = 0.1429$	$d_{1e} = 1.96$; $d_{2e} = 2.84$; $d_{3e} = 1.68$ $d_{avg} = 2.16$

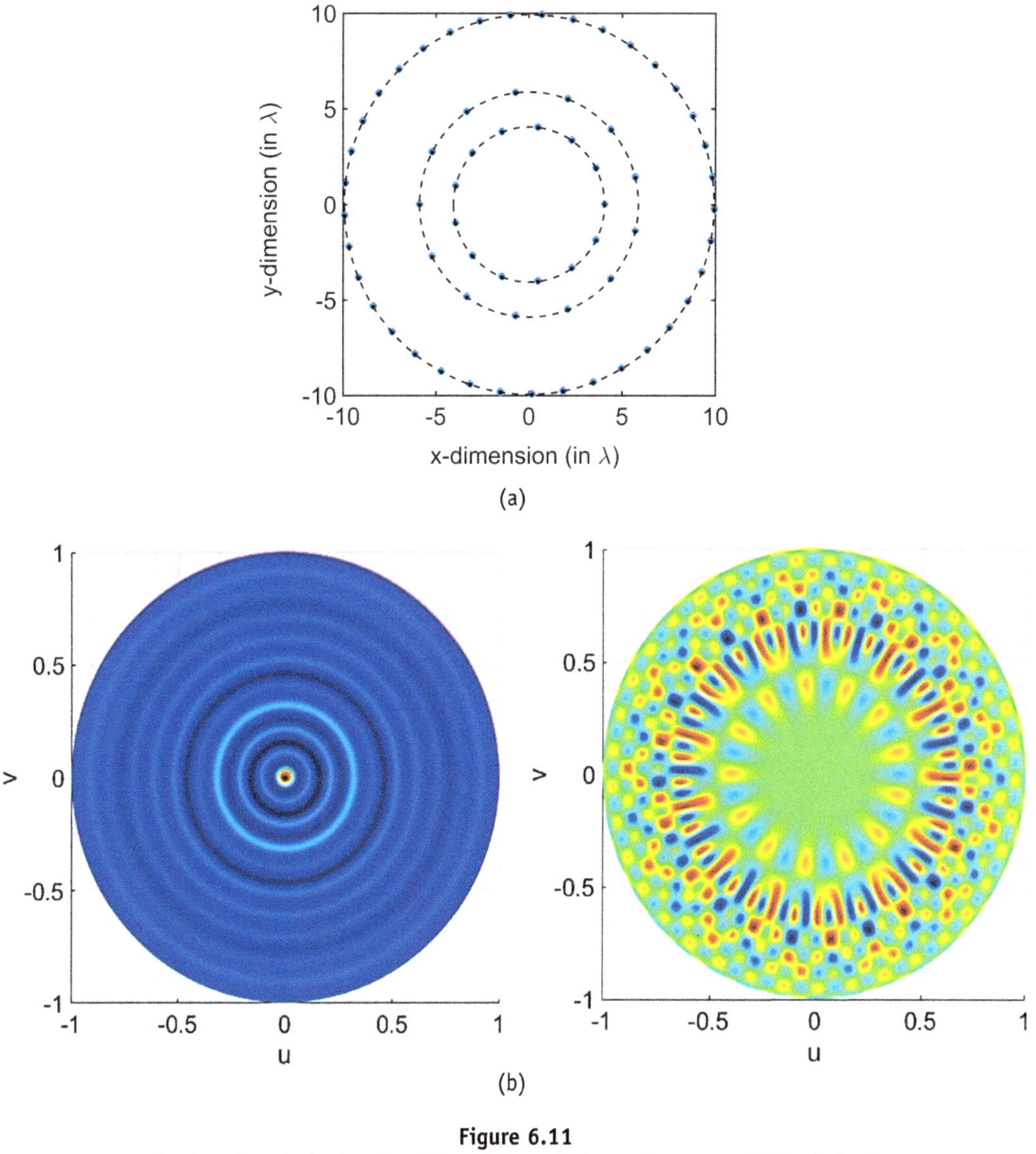

Figure 6.11
Design of optimized 3-ring CRA and its radiative performance. (a) Optimized geometrical layout of 63-element, 3-ring CRA. (b) Real and imaginary parts of AF, Re(F_A), and Im(F_A). (c) A 2-D AF plot with respect to different phi cuts. (Inset shows combined Re(F_A) and Im(F_A).) (d) A 3-D AF plot showing worst SLL over the hemisphere. (© 2019 IEEE. Reprinted with permission from [25].)

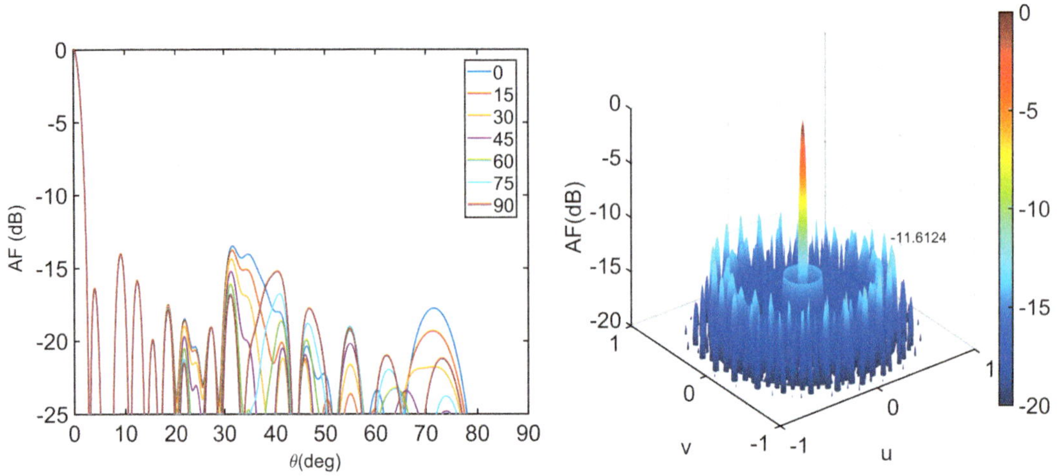

(c) (d)

Figure 6.11
(Continued)

gives D_o = 21.72 dBi including the effect of mutual coupling. Thus, optimized results as obtained are almost optimal.

6.6.1.2 Wide-Scan and Wideband Performance

Tables 6.3 and 6.4 show the simulated wideband and wide-scan performance of CRA using (4.15) to account for the effect of the MC on its performance. The variation in MC due to variation in d_{ke} or d_{avg} with respect to the change in f leads to the variation in D_o. However, since $d_{avg} > 2\lambda$, the MC is very small in magnitude and, thus, variation in D_o is small as shown in earlier work [14]. Table 6.3 shows the performance of CRA over a wide FBW = 1.55. The variations observed in D_o is 1.75 dB (with a maximum at $f/f_o = 1$) and in PSLL is 3.33 dB, reasonably good. Further, ASLL < –15.56 dB and beam broadening observed is in proportional to the variation in the electrical aperture $\propto f$. Table 6.4 shows the wide-scan performance of CRA simulated across 360° hemisphere in steps of 15° in φ_0 and 15° variation in θ_o. The variations observed in D_o is 1.43 dB and in PSLL is 1.67 dB, reasonably good over a wide-scan volume.

Table 6.3
Three-Ring CRA Wideband Performance

f/f_o	PSLL (dB)	ASLL (dB)	D_o (dBi)	HPBW (deg)
0.2	−13.98	−13.97	20.16	12.6
0.4	−13.98	−15.56	21.85	6.3
0.6	−12.69	−16.83	20.61	4.1
0.8	−11.13	−16.91	21.05	3.1
1.0	−11.13	−17.21	21.91	2.5
1.2	−11.13	−16.46	21.09	2.0
1.4	−11.12	−16.52	20.86	1.7
1.5	−10.65	−16.41	21.03	1.5

In order to summarize, a 3-ring CRA design with 63 elements is demonstrated a rotationally symmetric pattern with a sparsity of (1 – 63/217) ~ 70% and d_{avg} ~ 2.16λ, where 217 is the number of elements in the denser counterparts assuming three rings having same radii with elements separated by 0.5λ. It demonstrated a PSLL

Table 6.4
Three-Ring CRA Scan Performance at the Center Frequency

θ_o (deg)	φ_o (deg)	PSLL (dB)	ASLL (dB)	D_o (dBi)	HPBW (deg)
0	0	−11.13	−17.21	21.91	2.50
15	0	−11.59	−17.80	20.83	2.50
15	15	−11.14	−17.78	20.79	2.50
30	0	−10.01	−16.97	20.75	2.50
30	15	−10.04	−16.91	20.62	2.90
30	30	−10.07	−17.04	20.69	2.90
45	0	−9.83	−17.76	21.16	3.55
45	15	−9.85	−17.44	20.89	3.55
45	30	−9.84	−17.44	21.01	3.55
45	45	−9.74	−17.30	21.09	3.55
60	0	−9.62	−17.53	20.89	5.04
60	15	−9.46	−17.29	20.48	5.04
60	30	−9.46	−17.69	20.70	5.04
60	45	−9.56	−17.65	20.85	5.04

better than –10.65 dB and –9.56 dB over the scan volume and the operating bandwidth, respectively. The variation in the directivity is lesser than 1 dB over the scan and frequency ranges. Henceforth, a wideband and wide scan performance is illustrated.

6.6.1.3 Electromagnetic Simulation-Aided Validation of CRA

The simulated design of CRA is validated with an experiment carried out with realistic radiation modeled in Ansys HFSS based on the FEM technique. A practical radiator, U-slot fed microstrip patch antenna (UMSA), is modeled on a low loss dielectric with a superstrate, both realized from a 62-mil-thick Rogers 5880 substrate (ε_r = 2.2, loss tangent = 10^{-3}). UMSA operates over a wide bandwidth (more than 1 GHz) in X-band. The inset in Figure 6.12(a) shows the modeled geometry of UMSA. The antenna demonstrates a VSWR of 2:1 over a frequency range of 9 to 10.5 GHz (1.5-GHz bandwidth) with HPBW better than 90° in both E- and H-planes. A 63-element CRA is modeled using UMSA based on optimized design parameters listed in Table 6.2 as shown in Figure 6.12(a). Figure 6.12(b) shows the simulated reflection coefficient in decibels illustrating self-coupling and mutual coupling terms. The self-reflection coefficient is better than –10 dB over a bandwidth of 9.03 to 10.3 GHz in array environment consisting of all of the 63 elements being turned on, fed equally with unity amplitude. This demonstrates the wideband capabilities of the designed CRA. The worst-case coupling is better than –19.5 dB, demonstrating a low level of mutual coupling between the elements. Figure 6.12(c) shows the 3-D polar plot of the realized gain (in dBi) showing a maximum gain of 20 dBi with low SLL around the hemisphere as desired. The gain is lesser by 1.91 dB compared to simulations, which shows the effect of the coupling in the real scenario of inclusion of practical radiator modeling and the losses in the dielectric, conductor, and free space, included in the em simulations.

Figure 6.12(d) shows the 2-D plots of realized gain in three φ-planes, namely, 0° (red curve), 45° (blue curve), and 90° (black curve), respectively. It shows a worst PSLL better than –12.1 dB in agreement with Figure 6.11(c) (isotropic element assumption). However, it is seen from Figure 6.12(c) that a higher SLL is seen across the hemisphere better than –11 dB in agreement with Figure 6.11(d) (simulated with isotropic assumption of radiator). Figure 6.11(e)

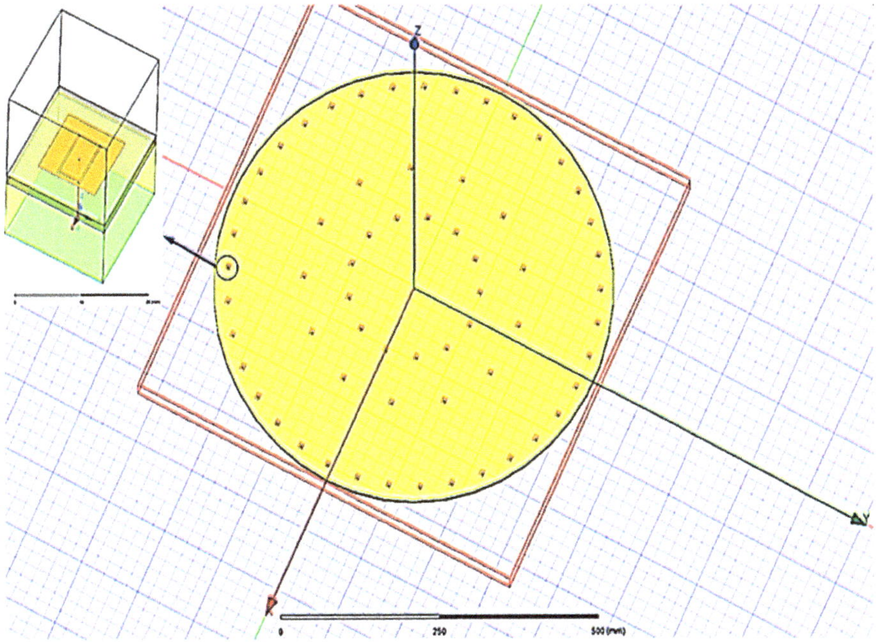

(a)

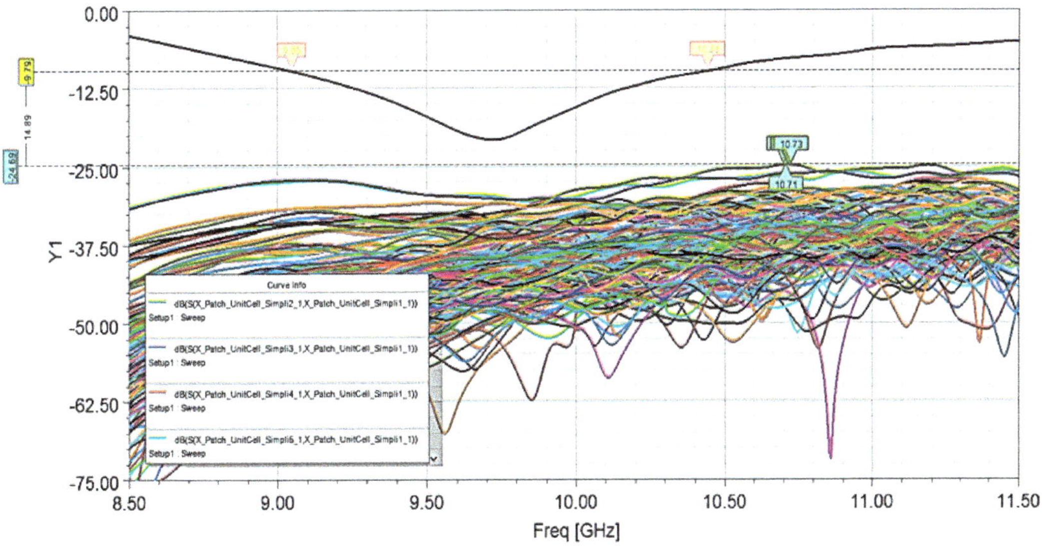

(b)

Figure 6.12

Full-wave modeling of a 63-element CRA and the results. (a) Geometrical model of CRA and UMSA (shown as an inset). (b) Reflection coefficient (in decibels): self-coupling and mutual coupling terms. (c) A 3-D polar plot of realized gain (in dB). (d) A 3-D polar plot of realized gain (in dB). (e) Scanned performance of array using HFSS.

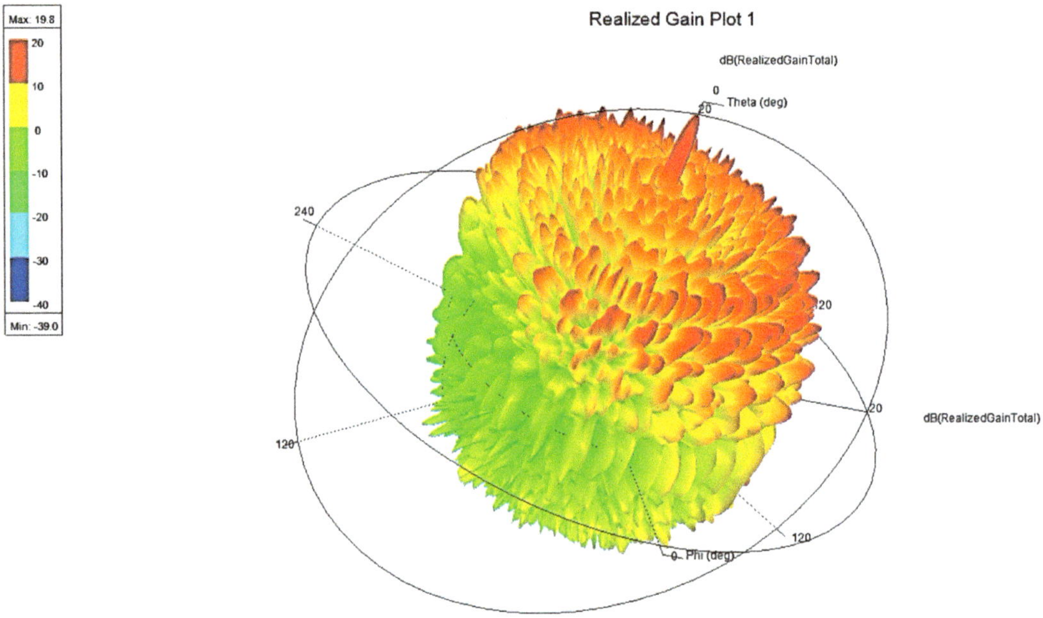

(c)

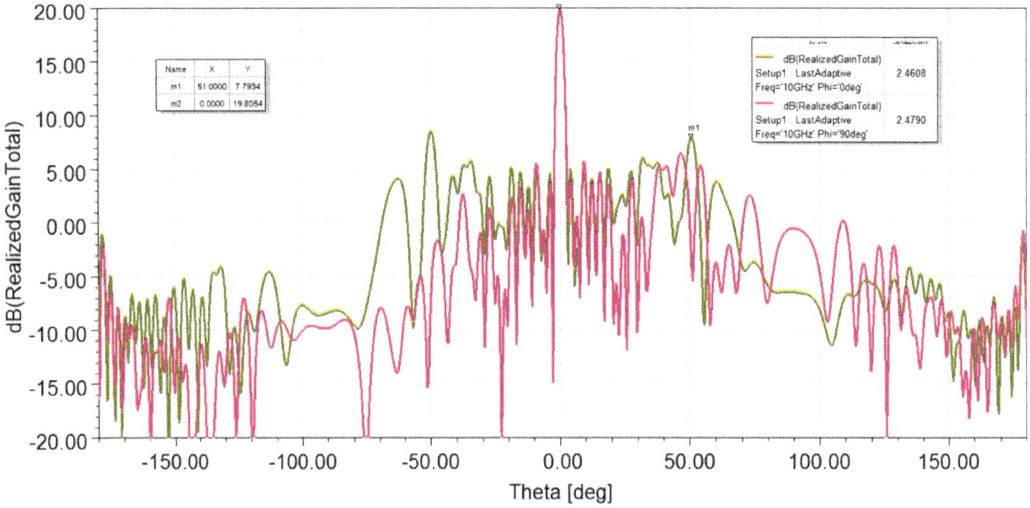

(d)

Figure 6.12
(Continued)

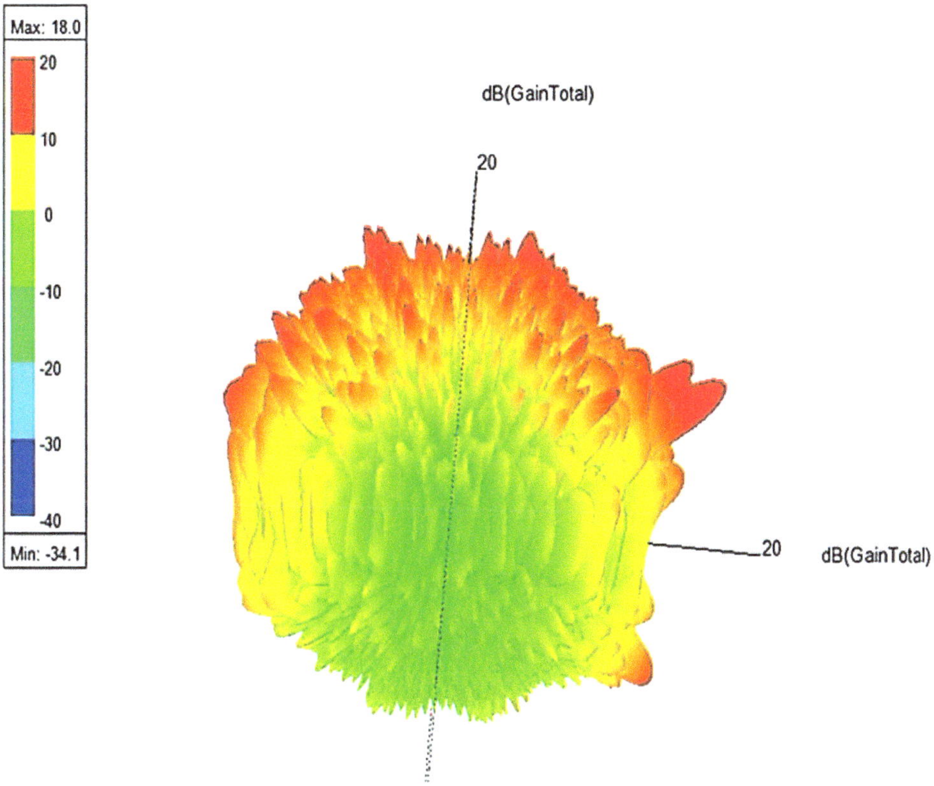

(e)

Figure 6.12
(Continued)

shows the scanned performance at a 60° scan angle showing no onset of any grating lobe, thus validating the wide scan performance using HFSS as well. The SLL is also in agreement with the MATLAB simulations. Hence, the accuracy of CRA sparse array design is validated with EM solver HFSS showing a very close agreement of results.

6.6.2 Benchmark Study of CRA with a Larger Aperture

The proposed design technique based on the Bessel function approach is validated with em solvers and, in the following sections, a few new designs are presented and validated with the existing literature.

6.6.2.1 Benchmark I

A 6-ring sparse CRA with $R_{ap} \sim 5\lambda$ is chosen as Benchmark I from an earlier reference by Zhao et al. [19]. A dense CRA with uniform interring and interelement separation (with $d_{ke} \sim 0.5\lambda$) consists of 341 elements distributed in 10 rings. Zhao et al. demonstrated a 6-ring CRA with 152 elements (one element at the center) [19]. Zhao's structure is resimulated here and it demonstrated PSLL ~ –23.1 dB, HPBW ~ 5.7°, D_o ~ –27.74 dBi, and η_{sp} = 55.4% (with respect to dense CRA). The center element is not desirable for monopulse tracking phased array radars requiring quadrant symmetry [1]; hence, sparse CRA without central element is synthesized using the present technique and presented below. Here, two designs of CRA with 5 and 6 rings are presented. Table 6.5 summarizes the comparison of the designed 6- and 5-ring CRA in with above CRA [19]. The 6-ring CRA demonstrates comparable performance, PSLL ~ –23.27 dB, HPBW ~ 5.9°, D_o ~ 25.56 dBi with only 142 elements (η_{sp} = 58.4%), lesser by 10 numbers of elements in comparison [19]. The 5-ring CRA aims at a higher η_{sp} = 69.2% ~ 70% and it demonstrates PSLL ~ –19.7 dB, HPBW ~ 5.8°, D_o ~ 22.61 dBi with only 105 elements. Figures 6.13(a, b) show the 3-D plots of AF for 6-ring and 5-ring CRAs (geometries shown as an inset) at boresight. A

Table 6.5
Benchmark I

CRA	a_k (λ)	N_k	d_{ke} (λ)	PSLL/ASLL (dB)
Zhao et al. [22]	[0.77 1.44 2.37 3.04 4.03 4.98]	[10 16 23 27 34 41] (N_{TOT} = 152) η_{sp} = 55.4%	[0.4838 0.5655 0.6474 0.7074 0.7447 0.7632] d_{avg} = 0.6520	–23.1/–23.2
M = 6	[0.8 1.5 2.35 2.90 3.98 4.9]	[11 17 19 23 43 29] (N_{TOT} = 142) η_{sp} = 58.3%	[0.4570 0.5544 0.7771 0.7922 0.5816 1.0616] d_{avg} = 0.704	–23.3/–23.1
M = 5	[0.77 1.56 2.53 3.58 4.83]	[7 13 21 33 31] (N_{TOT} = 105) η_{sp} = 69.2%	[0.69 0.75 0.73 0.68 0.97] d_{avg} = 0.767	–19.7/–23.5

comparable sidelobe level envelope with much higher sparsity is achieved in the designs in comparison [19].

6.6.2.2 Benchmark II

A sparse CRA with an aperture diameter of 8.73λ is considered from [22] as the next benchmark. Jiang et al. [29] synthesized a 7-ring CRA using GA initially, and later Zhao et al. [22] synthesized a 6-ring CRA using IWO with a center element. They demonstrated a PSLL of −24.58 dB and −24.7 dB, respectively, and achieved d_{avg} not better than 0.55λ (see Table 6.6), indicating that CRAs are not sparse

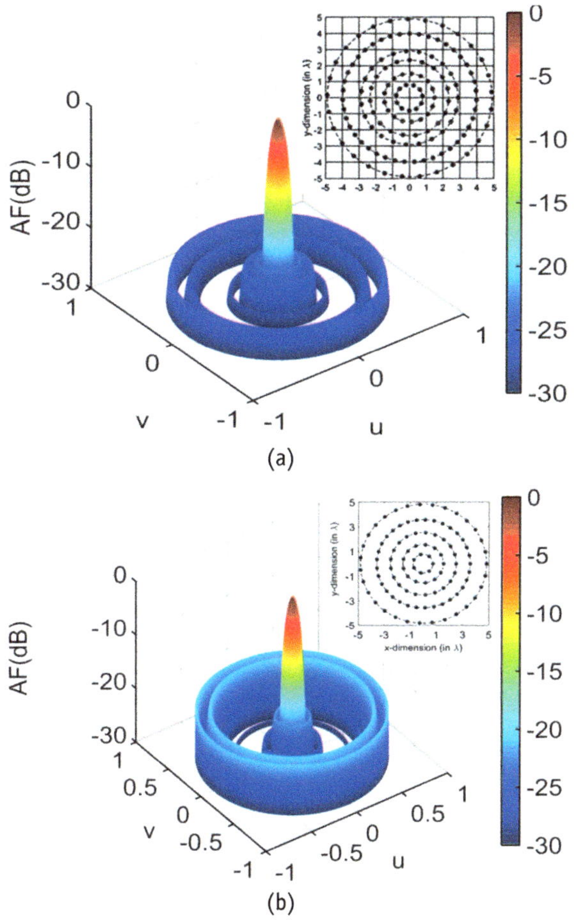

Figure 6.13
The 3-D AF plot of a CRA. (a) A 6-ring CRA. (b) A 5-ring CRA. (© 2019 IEEE. Reprinted with permission from [25].)

Table 6.6
Benchmark II

CRA	a_k (λ)	N_k	d_{ke} (λ)	PSLL/ASLL (dB)
[29] $M = 7$	[0.5 1 1.5 2 2.61 3.5 4.365]	[6 12 18 25 32 43 54] $N_{TOT} = 191$	[0 0.524 0.524 0.524 0.503 0.513 0.511 0.508] $d_{avg} = 0.451$	−24.6/−26.2
[22] $M = 6$	[0.75 1.39 2.06 2.64 3.62 4.365]	[10 15 20 24 31 36] $N_{TOT} = 137$	[0.471 0.582 0.647 0.691 0.734 0.762] $d_{avg} = 0.555$	−24.7/−25.2
$M = 6$	[0.55 0.965 1.5 1.96 2.72 4.43]	[5 7 13 19 31 29] $N_{TOT} = 104$	[0.6912 0.8661 0.7250 0.6482 0.5513 0.9598] $d_{avg} = 0.741$	−21.5/−21.7

and a strong mutual coupling definitely exists. Using the present technique, a 104-element, 6-ring sparse CRA is designed with PSLL < −21.54 dB and ASLL ~ −21.75 dB. Figure 6.14 shows a 3-D plot of AF. The number of elements is lesser by 87 and 33 in comparison to 191-element and 137-element CRA, respectively, and the geometry (shown as an inset in Figure 6.14) shows $d_{avg} = 0.741\lambda > 0.5\lambda$,

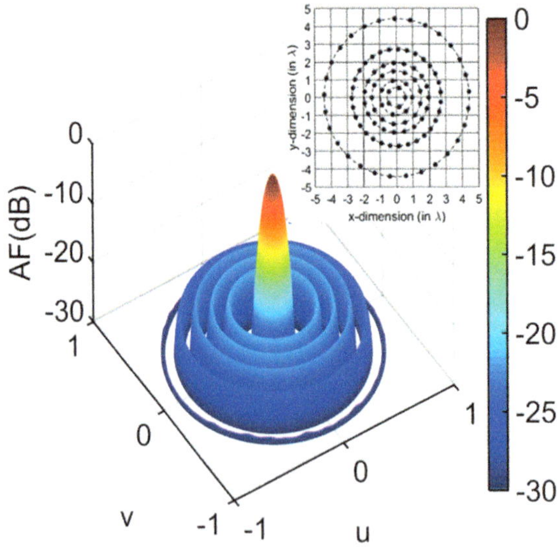

Figure 6.14
The 3-D AF plot for 6-ring CRA. (© 2019 IEEE. Reprinted with permission from [25].)

indicating medium sparsity in the design [14]. Table 6.6 summarizes the comparative performance of these arrays. Due to smaller R_{ap}, the effect of higher-order Bessel functions is observed in the FSLL region (nonuniformity in pattern) in Figure 6.14. Although the SLL in FSLL region is lower than –28 dB.

6.7 Fractal Concentric Ring Circular Arrays

Fractal antenna arrays are naturally sparse in nature leading to a lowering of the sidelobe level and narrow angular resolution. I find it interesting and worth discussing in this book as another approach for the realization for the concentric ring antenna arrays. Fractal is an object demonstrating self-similarity based on a recursively generated object. The term fractal was originally introduced by Mandelbrot [30], describing few complex shapes possessing an inherent self-similarity in the geometrical layout. Mandelbrot defined fractal as a set for which the Hausdorff dimension strictly exceeds its topological dimension. A fractal can have integer as well as noninteger dimension [30–34]. Jaggard et al. [31, 33] introduced fractal electrodynamics based on the fractal concept useful for studying a novel class of radiation, propagation, and scattering problems. A lot of fractal-shaped antennas and fractal antenna arrays geometries have been introduced in the literature. The self-similarity and self-scaling features incorporate the wideband nature inherently within the fractal shapes. The study of these two is called fractal antenna engineering [32]. The dimension of fractal, or fractal dimension, D, is defined as

$$D = \frac{\log n}{\log 1/f} \tag{6.32}$$

In (6.32), n is the copies of the original dimension scaled down by a factor, f. For instance, a square can be divided into n^2 copies of scale $1/n$. Fractal shapes have other properties such as scale-invariance, plane-filling or space-filling nature, and lacunarity. Lacunarity is the area of fractal having gaps (i.e., not filled or occupied).

Here, the focus is upon the design of CRAs using fractal engineering principles. The cantor fractal shapes are utilized to design one such antenna. For a cantor set, $D = 0.6309$.

6.7.1 Cantor Fractal Set

A basic cantor set [31] is an infinite set of points in the unit interval [0,1]. The points are 0, 1, 1/3, 2/3, 1/9, 2/9, 7/9, 8/9, 1/27, 2/27, ... The construction of cantor set may be described as:

1. Divide the unit interval [0,1] into three equal parts of length 1/3. Take away the middle interval [1/3, 2/3] but not the numbers 1/3 and 2/3.
2. This leaves the two intervals [0,1/3] and [2/3 and 1] of length 1/3 each.
3. Repeat step 1 for these two intervals leading to four intervals of length 1/9 each. Again, the middle section of each group is removed and the intervals remaining are [0, 1/9], [2/9, 3/9], [7/9, 8/9], and [8/9,1].
4. These steps can be repeated infinite times to generate a cantor set.

Thus, 2^n intervals of the length $1/3^n$ are generated after the nth iteration step. Figure 6.15 shows the cantor set generated after 4 iterative stages, S = 0, 1, 2, and 3. The fractal dimension for the set is $D = 0.6903$.

6.7.1.1 Design of Cantor Fractal Set-Based CRA

Figure 6.15 shows an 8-ring CRA (with radii of the rings as 5.0, 4.63, 4.26, 3.89, 2.78, 2.41, 2.04, 1.67 (in λ)) where the radii are chosen from the coordinates of the removed sections shown by the dotted lines. The elements are uniformly distributed in the rings with equal interelement spacing of 0.5λ. Therefore, the kth ring consists of $N_k = 2\pi R_k/d_k$ elements, where R_k is the radii and d_k is the interelement separation. Figure 6.16 shows the designed CRA with 10λ aperture with the array factor plot in azimuth. It is seen that except the first near SLL (~–16.4 dB), the worst SLL in the complete angular region is below –20 dB.

This is an example of regular self-similar cantor array. A further reduction in the SLL is possible by employing the optimization techniques based on the stochastic techniques such as GA and PSO, to introduce randomization in the CRA leading to the design of random fractal CRAs [32]. The technique can be combined with the Bessel function approach discussed previously to generate sparser CRAs with better SLL performance.

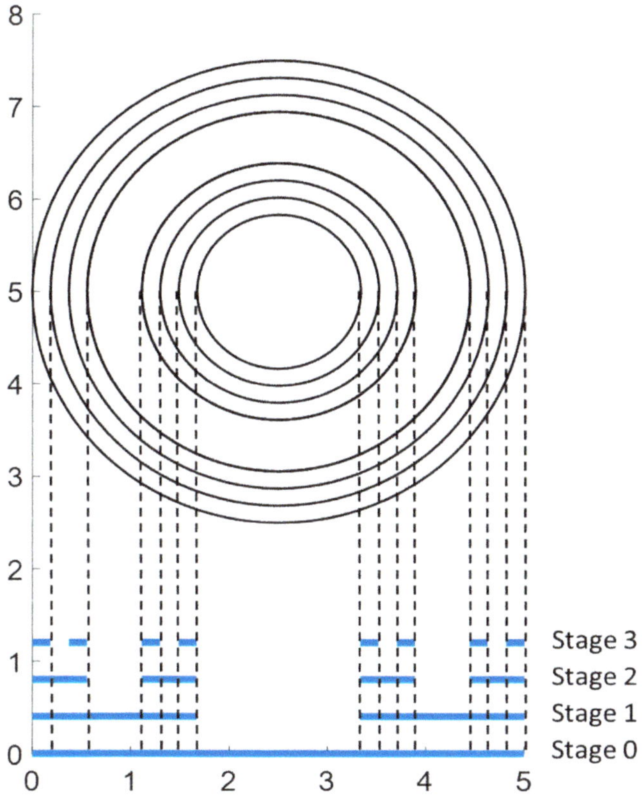

Figure 6.15
Cantor fractal set-based CRA.

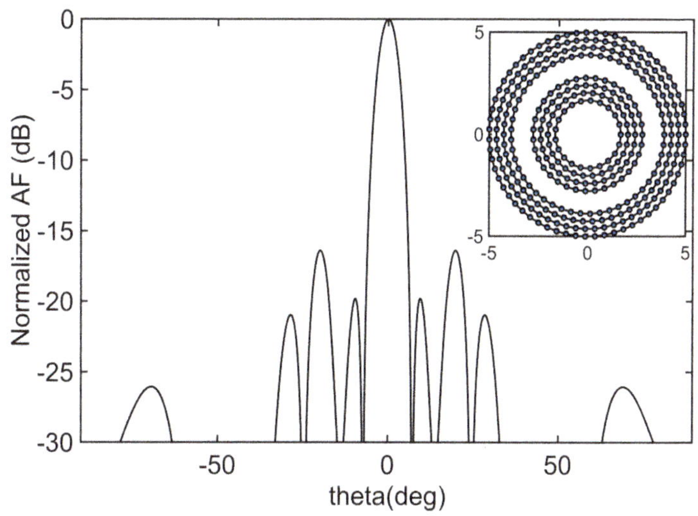

Figure 6.16
Cantor CRA and array factor plot in azimuth plane.

6.8 Spiral Concentric Ring Circular Arrays

The spiral concentric ring arrays is another form of CRAs that again inherit the properties of fractal antenna arrays. Spirals are the most common form visible in nature in most of the cases such as snail shell, milky way arms, sunflower seed, DNA double helix, hurricanes, vine tendrils, sea shells, pine fruit, and cactus (see Figure 6.17). Most of these shapes can be described by using the Fermat's spiral mathematically [35–38]. Spirals offers great potential to the antenna arrays' design by offering a good distribution of energy due to sidelobes and the grating lobes and providing good space filling of antenna aperture by the antenna elements.

A Fermat spiral encloses an equal area within each curve and can be described mathematically using (6.32) and (6.34).

$$\xi_n = d_0\sqrt{\frac{n}{\pi}};\ n = 1,\ldots,N \tag{6.33}$$

$$\varphi_n = 2\pi n \bullet GR;\ n = 1,\ldots,N \tag{6.34}$$

N is the total number of elements in an antenna array, d_0 is the distance between two adjacent elements in the xOy plane, GR is the

Figure 6.17
Illustrations of spirals existing in nature. (Source: Internet.)

golden ratio, ξ_n represents the distance of the nth element from the center of the spiral, and φ_n is the angular location of the element from the horizontal axis of the xOy plane. This may also be referred to as a uniform spiral antenna array.

Considering a circular aperture with diameter 15λ filled with 200 elements, using the above equations, a spiral antenna array is drawn as shown in Figure 6.18. Figure 6.18(a) shows the geometry of the spiral antenna array, and Figure 6.18(b) shows the radiation pattern assuming that the array is generated with isotropic antenna radiators with the ground plane. The computed array parameters are $D_o = 25.7$ dBi, HPBW = 2°, PSLL < –17.2 dB, and ASLL = –24.2 dB.

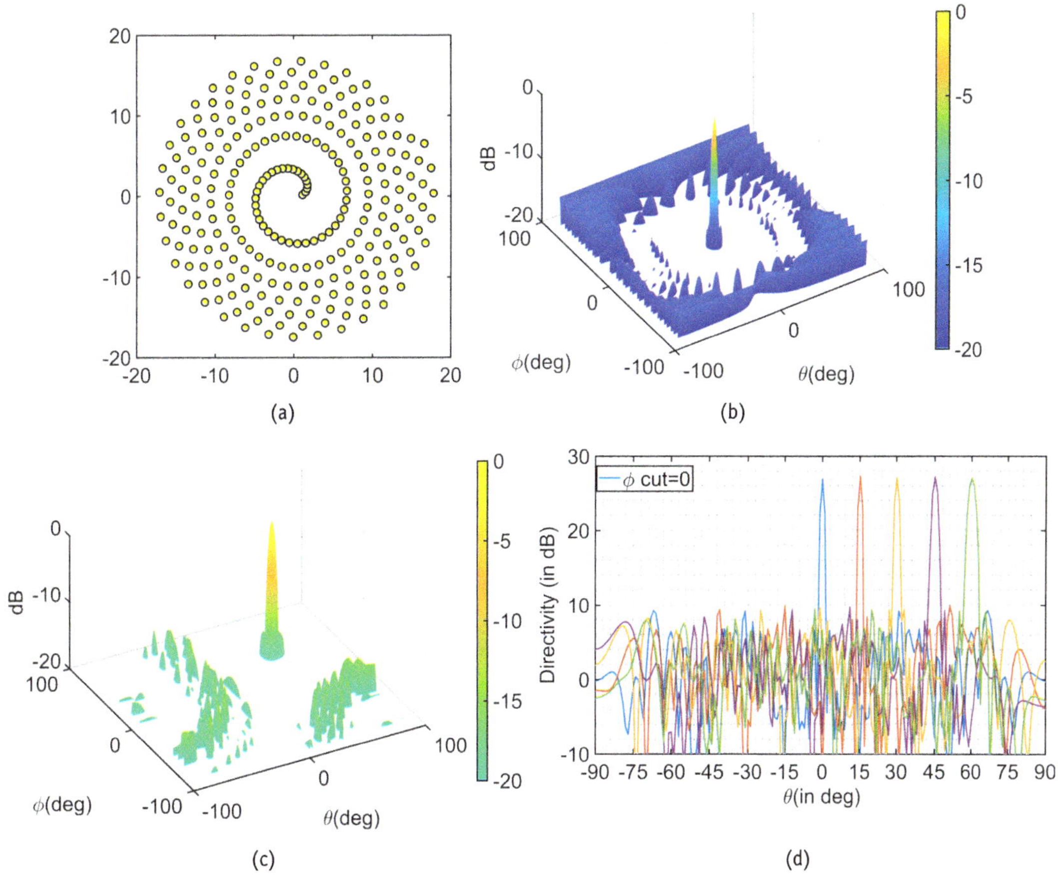

Figure 6.18
Spiral antenna array performance. (a) Geometry of spiral array. (b) Normalized radiation pattern of spiral array. (c) Scanned radiation pattern at (45°, 45°). (d) A 2-D scanned radiation pattern.

Figure 6.18(c) shows the scanned performance with PSLL < –15.18 dB at scan angle (45°, 45°). Thus, a wide scan performance is shown with 2-dB degradation in PSLL. Figure 6.18(d) further shows the wide scan performance of array in terms of the plot of directivity computed using the exact equation from Chapter 4. The variation in the directivity is observed to be less than –1 dB until 60° showing a wide scan performance. A further minimization of SLL is also possible by employing spatial density distribution dictated by a prescribed amplitude distribution such as Taylor [35, 39–41].

Vigano et al. [35] proposed the spatial density tapering based on the equation given in (6.35) and readers are encouraged to read the reference in detail.

$$2\pi \int_{R_{n-1}}^{R_n} A(r) r \, dr = \frac{2\pi}{N} \int_{0}^{R_{ap}} A(r) r \, dr \tag{6.35}$$

R_n and R_{n-1} represents the radii of the two rings enclosing the nth element, N is the total number of elements in the array, R_{ap} is

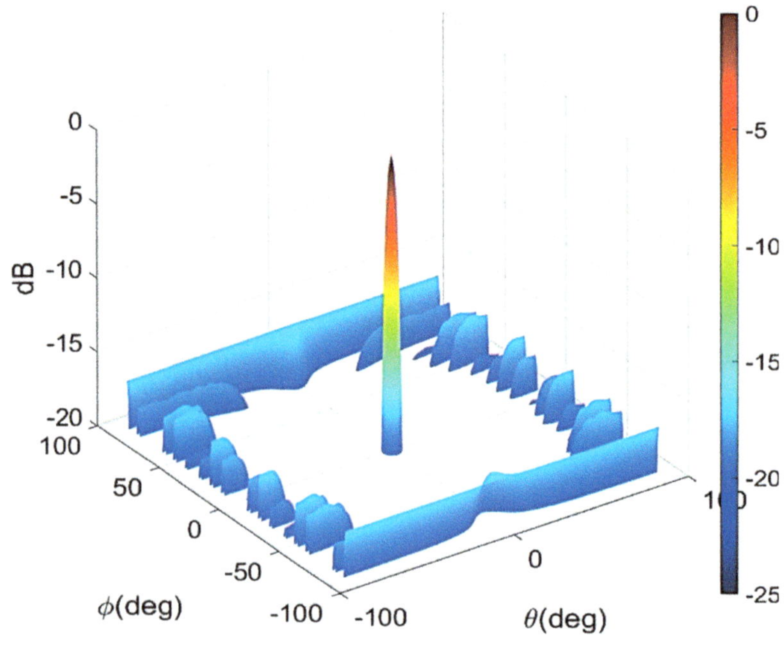

Figure 6.19
Radiation pattern of a spatially distributed antenna array.

the radius of the circular aperture of array, r is denoted the distance for the center of array, and $A(r)$ represent the desired amplitude distribution. Thus, (6.28) emulates the required aperture current distribution by equating the current distribution in the nth annular ring defined be two radii, R_n and R_{n-1}, equal to the Nth part of the total aperture current. Thus, (6.27) and (6.28) together give the number of elements in the nth annular ring and their angular location, respectively.

The number of elements is 274 in aperture with [8, 23, 35, 42, 45, 39, 36, 46] elements distributed in eight rings, respectively. Figure 6.19 shows the normalized radiation pattern in decibels. However, an increase in not near SLL and far SLL is observed. The near SLL is better than –29 dB and the worst SLL in the far SLL region is –16.9 dB. However, referring to the Bessel function approach, if it is combined with the present approach, a much lower far SLL can be achieved as well deterministically.

References

[1] Balanis, C. A., *Antenna Theory: Analysis and Design*, Second Edition, New York: John Wiley & Sons, 1997.

[2] Stearns, C., and A. Stewart, "An Investigation of Concentric Ring Antennas with Low Sidelobes," *IEEE Transactions on Antennas and Propagation*, Vol. 13, No. 6, 1965, pp. 856–863.

[3] Kaifas, T. N., et al., "Multibeam Antennas for Global Satellite Coverage: Theory and Design," *IET Microwaves, Antennas & Propagation*, Vol. 10, No. 14, 2016, pp. 1475–1484.

[4] Bianchi, D., S. Genovesi, and A. Monorchio, "Constrained Pareto Optimization of Wide Band and Steerable Concentric Ring Arrays," *IEEE Transactions on Antennas and Propagation*, Vol. 60, No. 7, 2012, pp. 3195–3204.

[5] El-Kamchouchi, H., "Ultra-Low-Sidelobe-Level Concentric-Ring-Array Pattern Synthesis Using Bessel Neural Networks [Antenna Designer's Notebook]," *IEEE Antennas and Propagation Magazine*, Vol. 52, No. 4, 2012, pp. 102–105.

[6] Salas-Sánchez, A. Á., et al., "Synthesis of Taylor-Like Patterns with Uniformly Excited Multi-Ring Planar Antennas," *IEEE Transactions on Antennas and Propagation*, Vol. 62, No. 4, 2014, pp. 1589–1595.

[7] Kumar, B. P., and G. R. Branner, "Design of Low Sidelobe Circular Ring Array by Element Radius Optimization," *IEEE Antennas and Propagation Society International Symposium. 1999 Digest*, Orlando, FL, July 11–16, 1999, pp. 2032–2035.

[8] Das, R., "Broadbanding of Concentric Planar Ring Arrays by Space Tapering," *Radio and Electronic Engineer*, Vol. 33, No. 4, 1967, pp. 211–222.

[9] Milligan, T. A., "Space-Tapered Circular (Ring) Array," *IEEE Antennas and Propagation Magazine*, Vol. 46, No. 3, 2004, pp. 70–73.

[10] Haupt, R. L., "Optimized Element Spacing for Low Sidelobe Concentric Ring Arrays," *IEEE Transactions on Antennas and Propagation*, Vol. 56, No. 1, 2008, pp. 266–268.

[11] Jiang, Y., et al., "Synthesis of Uniformly Excited Concentric Ring Arrays Using the Improved Integer GA," *IEEE Antennas and Wireless Propagation Letters*, Vol. 15, 2016, pp. 1124–1127.

[12] Guo, Q., C. Chen, and Y. Jiang, "An Effective Approach for the Synthesis of Uniform Amplitude Concentric Ring Arrays," *IEEE Antennas and Wireless Propagation Letters*, Vol. 16, 2017, pp. 2558–2561.

[13] Special Issue, "Innovative Phased Array Antennas Based on Non-Regular Lattices and Overlapped Subarrays: Part 1," *IEEE Transactions on Antennas and Propagation*, Vol. 62, No. 4, 2014.

[14] Kedar, A., and L. P. Ligthart, "Wide Scanning Characteristics of Sparse Phased Array Antennas Using an Analytical Expression for Directivity," *IEEE Transactions on Antennas and Propagation*, Vol. 67, No. 2, 2018, pp. 905–914.

[15] Bucci, O. M., and D. Pincherra, "A Generalized Hybrid Approach for the Synthesis of Uniform Amplitude Pencil Beam Ring-Arrays," *IEEE Transactions on Antennas and Propagation*, Vol. 60, No. 1, 2012, pp. 174–183.

[16] Angeletti, P., G. Toso, and G. Ruggerini, "Array Antennas with Jointly Optimized Elements Positions and Dimensions Part II: Planar Circular Arrays," *IEEE Transactions on Antennas and Propagation*, Vol. 62, No. 4, 2014, pp. 1627–1639.

[17] Bucci, O., S. Perna, and D. Pinchera, "Advances in the Deterministic Synthesis of Uniform Amplitude Pencil Beam Concentric Ring Arrays," *IEEE Transactions on Antennas and Propagation*, Vol. 60, No. 7, 2012, pp. 3504–3509.

[18] Xu, K., et al., "Versatile Beam Forming with Concentric Excitations Based on Multiple Weighted Sinc or Bessel Function Distribution," *IEEE Transactions on Antennas and Propagation*, Vol. 61, No. 8, 2013, pp. 4082–4090.

[19] Zhao, X., Y. Zhang, and Q. Yang, "Synthesis of Sparse Concentric Ring Arrays Based on Bessel Function," *IET Microwaves, Antennas & Propagation*, Vol. 11, No. 11, 2017, pp. 1651–1660.

[20] Sun, G., et al., "Radiation Beam Pattern Synthesis of Concentric Circular Antenna Arrays Using Hybrid Approach Based on Cuckoo Search," *IEEE Transactions on Antennas and Propagation*, Vol. 66, No. 9, 2018, pp. 4563–4576.

[21] Carlin, M., G. Oliveri, and A. Massa, "Hybrid BCS-Deterministic Approach for Sparse Concentric Ring Isophoric Arrays," *IEEE Transactions on Antennas and Propagation*, Vol. 63, No. 1, 2015, pp. 378–383.

[22] Zhao, X., Q. Yang, and Y. Zhang, "A Hybrid Method for the Optimal Synthesis of 3-D Patterns of Sparse Concentric Ring Arrays," *IEEE Transactions on Antennas and Propagation*, Vol. 64, No. 2, 2016, pp. 515–524.

[23] Jiang, Y., et al., "A Hybrid Strategy Based on Weighting Density and Genetic Algorithm for the Synthesis of Uniformly Weighted Concentric Ring Arrays," *IEEE Antennas and Wireless Propagation Letters*, Vol. 16, 2017, pp. 186–189.

[24] Wolpert, D. H., and W. G. Macready, "No Free Lunch Theorems for Optimization," *IEEE Transactions on Evolutionary Computation*, Vol. 1, No. 1, 1997, pp. 67–82.

[25] A. Kedar, "Deterministic Synthesis of Wide Scanning Sparse Concentric Ring Antenna Arrays," IEEE Transactions on Antennas and Propagation, Vol. 67, No. 12, Dec. 2019, pp. 7387-7395.

[26] Abramowitz, M., and I. A. Stegun, *Handbook of Mathematical Functions with Formulas, Graphs, and Mathematical Tables*. Washington, D.C.: United States Department of Commerce National Bureau of Standards, 1964.

[27] Elliott, R. S., *Antenna Theory and Design*, Upper Saddle River, NJ: Prentice-Hall, 1981.

[28] Hansen, R. C., "Tables of Taylor Distributions for Circular Aperture Antennas," *IRE Transactions on Antennas and Propagation*, Vol. 8, No. 1, 1960, pp. 17–22.

[29] Jiang, Y., and S. Zhang, "An Innovative Strategy for Synthesis of Uniformly Weighted Circular Aperture Antenna Array Based on the Weighting Density Method," *IEEE Antennas and Wireless Propagation Letters*, Vol. 12, 2013, pp. 725–728.

[30] Madelbrot, B. B., *The Fractal Geometry of Nature*, New York: W. H. Freeman and Co., 1983.

[31] Peitgen, H. O., H. Jurgens, and D. Saupe, *Chaos and Fractals: New Frontiers of Science*, New York: Springer, 1992.

[32] Werner, D. H., et al., "Genetically Engineered Dual-Band Fractal Antennas," *IEEE Antennas and Propagation Society International Symposium*, Boston, MA, July 8–13, 2001, pp. 628–631.

[33] Jaggard, D. L., and A. D. Jaggard, "Cantor Ring Arrays," *Microwave and Optical Technology Letters*, Vol. 19, No. 2, 1998, pp. 121–125.

[34] Vinoy, K. J., et al., "Hilbert Curve Fractal Antenna: A Small Resonant Antenna for VHF/UHF Applications," *Microwave and Optical Technology Letters*, Vol. 29, 2001, pp. 215–219.

[35] Vigano, M. C., et al., "Sunflower Array Antenna with Adjustable Density Taper," *International Journal of Antennas and Propagation*, Vol. 2009, Article ID 624035, 2009.

[36] Lockwood, E. H., *A Book of Curves*, Cambridge, England: Cambridge University Press, 1967.

[37] Gray, A., *Modern Differential Geometry of Curves and Surfaces*, Boca Raton, FL: CRC Press, 1993, pp. 69–70.

[38] Lee, X., "Equiangular Spiral," http://xahlee.info/SpecialPlaneCurves_dir/EquiangularSpiral_dir/equiangularSpiral.html.

[39] Caratelli, D., G. Toso, and P. Angeletti, "On the Deterministic Synthesis of Aperiodic Ring Antenna Arrays," *8th European Conference on Antennas and Propagation (EuCAP 2014)*, The Hague, Netherlands, April 6–11, 2014, pp. 1250–1254.

[40] Angeletti, P., G. Toso, and G. Ruggerini, "Sparse Array Antennas with Optimized Elements Positions and Dimensions," *8th European Conference on Antennas and Propagation (EuCAP 2014)*, The Hague, Netherlands, April 6–11, 2014, pp. 3142–3145.

[41] Kedar, A., "Wide Bandwidth and Wide Scanning Phased Array Antennas," *2018 IEEE Indian Conference on Antennas and Propagation (InCAP)*, Hyderabad, India, December 16–19, 2018, pp. 1–4.

7

Planar Sparse Array Antennas

7.1 Introduction

This chapter focuses on the synthesis of planar sparse antenna arrays (PSAAs) in general. The origin of first concept on planar sparse antenna arrays dates back to the 1960s [1]. Sandler [1] gave the general analytical expressions for unequally spaced arrays and analyzed them in terms of the equivalent uniformly spaced arrays. He established the broadband qualities of these and demonstrated some equivalence between the amplitude and spatial variation in the uniformly and nonuniformly spaced arrays. Lindsay [2] studied the far-field distribution of nonuniform current distribution in the circular array. Ishimaru [4] was the first to conceptualize the design process of nonuniform arrays. Maffett [5] formulated the array factor for nonuniform arrays. Lo [6, 7] gave the concept of the space weighting in the antenna arrays to lower down the sidelobe level (SLL) of a uniformly fed antenna array (UFAAs) with nonuniform inert-element spacing. The first concept of random array thinning to improve the array performance was statistically shown by Maher and Cheng in 1962 [8]. A number of papers were published on the various new concepts being presented for the first time related to this class of nonuniform or SAAs [2–23]. Most of these techniques were based on deterministic techniques utilizing the available computer resources. With the advent of computationally efficient and faster computers, the iterative techniques gained popularity for the design of antenna arrays [23–48]. However, still these techniques have limitations and prove to be costly in terms of time and resource utilization if the size of an array is very large. Thus, the deterministic techniques are very useful to design SAAs and gain a physical insight into their characteristics.

This chapter discusses a succinct overview of various techniques available for the design of sparse antenna arrays. Further, the design methodologies that I developed based on the deterministic approach will be discussed in detail supported with the numerical case studies.

7.2 Design Techniques for Sparse Antenna Arrays

A flowchart summarizing the various techniques for the design of SAAs has been already given in Chapter 5. This section summarizes the various techniques for the design of SAAs based on my survey as:

- *Gradient-based optimizers* [45, 49–54]: They comprise most of the local optimization techniques and, as the name suggests, depend solely upon the gradient information to solve the problems. They can handle large design parameters and require customized tuning of the variables specific to the problem at hand. The limitations of these techniques are the readily available information about the gradient and the optimization cycle frequently trapping into the local optima. However, these techniques do not limit the popularity and usefulness of these procedures and the alternate ways to handle their limitations are suggested in the vast literature. A few of the examples for these techniques are conjugate gradient, least square, Fletcher-Reeves, compressive sensing (CS), and Bayesian CS (BCS).
- *Global or metaheuristic or stochastic techniques* [46, 47]: These techniques comprise mainly nature-based heuristic techniques such as the genetic algorithm (GA), particle swarm optimizer (PSO), gray wolf optimizer (GWO), invasive weed optimizer (IWO), and simulated annealing (SA). These techniques search for a global optimum and are very efficient in finding one. However, these techniques require larger

computation time, which increases proportionately with the size of the solution space. The solution space is dependent upon the size of the antenna array, which delimits its applications for very large antenna arrays.

- *Fourier transform-based techniques* [55–60]: Iterative Fourier transform (IFT) or iterative fast Fourier transform (IFFT)-based optimization is another computationally efficient technique. They utilize the concept that the aperture illumination and array factor of an antenna array forms a Fourier pair. The limitation of these techniques is the requirement of a regular grid to apply such a technique. However, nonuniform FFT is also proposed [60] to handle nonuniform array grid, but with its own limitations. These techniques are more useful for array thinning (removing elements from a regular grid) rather than sparse arrays where underlying grid are nonuniform.
- *Deterministic methodologies* [31–34, 36, 40, 41]: Deterministic methods form an interesting category of the synthesis techniques for sparse antenna array as they do not require any sort of iterations and in minimal steps may give a design of sparse antenna array as per the desired specifications. Most of these techniques are based upon space-tapering or density-tapering techniques. These techniques work efficiently irrespective of the size of an antenna. They give a very crisp and clear picture about the behavior of SAAs and their salient characteristics. However, they may sometimes give a suboptimal solution.
- *Hybridization of the above techniques* [24, 25, 44]: This category overcomes the limitations of all the above categories by combining few of the above categories leading to a computationally efficient and accurate methodology to arrive at a globally optimum solution. For instance, deterministic techniques may be used to provide a suboptimal solution that may provide an initial cue to stochastic techniques to arrive at a global optimum in minimal computational time.

7.3 Design Fundamentals

A PSAA may be characterized for its effectiveness for the various applications, especially in phased array antennas (i.e., sparse phased array antennas) based on its particular characteristic parameters such directivity (D_o), sidelobe level (SLL) envelope (PSLL and ASLL), and mean (or average) interelement spacing (d_{avg}). Chapter 4 presented an exact and approximate expression for the calculation of D_o and (4.14) is used in this chapter for the computation of the directive characteristics while accounting for the effect of the mutual coupling on the performance. The next section describes the projection technique (PT) or method and its use to estimate d_{avg}.

7.3.1 Mean or Average Distance and Projection Method

In this section, the concept of mean or average interelement spacing, d_{avg}, is introduced, which is an important parameter in predicting the performance especially of nonuniform (or sparse) antenna array; however, it is applicable for uniform (dense) arrays as well. Figure 7.1 shows a planar array lying in the Oxy plane and its associated coordinate system.

Referring to Figure 7.1, assume that all of the elements are uniformly fed (isophoric) and are radiating in the upper half space only

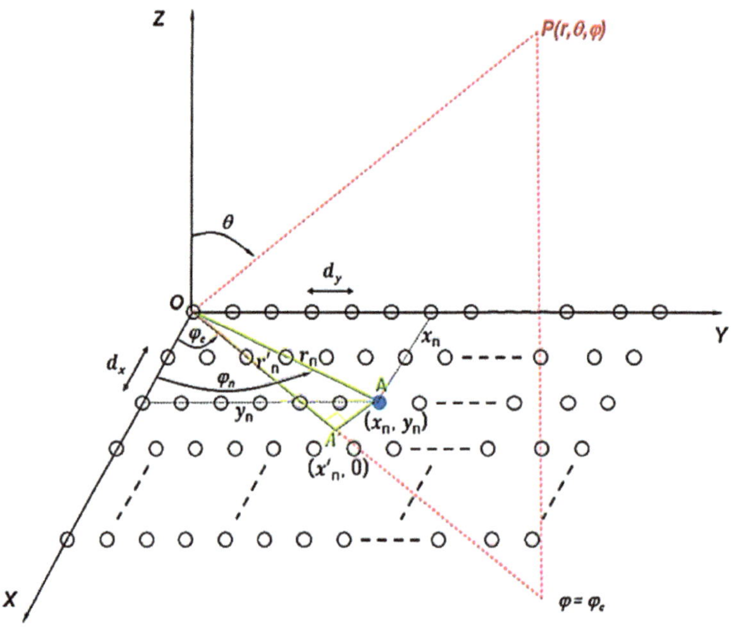

Figure 7.1
Arbitrary planar array and the reference coordinate system.

(assuming that isotropic antenna elements are backed by a ground plane, a realistic assumption for the phased array antennas). The Oxy plane accommodates the planar antenna array consisting of $N (= L \times M)$ elements, where L is the number of rows and M is the number of columns. The nth element is represented by $A(x_n, y_n)$ or $A(r_n, \varphi_n)$ in Cartesian or spherical coordinates, respectively, where $n = mL + \ell + 1$ ($m = 0, 1, 2, \ldots, M-1$; $\ell = 0, 1, 2, \ldots, L-1$). This way of numbering the antenna elements helps in considering any arbitrary array geometry for the analysis.

$$x_n = \ell d_x,\ y_n = md_y \tag{7.1}$$

$$r_n = \sqrt{x_n^2 + y_n^2} \tag{7.2}$$

$$\varphi_n = \operatorname{atan}\left(\frac{y_n}{x_n}\right) \tag{7.3}$$

The $\varphi = \varphi_C$ represents the φ-cut at which the far-field pattern of the array at $P(r, \theta, \varphi)$ is observed.

The PT is based on the fact that the distance from the point of the observation of the far field radiated from all of the scatterer/transmitter points on the radiating aperture is exactly the same (i.e., the far-field approximation [61, 68].) It is an established fact that the projection of a planar array on a particular $\varphi = \varphi_C$ cut is same as of a corresponding virtual linear antenna array of the same size as the projection [61–63]. It is possible to project all of the elements on a line corresponding to a particular $\varphi = \varphi_C$ cut, that is, shifted $x = x'$ axis (see Figure 7.1). Considering the nth element at position $A(x_n, y_n)$ or $A(r_n, \varphi_n)$ in the Oxy plane, and projecting it along the $\varphi = \varphi_C$ cut, the new location of that element becomes $A'(x_n', 0)$. The angle AOA′ becomes $(\varphi_n - \varphi_c)$. Hence, the shifted or projected distance of the nth element on a line corresponding to $\varphi = \varphi_C$ cut with respect to the origin, O, becomes:

$$r_n' = r_n \bullet \cos\left(\left|\varphi_n - \varphi_c\right|\right) \tag{7.4}$$

Thus r_n' corresponds to the new projected location of the array elements in a planar array, which is collapsed into a linear antenna array along the particular $\varphi = \varphi_C$ cut. The virtual amplitude of each of the elements in the projected linear antenna array will be decided by the fact that how many elements of a planar array have the same shifted location, r'_n (i.e., coincide) in the particular $\varphi = \varphi_C$ cut. This explains the reason why SLL as observed in the diagonal intercardinal planes is lower than the principal planes ($\varphi_c = 0°, 90°$) [61] for a uniformly fed array. This fact will be illustrated with the numerical test cases in the subsequent discussions.

The mean or average distance of the elements of a virtual linear antenna array along a particular $\varphi = \varphi_C$ cut is represented as:

$$d_{\text{avg}}(\varphi_c) = \frac{1}{N}\sum_{n=1}^{N}\left[r'_n - r'_{n-1}\right] = \frac{L(\varphi_c)}{N} \tag{7.5}$$

$L(\varphi_c)$ is the total length of the collapsed (or projected) linear array antenna along the φ_c cut. Thus, in a similar fashion, d_{avg} can be calculated at various φ_C-cuts and averaging these values will give the average or mean interelement spacing of the planar antenna array. The next sections present a numerical case study illustrating qualification of SAAs in terms of d_{avg} using the PT.

Figure 7.2 illustrates the mechanism of the formation of the projected (collapsed) linear antenna arrays along a φ_c-cut from a planar antenna array lying in the XY plane with an arbitrary-shaped boundary. A projection axis is shown at angle φ_c with respect to the x-axis where green dots show the projected location at which collapsed projections from array elements (in blue) are considered along the light blue lines as shown. For instance, it can be seen that A_1 and A_2 elements' projections collapse at the same projection location leading to a projected amplitude of 2. Similarly, it is seen that A_4 and A_6 elements' projections lead to the projected amplitude of 2 and A_3 leads to the projected amplitude of 1. In this manner, all the projected amplitudes as a result of projections from different array elements may be considered forming an amplitude-tapered collapsed linear antenna array emulating the performance of the PAA at φ_c-cut.

7.3.2 A Triangular Antenna Array

A 28- (4 × 7) elements, isophoric (equally fed), triangular array antenna is considered (see Figure 7.3(a)) as a numerical test case illustrating the application of the projection technique. It is assumed

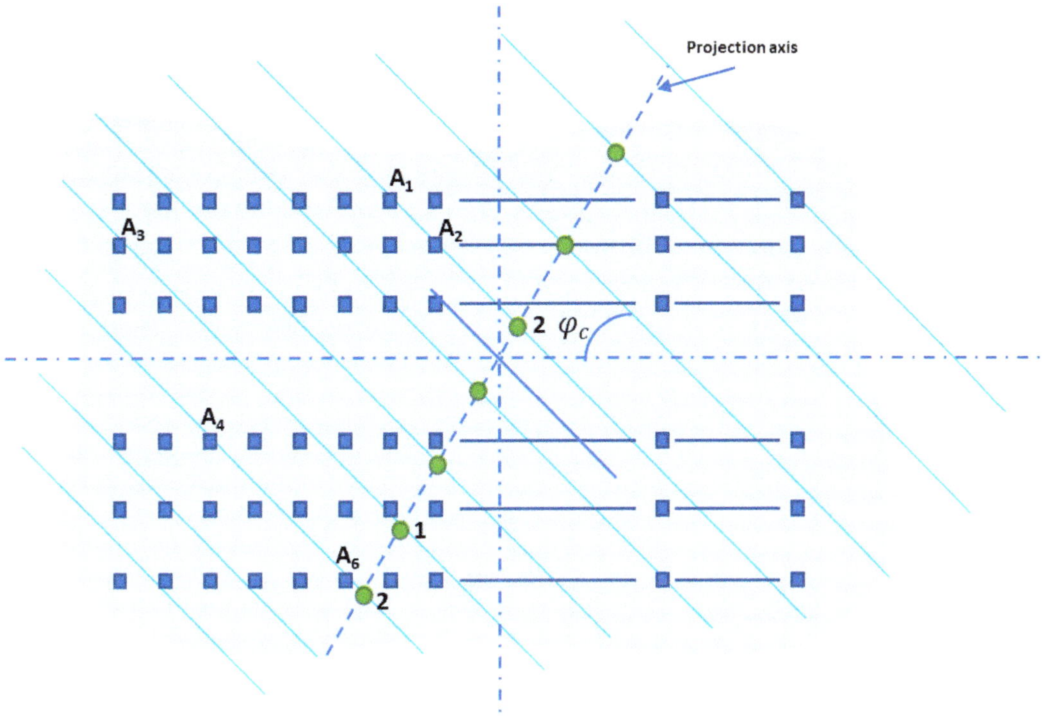

Figure 7.2
Illustration of the projection mechanism.

to radiate only in the upper half space, where $d_y = 0.25$ (along the y-axis) and $d_x = \sqrt{3}d_y$ (~0.433λ) (along the x-axis). Here, only 10 isotropic elements with an element directivity of 2 (isotropic radiator backed by a PEC ground plane) are excited uniformly and the arrows show the direction of E-field vector. Figure 7.3(b) shows the computed AF plot utilizing (4.14) for the antenna array and compares it with the collapsed LAA formed using the PT at $\varphi_c = 0°$. The two plots are in exact agreement except for different peak magnitudes. The difference is well-understood as PT assumes a virtual LAA with projected locations and does not consider the total number of elements in the actual PAA. The collapsed LAA at $\varphi_c = 0°$ (along the x-direction) consists of only 4 elements with equivalent amplitudes, namely, [4, 3, 2, 1] and maximum directivity, $D_o = 10\log(4) \sim 6.02$ dBi. Although the directivity of the planar array as calculated using the exact expression is 22.65 (13.5 dBi), the PT can used to analyze the array performance along different φ_c cuts and this will save the computation time in the case of large antenna arrays with a greater number of elements. Figures 7.3(c, d) show the array's scanned performance in two principal planes showing a wide scan performance with no onset of grating lobes (GLs).

Table 7.1 summarizes the array's performance at different φ_c cuts using the PT. The variables, NN, a_n, and x_n, represent the number of elements, effective amplitudes, and the respective locations in the virtual LAA at the respective φ_c-cuts. The array performance parameters (APMs) (i.e., d_{avg}, HPBW, ASLL, and PSLL) are also listed. It

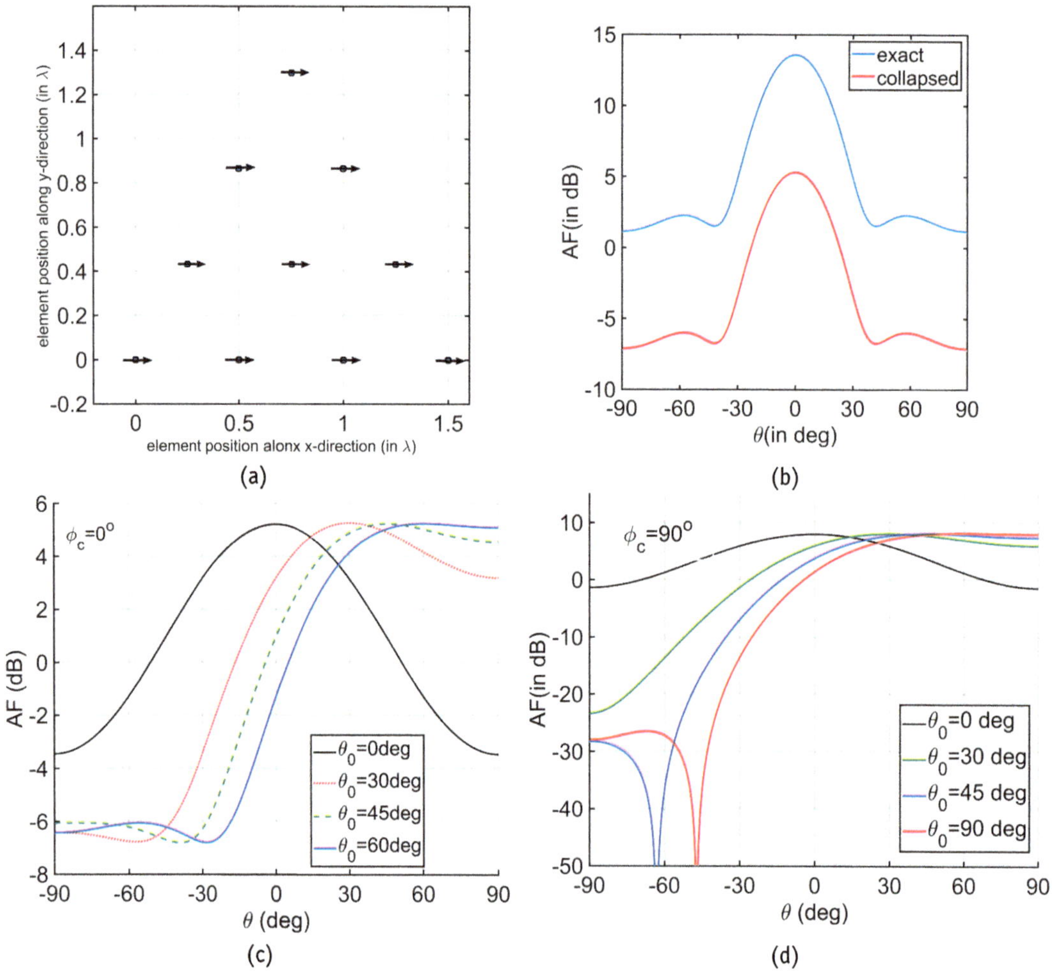

Figure 7.3
Triangular array performance. (a) A triangular array. (b) Array factor plot. (c) Scanned array performance ($\varphi_c = 0°$). (d) Scanned array performance ($\varphi_c = 90°$).

Table 7.1
Triangular Array Performance at Different φ_c-Cuts Using the PT

φ_c (deg)	*NN*	a_n	x_n (in λ)	d_{avg} (in λ)	HPBW (deg)	ASLL (dB)	PSLL (dB)
0	4	4 3 2 1	0.0000 0.4330 0.8660 1.2990	0.43	34.8	−11.7	−11.3
30	7	1 1 2 2 2 1 1	0 0.25 0.5 0.75 1.0 1.25 1.5	0.25	34.72	−38.7	−34.7
45	10	1 1 1 1 1 1 1 1 1 1	0 0.3536 0.4830 0.7071 0.8365 0.9659 1.0607 1.1901 1.3195 1.4489	0.16	34.76	−15.2	−14.4
60	4	1 2 3 4	0.0000 0.4330 0.8660 1.2990	0.43	34.8	−11.7	−11.3
90	7	1 1 2 2 2 1 1	0 0.25 0.5 0.75 1.0 1.25 1.5	0.25	34.72	−38.7	−34.7

is clearly visible that, even in uniformly fed dense antenna arrays, a lower SLL is achieved at intercardinal planes, such as −34.7 dB at $\varphi_c = 30°$. Thus, due to the varying sizes of the virtual LAAs with different lengths depending upon the respective φ_c-cuts, there is a nonuniform SLL envelope and the scanning performance.

Table 7.2 shows the d_{avg} for different interelement separation combinations with respect to the different φ_c-cuts. Thus, using the concept of mean spacing the sparsity in the antenna array can be quantified and the observed SLL variation along the hemisphere can be explained logically. It will be shown that the PT can be used to design wide scan, wideband sparse antenna arrays with low PSLL deterministically further in this chapter.

7.4 Array with Random Isotropic Element Allocation

It has been an established fact that the onset of GLs is inevitable while studying the characteristics of dense arrays and regular sparse array antennas. SAAs have an upper edge over their denser counterpart due to their random elements' locations leading to a GL-free wide scan behavior. The problem of GLs is tackled by disrupting the periodicity and introducing the randomness or aperiodicity in the array grid leading to nonuniform or aperiodic SAAs. Due to

Table 7.2
Average Interelement Spacing at Different φ_c-Cuts Using the PT

φ_c (deg)	0	30	45	60	90
$d_y = 0.125\lambda;\ d_x = \sqrt{3}d_y = 0.2165\lambda$	0.2165	0.125	0.08	0.2165	0.125
$d_y = 0.25\lambda;\ d_x = \sqrt{3}d_y = 0.433\lambda$	0.433	0.25	0.16	0.433	0.25
$d_y = 0.375\lambda;\ d_x = \sqrt{3}d_y = 0.6495\lambda$	0.6495	0.375	0.2415	0.6495	0.375
$d_y = 0.5\lambda;\ d_x = \sqrt{3}d_y = 0.866\lambda$	0.866	0.5	0.322	0.866	0.5
$d_y = 0.75\lambda;\ d_x = \sqrt{3}d_y = 1.299\lambda$	1.299	0.75	0.483	1.299	0.75
$d_y = 1.0\lambda;\ d_x = \sqrt{3}d_y = 1.732\lambda$	1.7321	1	0.644	1.7321	1
$d_y = 2.0\lambda;\ d_x = \sqrt{3}d_y = 3.46\lambda$	3.4641	2	1.2879	3.4641	2

the distribution of element locations randomly over the aperture, it requires adopting the statistical techniques to analyze these antennas and introduce statistical APMs for their performance characterization. This section discusses the characteristics of SAAs with random element allocation and calculate the normalized directivity as a function of the phasing angle and the mean distance, d_{avg}, between the elements. The assumptions made are:

- A square planar aperture array of the size $L \times L$ is considered.
- The elements have coordinates x_n and y_n ($n = 1, \ldots, N$) and are independent and uniformly distributed within the interval $(-L/2, L/2)$ with a probability density of, say:

$$W(x_n) = W(y_n) = \begin{cases} \dfrac{1}{L} & \text{for } |x_n| \le \dfrac{L}{2} \quad |y_n| \le \dfrac{L}{2} \\ 0 & \text{for } |x_n| > \dfrac{L}{2} \quad |y_n| > \dfrac{L}{2} \end{cases} \tag{7.6}$$

- The mean or average directive gain can be calculated as described before in Chapter 4. First, calculate the average of the expression in the denominator of (4.10) (to simplify, assume $\varphi_0 = 0$),

$$\begin{aligned}\overline{P_{n,m}} &= \overline{\cos\left[k\sin\theta_0\left(x_n - x_m\right)\right]\frac{\sin\left(kr_{nm}\right)}{kr_{nm}}} \\ &= \int_{-L}^{L}\int_{-L}^{L}\frac{\sin\left(k\sqrt{x^2+y^2}\right)}{k\sqrt{x^2+y^2}}\cos\left(k\sin\theta_0\right)W^1(x,y)\,dx\,dy\end{aligned} \tag{7.7}$$

In (7.7), quantities are average values and $W^1(x, y)$ is the probability density with $x = x_n - x_m$ and $y = y_n - y_m$ $(m \neq n)$, that is,

$$\begin{aligned}W^1(x,y) &= \frac{1}{L^2}\left(1-\frac{|x|}{L}\right)\left(1-\frac{|y|}{L}\right) \quad \text{for } |x| \le L \text{ and } |y| \le L \\ &= 0 \quad\quad \text{for } |x| > L \text{ and } |y| > L\end{aligned} \tag{7. 8}$$

Further, it may be assumed in the integral that $(1 - |x|/L)(1 - |y|/L) \approx 1$, the integration limits to be $\pm\infty$ and $\cos\theta_0 > 3/4\sqrt{2\lambda/L}$. Henceforth, (7.7) becomes:

$$\begin{aligned}\overline{P_{n,m}} &= \frac{1}{L^2}\int_{-\infty}^{\infty}\int_{-\infty}^{\infty}\frac{\sin\left(k\sqrt{x^2+y^2}\right)}{k\sqrt{x^2+y^2}}\cos\left(kx\sin\theta_0\right)dx\,dy \\ &= \frac{\pi}{kL}\int_{+\infty}^{\infty} I_0(kx)\cos\left(kx\sin\theta_0\right)dx = \frac{2\pi}{k^2L^2\cos\theta_0}\end{aligned} \tag{7.9}$$

Thus, the normalized value of the mean directivity becomes

$$\overline{g} = \frac{1}{1+\dfrac{2\pi N}{(kL)^2\cos\theta_0}} = \frac{1}{1+\dfrac{1}{2\pi\left(\dfrac{d_{\text{avg}}}{\lambda}\right)^2\cos\theta_0}} \tag{7.10}$$

with $d_{\text{avg}} = L/\sqrt{N}$ being the mean interval between the elements.

The different realizations of element allocations over the aperture lead to the mean value $\bar{g}$, valid for $\cos\theta_0 > 3/4\sqrt{2\lambda/L}$. This means that the antenna beam does not touch the aperture plane. The expression (7.10) for the mean $\bar{g}$ forms the basis for the analysis of sparse arrays with random element allocation.

In the case of high density of elements (dense array), if

$$\frac{1}{2\pi\left(\frac{d_{avg}}{\lambda}\right)^2 \cos\theta_0} \gg 1$$

then

$$\bar{g} = 2\pi\left(\frac{d_{avg}}{\lambda}\right)^2 \cos\theta_0 \quad (7.11)$$

Hence, the directivity varies as function of phasing angle in accordance with the cosine law. The mean normalized directive gain, $\bar{g}$, is proportional to the projection of the planar aperture in the phasing direction. In the case of the low density of element (sparse array), if $d_{avg}/\lambda \gg 1$, then:

$$\bar{g} \sim 1 \quad (7.12)$$

or

$$\bar{G} = N \quad (7.13)$$

Hence, the directivity does not depend upon the beam direction and equals approximately the total number of elements if mutual coupling is neglected, which is a valid assumption in the case of SAAs showing their superiority over the denser counterparts.

The mean normalized directive gain, $\bar{g}$, is proportional to the projection of the planar aperture along the phasing direction. The variation of $\bar{g}$ response as a function of the beam direction for different values of the mean distance between the elements is shown in Figure 7.4. Figure 7.5 shows the $\bar{g}$ as a function of the mean distance, d/λ, between the elements and the antenna beam direction as a parameter. The evaluation of the directivity for $\cos\theta_0 > 3/4\sqrt{2\lambda/L}$

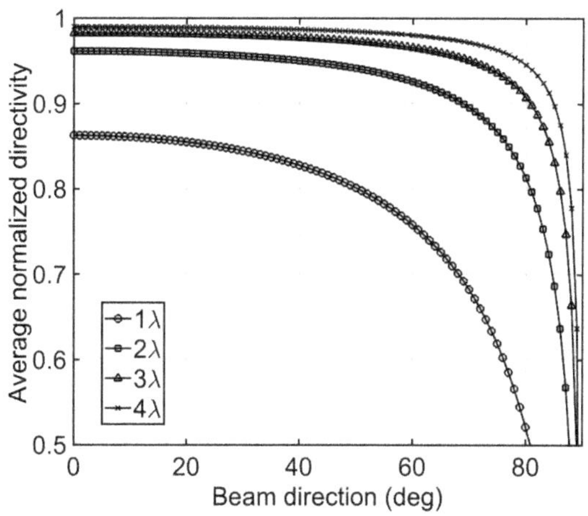

Figure 7.4
Normalized directivity versus the scan angle at different mean distances, d_{avg}.

is needed because the main beam is touching the aperture. The above condition is needed to avoid unwanted tangential beams along the aperture plane.

The evaluation of tangential beams can be done for $\theta_0 = 90°$, that is, when the main beam is touching the aperture.

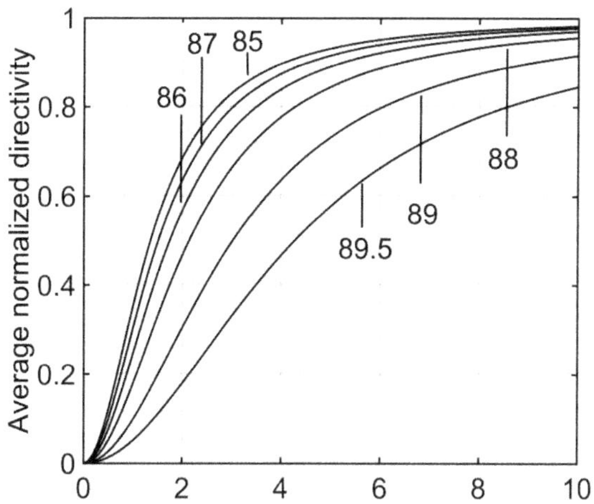

Figure 7.5
Normalized directivity versus d_{avg}/λ between the elements and the antenna beam direction.

$$\begin{aligned}\overline{P_{n,m}} &= \frac{2\pi}{kL^2}\int_{-\infty}^{\infty} I_0(kx)\cos(kx)\left(1-\frac{|x|}{L}\right)dx \\ &\simeq \frac{2\pi}{kL^2}\left[\begin{aligned}&I_0(kL)\cos(kL)+\frac{2}{3}I_1(kL)\sin(kL)\\ &-\frac{1}{3}I_2(kL)\cos(kL)-\frac{I_1(kL)}{kL}\cos(kL)\end{aligned}\right] \\ &\simeq \frac{4\sqrt{\pi}}{3(kL)^{\frac{3}{2}}}\end{aligned} \tag{7.14}$$

$$\overline{g}_{\theta_0=90^\circ} = \frac{1}{1+\dfrac{\sqrt{2\lambda}\sqrt{L\lambda}}{3\pi d^2}} \tag{7.15}$$

Hence, for $\theta_0 = 90°$, (7.15) needs to be applied and also to be applied for the angles for which $\cos\theta_0 > 3/4\sqrt{2\lambda/L}$. The value of $\overline{g}$ for angles nearer to $\theta_0 = 90°$ may be given as:

$$\overline{g}_{\theta_0\le 90^\circ} = \frac{1}{1+\dfrac{\sqrt{2\lambda}\sqrt{L\lambda}}{3\pi d^2}+\dfrac{\sqrt{2L}\sqrt{L\lambda}}{15d^2}\left(\dfrac{\pi}{2}-\theta_0\right)^2} \tag{7.16}$$

This results into a smaller (in general negligible) dependence on the angle, $\theta_0 = 90°$, and for the angles deviating more than HPBW. As seen in Figure 7.4, the directive gain pattern for $\theta_0 = 90°$, goes to zero for large L. This is in agreement with theory, if we assume, $L \to \infty$. For an LAA, $L_x \gg L_y$ ($L_y = 0$), and

$$\overline{g} = \frac{1}{1+\dfrac{N\lambda}{2L_x}} \tag{7.17}$$

It can be clearly inferred from (7.17) that the directivity of linear random antenna array does not depend on the beam angle for any value of the mean distance between the elements.

For a rectangular array with $L_x > L_y$, we calculate the average of the denominator of the directive gain as:

$$\overline{P_{n,m}} \simeq \frac{4}{L_x L_y} \int_{-\infty}^{\infty}\int_{-\infty}^{\infty} \frac{\sin\left(k\sqrt{x^2+y^2}\right)}{k\sqrt{x^2+y^2}} \cos\left(kx\sin\theta_0\right)\left(1-\frac{y}{L_y}\right)dx\,dy$$

$$\simeq \frac{2\pi}{L_x L_y k}\int_0^{L_y} I_0\left(ky\cos\theta_0 L\right)\left(1-\frac{y}{L_y}\right)dy$$

$$\simeq \frac{2\pi}{L_x L_y k}\left\{\frac{\pi L_y}{2}\left[H_0\left(kL_y\cos\theta_0\right)I_0\left(kL_y\cos\theta_0\right)+I_0\left(kL_y\cos\theta_0\right)H_{-1}\left(kL_y\cos\theta_0\right)\right.\right.$$

$$\left.\left.-\frac{1}{\cos\theta}I_1\left(kL_y\cos\theta_0\right)\right]\right\} \tag{7.18}$$

where $H_0(\xi)$ and $H_{-1}(\xi)$ are the Strouve functions [64, 65]. If

$$kL_y\cos\theta_0 \gg 1 \Rightarrow \overline{P}_{nm} \simeq \frac{2\pi}{k^2 L_x L_y \cos\theta_0} \tag{7.19}$$

(the square array $L_x = L_y$; the results are same as before);

$$\overline{P}_{nm} \simeq \frac{\pi}{kL_x} \text{ for } L_y = 0 \text{ (for a linear antenna array)} \tag{7.20}$$

The value of average g is given by

$$\overline{g} = \frac{1}{1+\dfrac{2\pi N}{(kL)^2\cos\theta_0}} \tag{7.21}$$

The dispersion in $\overline{g}$ is defined as:

$$\overline{|g-\overline{g}|^2} = \overline{g^2} - \overline{g}^2$$

$$= \frac{N}{\sum_{m,n,p,q=1}^{N}\cos\left[k\left(x_n - x_m\right)\sin\theta_0\right]\cos\left[k\left(x_p - x_q\right)\sin\theta_0\right]} \tag{7.22}$$

$$\times \frac{1}{\dfrac{\sin\left(kr_{nm}\right)}{kr_{nm}}\dfrac{\sin\left(kr_{pq}\right)}{kr_{pq}}} - \overline{g}^2$$

For dense or sparse arrays, with a large number of elements and $\sqrt{|g - \bar{g}|^2} \ll \bar{g}$ yields that, for any element allocation (also random), the fluctuations of g are negligible with respect to the mean value of g. When both terms in the denominator of $\bar{g}$ are close to each other, the dispersion increases and the array cannot be described by averaging of neither dense nor SAA.

Thus, an SAA is easy to characterize by the probabilistic theories and any sparse array with randomly allocated element positions can be characterized by the average interelement separation in the array grid. The characteristics of randomly spaced SAA may be summarized as (assuming large average interelement spacing):

- The directivity is proportional to the total number of elements.
- The sidelobe level (SLL) is decided by the number of elements or vice versa.
- The angular resolution mainly depends upon the electrical aperture size.
- If the number of elements is fixed, the angular resolution can be increased by increasing the aperture size.
- The average SLL (ASLL) is inversely proportional to the number of elements ($\sim 1/N_{tot}$).
- They demonstrate the diminishing effect of the mutual coupling, which leads to the wide bandwidth and wide scan volume capabilities.

The next few sections describe the design and characterization of sparse array antennas based on a few techniques that I chose based on their efficacy and computational effectiveness in picturizing the characteristics of SAAs of any electrical size.

7.5 Statistical Density Tapering (SDT) Method

In this deterministic synthesis method [66], the density distribution of the elements within SAA's aperture is made proportional to the amplitude distribution in a regular dense antenna array. The density tapering is decided statistically by choosing the model amplitude taper (like Taylor and Chebyshev) as the probability density function (PDF), henceforth helping to choose the element locations. All of the fixed elements are applied equal excitation (isophoric array) and, hence, an isophoric and unequally spaced sparse antenna array is designed using the SDT.

SDT performance depends upon the chosen modal amplitude taper and predefined selection probability factor, q, to achieve SAA design with low SLL. It demands allocating discrete amplitude value, 1 or 0, based on the product of the normalized amplitude, a_n ($= I_n/\max\{I_n\}$) and q ($0 \le q \le 1$), that is,

$$F_n = q \bullet a_n \tag{7.23}$$

A random number, $0 < ran_n < 1$, is generated. If $F_n < ran_n$, then the element is allocated the amplitude 1 or else 0. The coefficients, I_n, are obtained from the modal amplitude distribution (i.e., either Taylor's or Chebyshev's distribution [61].) In SDT, q decides the sparsity in the array, that is, $\eta_{sp} \propto (1 - N_{EL}/K)$ (N_{EL} is the total number of elements left in SAA and K is the number of elements in the dense array counterpart). The average power pattern [66] may be represented as:

$$\overline{|F(\theta,\varphi)|^2} = q^2|F(\theta,\varphi)|^2 + \sum_{n=1}^{N_{EL}} qI_n(1 - qI_n) \tag{7.24}$$

In (7.24), the first part of the RHS expression denotes the modal amplitude-tapered antenna array and the second part of the expression depicts the statistical average SLL envelope outside the main lobe region. For, $q = 1$, the first part is equal to the amplitude-tapered array and, henceforth, it is called a naturally thinned array. It is observed that the PSLL $\propto 2/N_{EL}$ leading to PSLL $\propto 2/\eta_{sp}$ as will be shown with the numerical test cases in the following studies.

As a demonstrator a 625- (25 × 25) element dense antenna array (DAA) is considered with the underlying grid of a half-wavelength ($\sim\lambda/2$). Figure 7.6 shows the DAA performance with uniform and nonuniform excitations. A beam broadening by the virtue of the nonuniform excitations is seen while lowering the PSLL. SDT is employed to design a thinned sparse antenna array, naturally thinned (i.e., $q = 1$), and 100 iterations are employed to statistically design an array with lowest SLL envelope. The modal amplitude taper chosen is Taylor's distribution (SLL = −30 dB, $\bar{n} = 7$).

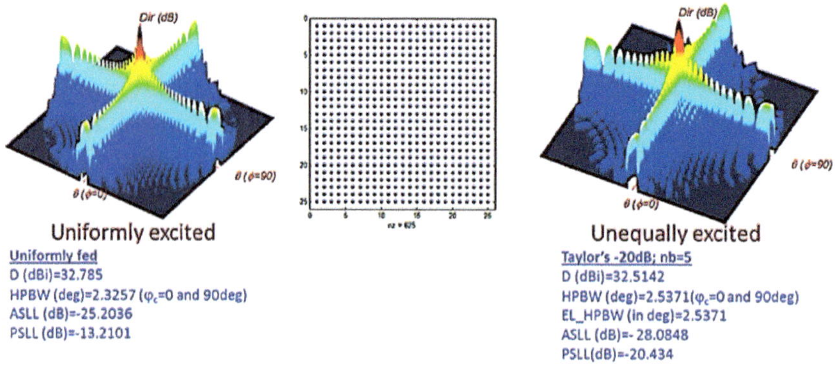

Figure 7.6
Radiation pattern performance of a 625-element DAA.

Figure 7.7 shows the performance of a thinned array using SDT. Figure 7.7(a) shows the thinned array consisting of 257 elements, that is, a sparsity of η_{sp} = 58.8% is achieved. Figure 7.7(b) shows the ASLL and PSLL achieved over each set of iterations and Figure 7.7(c) shows the normalized 3-D AF plot. The inset of Figure 7.7(c)

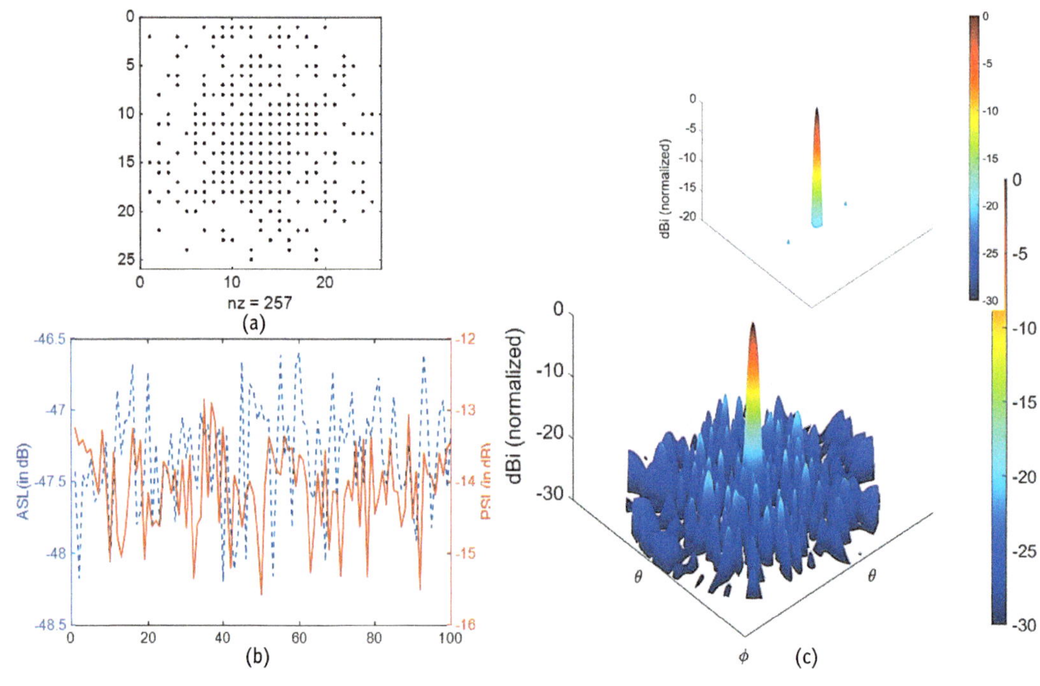

Figure 7.7
Array performance of a statistically tapered planar array (K = 625).

shows the worst PSLL < –19.2 dB at only two locations and the rest of all the locations have an SLL envelope better than –20 dB.

Figure 7.8 shows another statistically tapered SAA. Herein, a 10,000-element DAA is spatially tapered and a η_{sp} = 53.46% (more than 50% reduction in the number of elements) and PSLL < –23.9 dB is demonstrated, which is very impressive. Thus, the above two examples demonstrate the strength of SAAs where the spatial tapering emulates the amplitude-tapered DAAs performance leading to a lowered PSLL with equi-excitation of the array elements. It is observed that the maximum concentration of elements is in the middle region of SAA, which emulates the Taylor's amplitude taper behavior where the ratio of the center-to-edge element is very high, nearly 5.5 dB in the 625-element array. The maximum sparsity as demonstrated is near the edges of the arrays. The ASL demonstrated in both the cases is better than $1/N_{EL}$.

Figures 7.9(a, b) demonstrate that the results for ASLL and PSLL, respectively, converge better if the number of iterations is increased from 10 to 1,000 in the case of a 10,000-element array. Another observation made here is that *PSLL* = *ASLL* + (3 to 5) dB is typically achieved in SAAs. The limitation of the SDT technique is that it is very accurate for the large antenna arrays and it requires many

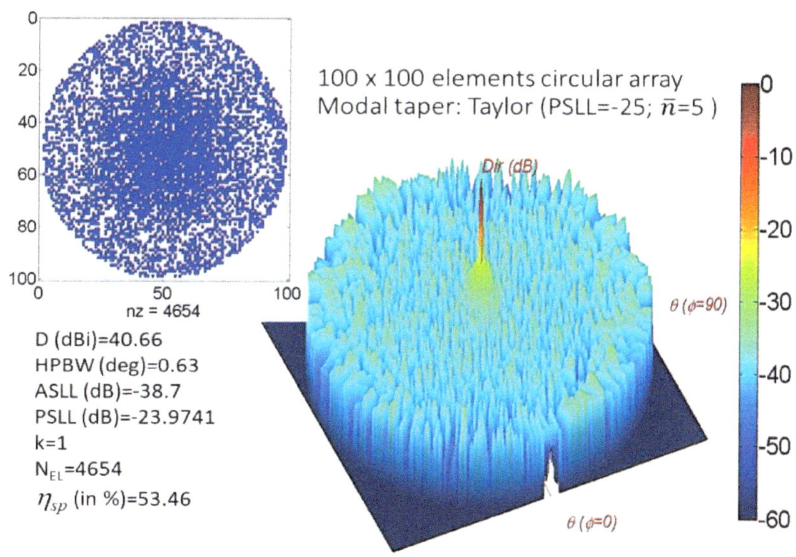

Figure 7.8
A 10,000-element statistically tapered antenna array.

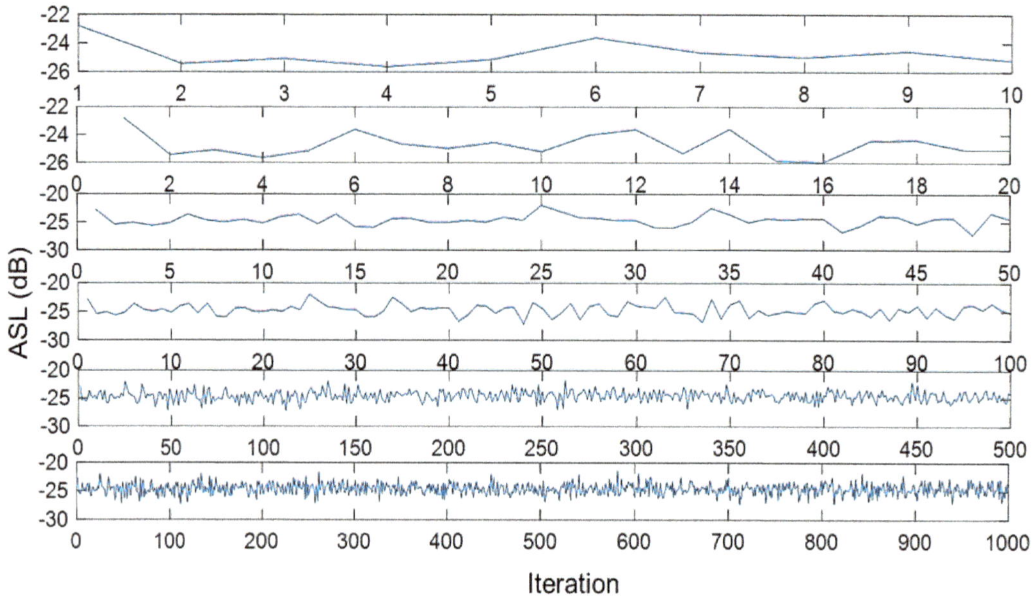

(a)

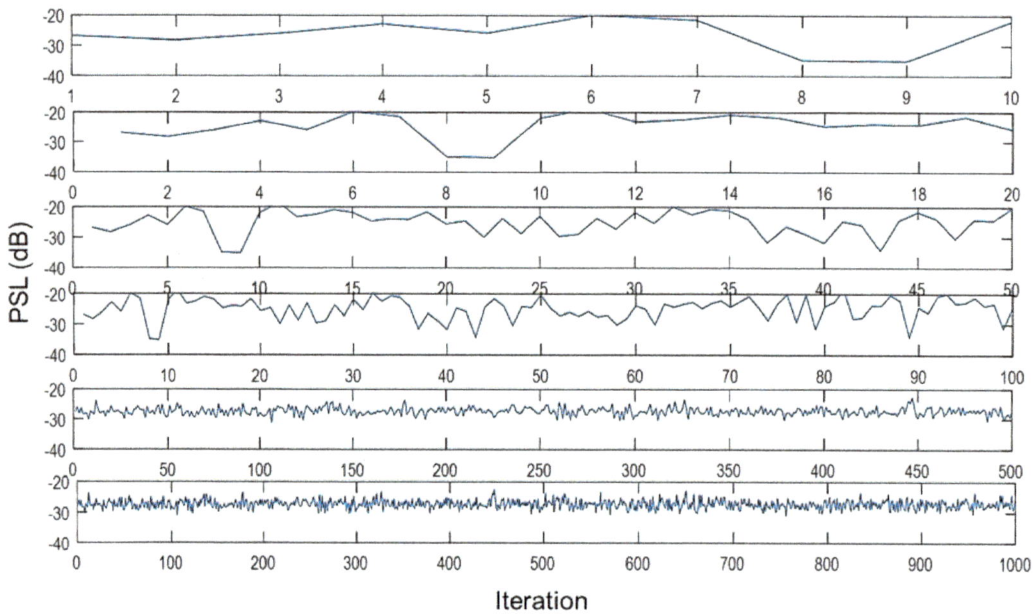

(b)

Figure 7.9

Statistical performance of an array in terms of ASLL. (a) ASLL variation. (b) SLL variation.

iterations to get a nearly optimal solution. Moreover, there is no control over the placement of the elements and the middle region with high density may lead to higher levels of the mutual coupling between the elements in the particular region.

7.6 Fractal Engineering Techniques

Fractal engineering is another interesting way to design an SAA with irregular array grids. These can also be used to generate interleaved arrays operating over different operating frequency bands. Sierpinski Fractal Carpet [67] is one of the simplest demonstrators used here to demonstrate the usefulness as stated above. Sierpinski was a professor at Lvov and Warsaw, and one of the most influential and reputed mathematicians in Poland. The basic geometric construction of the Sierpinski carpet starts from a square in the plane, divided into nine congruent squares of which the center one is removed. The process may be repeated as many times as required. Figure 7.10 shows three examples of fractal carpet geometries with

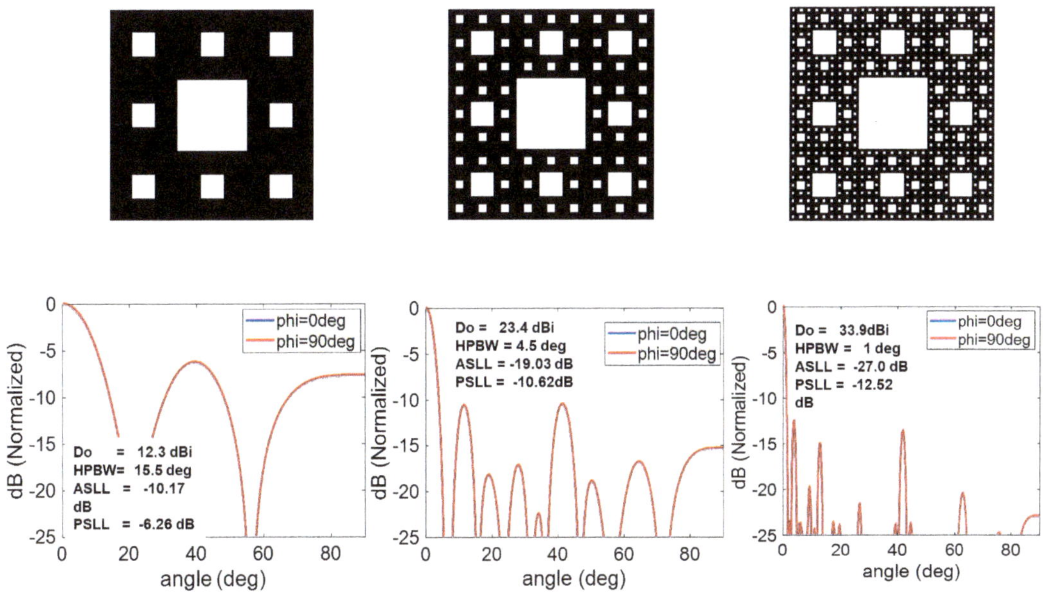

Figure 7.10
Sierpinski fractal carpet antennas.

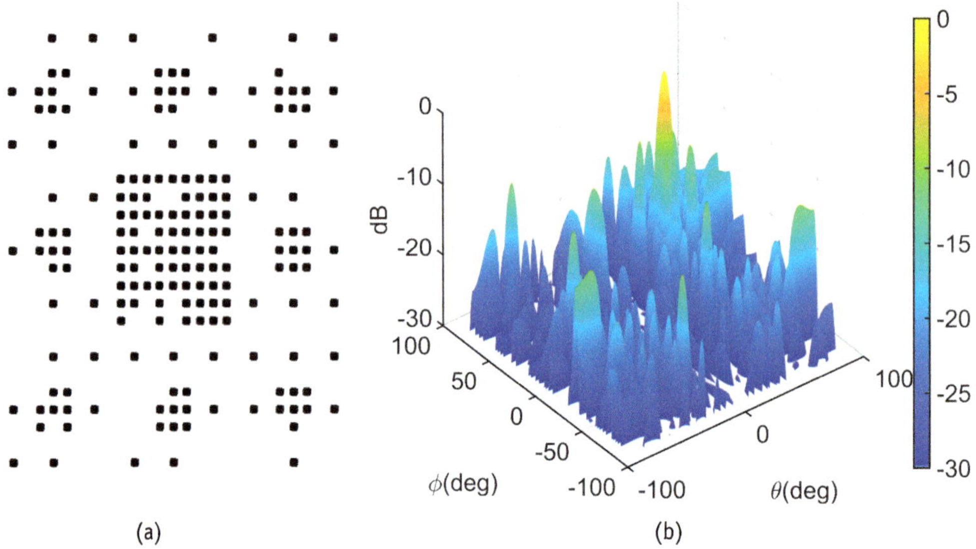

Figure 7.11
Sierpinski fractal carpet antennas (3 iterations) and the scanned 3-D AF plot.

iteration levels of 2, 3, and 4 and the corresponding AF plots in the two principal planes. The AF plots shown below are plotted assuming the respective fractal antennas operate in a single frequency band and the element locations are as shown in white spots. All the elements are uniformly fed with equal excitations.

Figure 7.11 shows the 3-D AF plot for the fractal antenna scanned at an angle (45°, 45°). The left side of the figure shows the array geometry where the black dots show the element locations. There are no GLs even at wide scan angle due to aperiodicity in the fractal antenna array, but the regularity in the subarray block leads to a periodic high SLL as seen in the figure. These can be further alleviated by destroying regularity in the subarrays by employing stochastic techniques such as GA. The next section shows a case study using heuristic techniques.

7.7 Hybridization of Statistical Density Tapering and Genetic Algorithm Techniques

In Section 7.5, it was mentioned that SDT has a few limitations and it is accurate only for the design of larger size antenna arrays, and the accuracy to achieve desired SLL improves with the large number of iterations. Moreover, larger sparsity, η_{sp} and low PSLL are difficult to achieve together with SDT. These limitations are overcome in this section by hybridizing SDT with the GA method (i.e., hybridization of deterministic and stochastic techniques). A numerical test case, a 400- (20 × 20) element antenna array with $\lambda/2$ array grid along x- and y-directed principal planes is considered to illustrate the process. On the application of SDT (number of iterations: 100; $q = 0.7$) and choosing a modal taper (Taylor's; PSLL = –25 dB, $\bar{n}$ = 6), a PSLL < –14.8 dB and ASLL < –17 dB were achieved along with a $\eta_{sp} = 49.7\%$. Figure 7.12 shows the array grid (Figure 7.12(a)), 3-D AF plot (normalized) (Figure 7.12(b)), and the variation of PSLL and ASLL with respect to the iterations (Figure 7.12(c)). It is observed that, although sparsity is achieved, PSLL is not lowered enough as desired.

The arrangements of 1s and 0s across the array aperture is taken as an initial cue for initiating the GA [68]. The parameters used for GA are: mutation rate = 0.2, population size = 8, number of generations = 300, and the applied fitness mask specified an SLL < –25 dB with maximum possible sparsity. Figure 7.13 shows the resultant array performance (the left side of the figure) and the array grid (the right side of the figure). It is seen that with the similar sparsity of 51% (a little better in comparison to the case when SDT is only applied), a much lower PSLL < –20 dB is achieved. Figure 7.14 shows the amplitude distribution over the array aperture using SDT alone and in conjunction with the GA method. It is seen using hybridization that there is a better control over distribution and a nearly quadrant symmetric distribution is achieved, which is good for a monopulse tracking operation.

Here, although sparsity is a high but average separation, $d_{avg} \leq 0.5\lambda$, indicating a strong mutual coupling between the array elements. Figures 7.15(a, b) show the wideband and wide scan performance of the array in X-band. The variation in the directivity observed is 1.86 dB from 8 to 12 GHz and 1.9 dB from boresight to a 60° scan angle. The SLL remain almost constant over the variation in frequency and scan angle and the worst PSLL< –18.9 dB is observed. Figure 7.15(c) shows the scan performance plots in the (u, v) coordinate system from (0°, 0°) to (60°, 50°) along the intercardinal φ_c-cuts. It is seen there is an increase in PSLL up to –16.5 dB in intercardinal planes. However, there is no visible onset of any

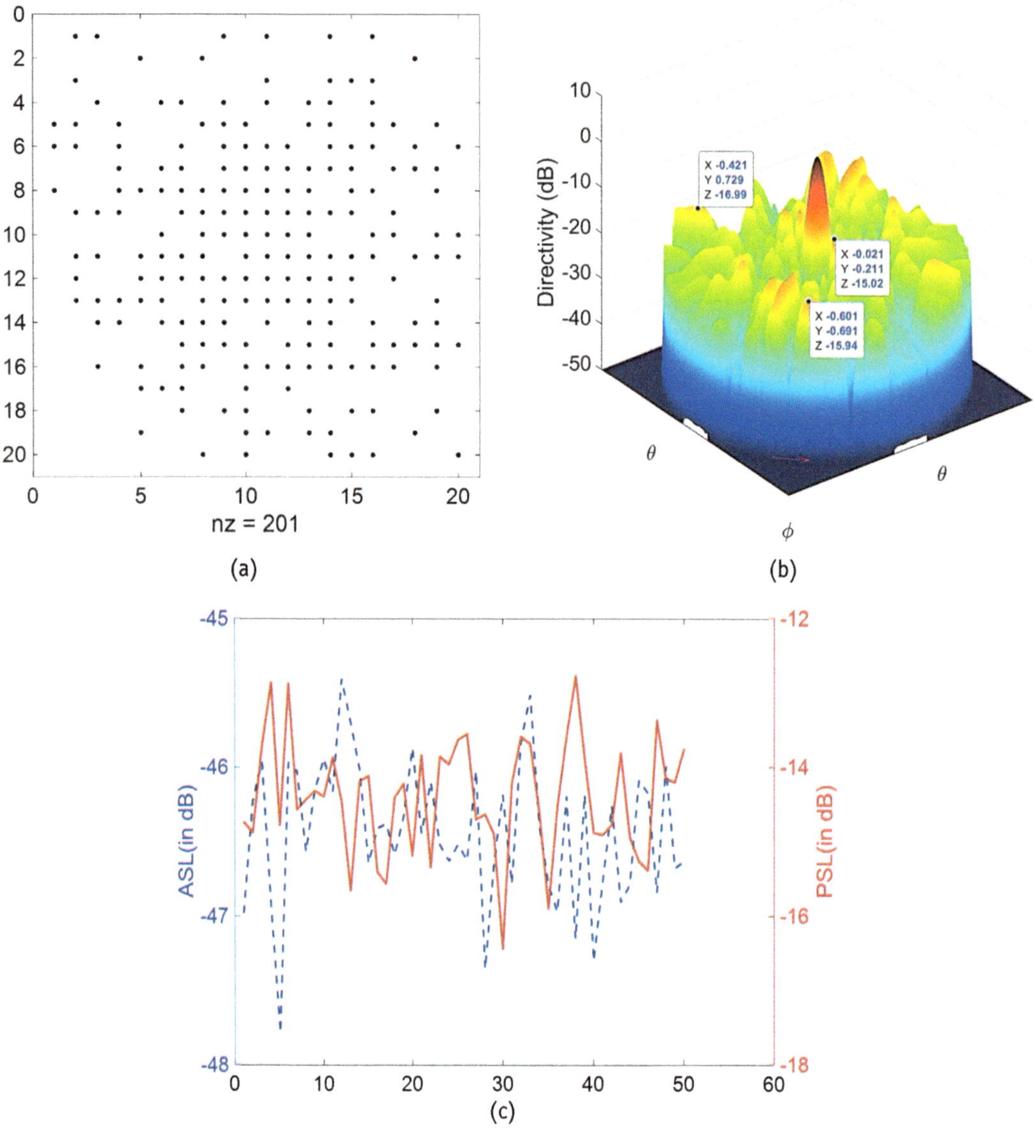

Figure 7.12
A 400-element antenna array performance using SDT: (a) array grid, (b) 3-D array factor plot, and (c) ASLL and PSLL versus iterations.

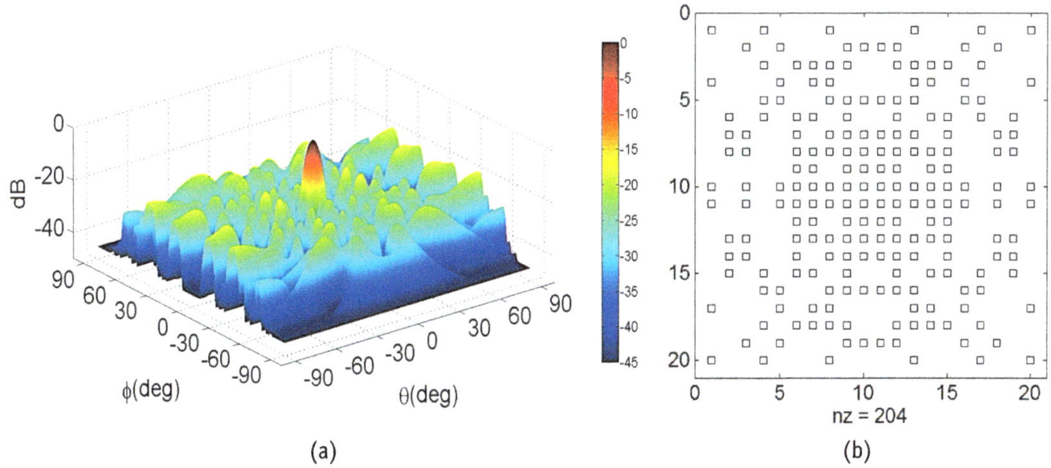

Figure 7.13
A 20 × 20-element antenna array performance using the GA.

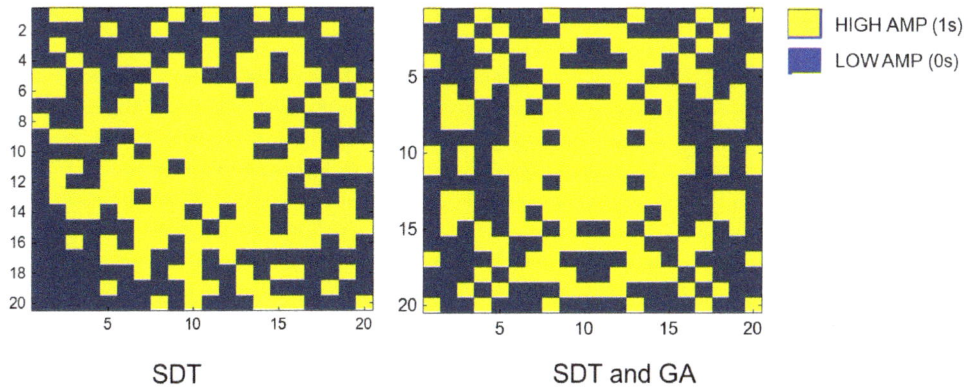

Figure 7.14
Amplitude distribution over the array aperture.

GL and a wide scan volume is ensured from the SAAs performance over the complete hemisphere.

A model of this geometry is made using an Ansys HFSS solver using a practical radiator, a U-slot microstrip patch antenna (US-MPA) operating in X-band for validation of the simulations using isotropic radiator. Figures 7.15(d, e) show the model and the simulated radiation pattern at 10 GHz. A gain of 28 dBi and PSLL < –16 dB is observed, which is in a very good agreement with the design simulations and the array parameters obtained using (4.14).

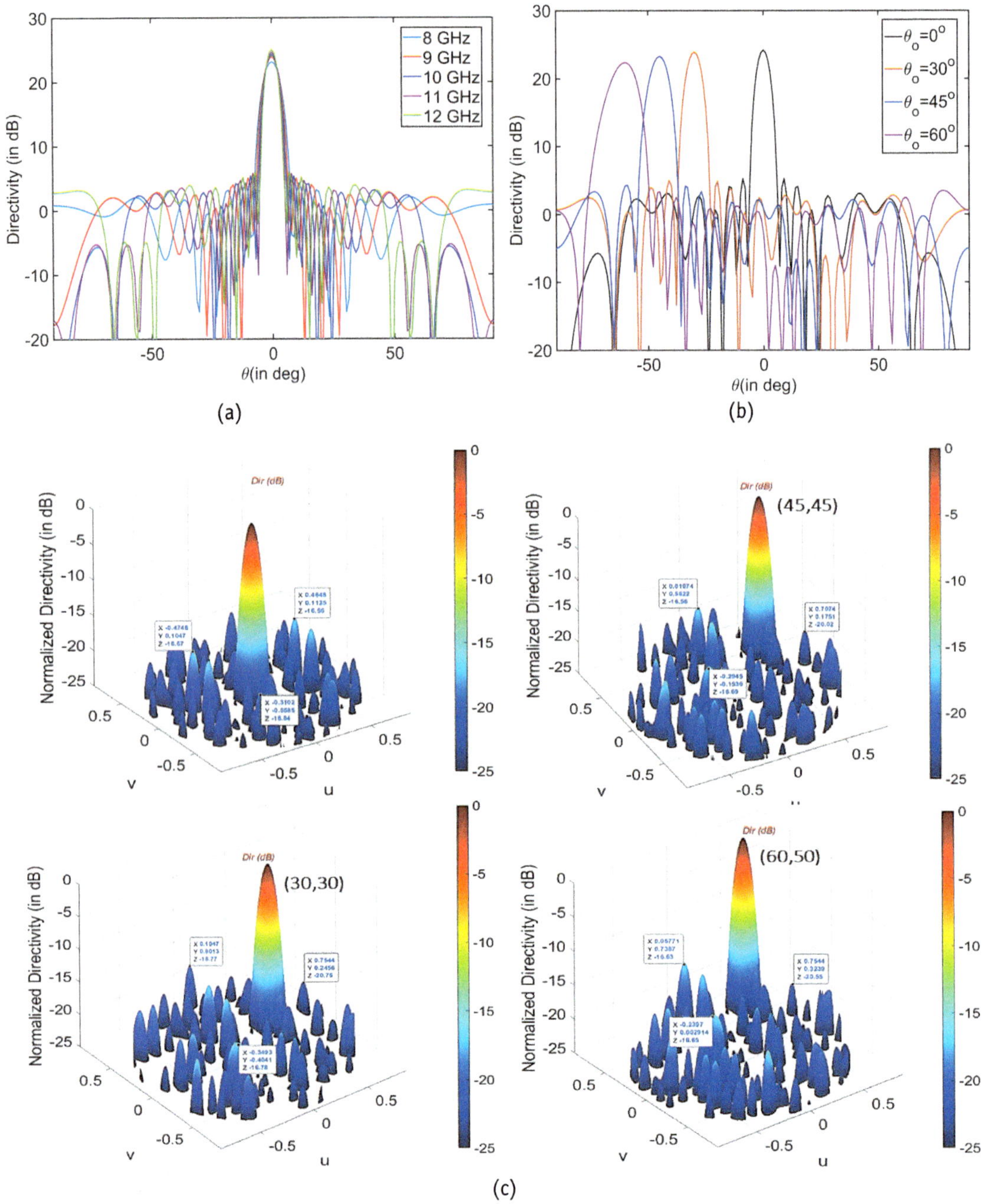

Figure 7.15

Array performance at 10 GHz along the $\varphi_c = 0°$ plane: (a) over the frequency bandwidth, (b) over the scan volume, (c) 3-D array patterns showing the intercardinal scan performance, (d) far-field pattern in two principal planes using HFSS model, and (e) HFSS model and 3-D array pattern simulations.

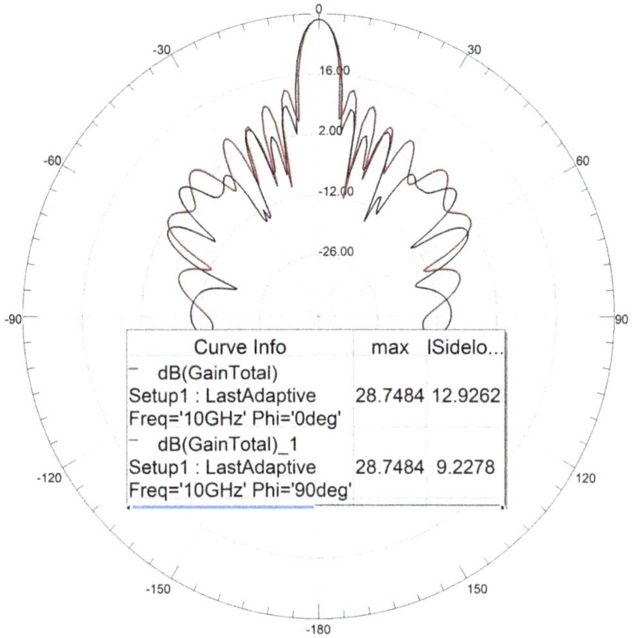

(d)

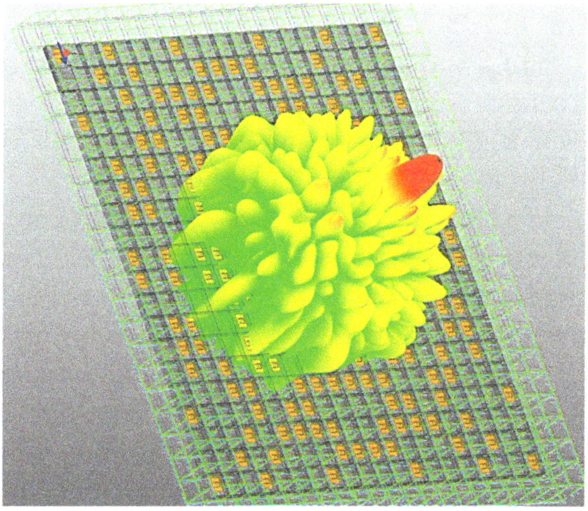

(e)

Figure 7.15
(Continued)

7.8 PT-Based Deterministic Design Methodology

In this section, an indigenous deterministic technique to design an SAA utilizing the PT or the theory of collapsed distributions (TCD) supported with illustrative examples is presented. Few deterministic guidelines are formulated based on the established mapping between the reference (or modal) taper and the actual (spatial) taper using TCD. Figure 7.16(a) presents an arbitrarily shaped array with boundary, S, in the XOY plane. In Figure 7.16(a), X′OY′ represents the rotated coordinate system about the z-axis at an angle φ_c with respect to the x-axis, it consists of N_r concentric rings, indexed, $n = 1, 2, \ldots, N_r$, sampling the desired continuous aperture current distribution. The outermost ring considered is ($n = N_r$) of length and width, L and W, respectively, approximates the array boundary, S, closely. The shaded portion represents the unaccounted portion worth neglecting while allotting the optimal locations to array elements strategically as presented later. $P(r, \theta, \varphi)$ and $A(x_n, y_n)$ represent the far-field observation point and the element location. The collapsed location of an array element along the OX′ line in the X′Oy′ plane is $a'(x'_n, 0)$ (no y-dependence). The array is characterized in terms of APCs, that is, D_o, −3-dB beamwidth, HPBW, SLL (ASLL and PSLL), and average or mean interelement spacing, d_{avg}, as defined below.

Figure 7.16(b) illustrates an example with $L = 10\lambda$; $W = 5\lambda$, $N_R = 20$, which helps in formulating the design guidelines and illustrate its efficacy simultaneously. The concentric rings control the variation of the surface current distribution from the center towards the edge. Each of these rings is split into a number of unit cells (UCs), each having an area, $\mathbf{a}_{\text{UC}} = \lambda^2/16$, having dimensions, either $\lambda/4 \times \lambda/4$ or $\lambda/2 \times \lambda/8$, such that no two UCs overlap each other as shown in Figure 7.16(b). Each of the UCs encapsulates a small portion of the total aperture current distribution while all of the rings are assumed to have uniform current distribution. The average amplitude of current in the nth ring, denoted as p_n. Assuming that $a = n\lambda/4$, the size of the nth ring is ($l_n \times w_n = 2a \times a$), if the rings are rectangular-shaped or ($2a \times 2a$), if the rings are square-shaped.

The total number of UCs in the nth ring is given by:

$$UC_n = \frac{\text{area of } n^{\text{th}} \text{ ring}}{\text{area of unit cell}} = \frac{l_n \times w_n}{a_{\text{UC}}} - \sum_{k=1}^{n-1} UC_k \tag{7.25}$$

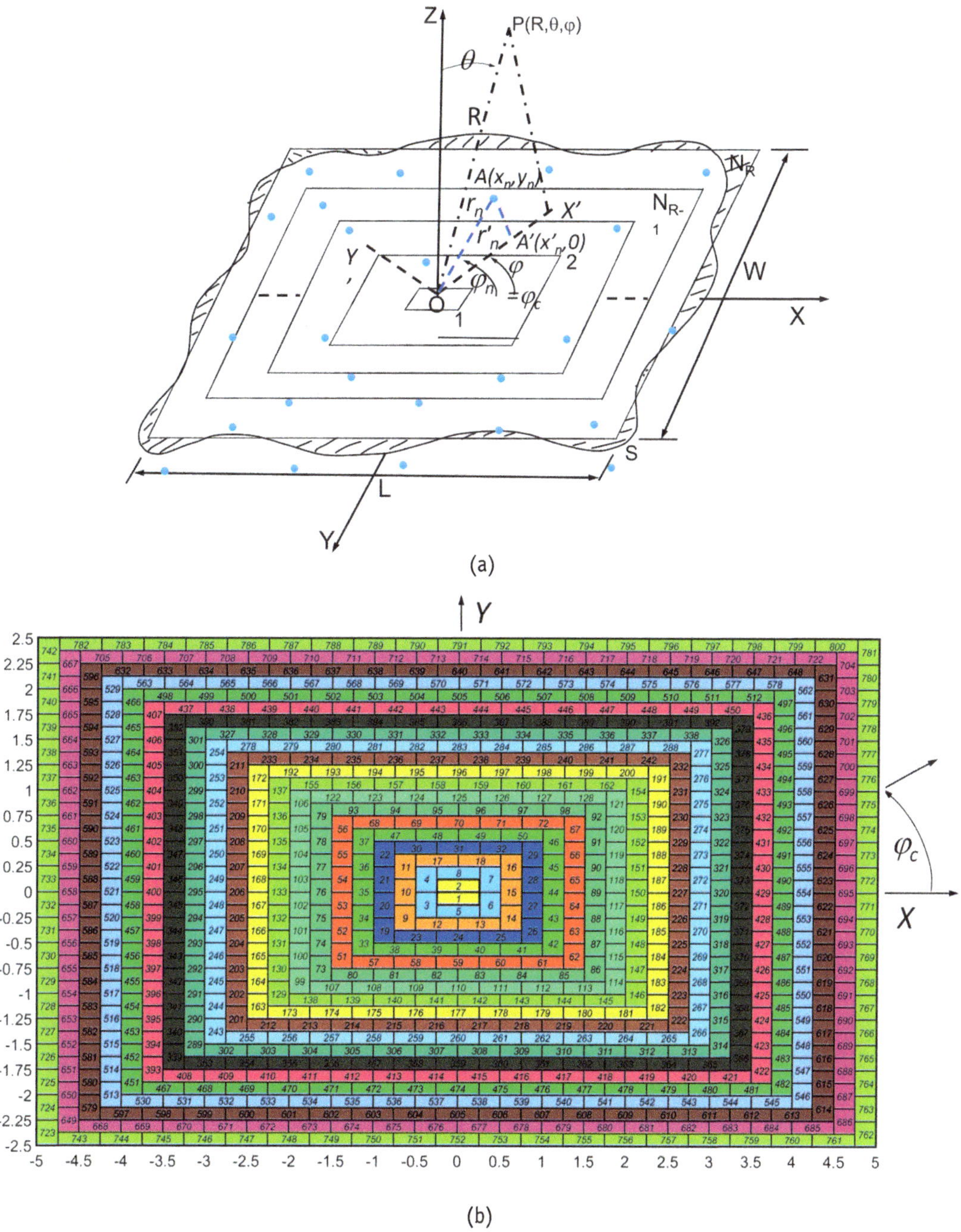

Figure 7.16

Array layout and discretization plan. (a) Array layout and associated coordinate system. (b) Arrangement of unit cells in the concentric rectangular rings.

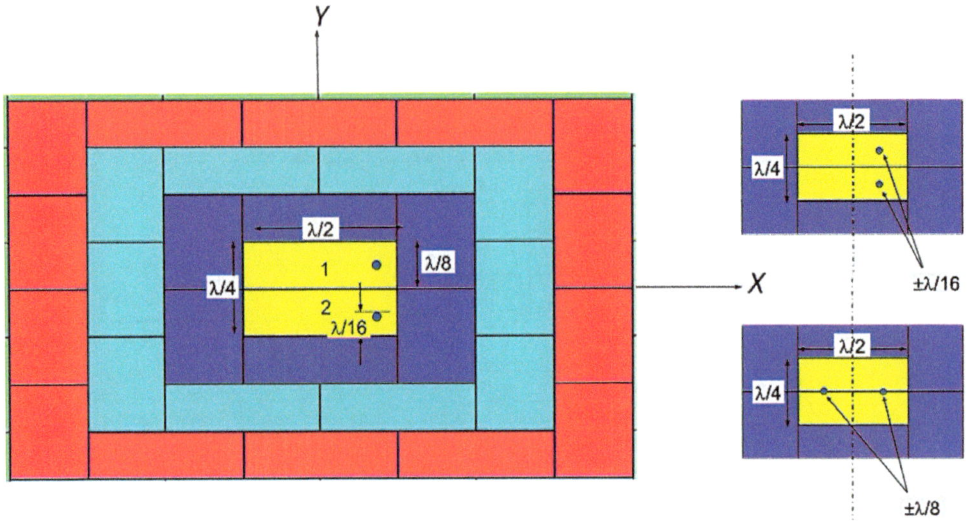

Figure 7.17
Innermost unit cells: (a) exploded view of inner rectangular rings, (b) split along the *x*-axis, and (c) split along the *y*-axis.

In (7.25), $UC_1 = 2$ ($n = 1$), that is, the ring has 2 UCs of the size $\lambda/2 \times \lambda/8$ or $\lambda/4 \times \lambda/4$ (see Figures 7.17(a–c)). The total number of UCs in an array equals $\sum_{n=1}^{N_R} UC_n$. The other possibility is splitting along the *y*-axis leading to the dimensions of the UCs as $\lambda/4 \times \lambda/4$.

TCD is used to compute collapsed amplitude distribution (CaD), p_n along a φ_c-cut. Having considered the projections along a principal plane, $\varphi_c = \varphi_1 = \pi/2$, (plane $y > 0$), ring 1 is split symmetrically along the *x*-axis and the center of UCs are shown as blue dots in Figure 7.13(a) at locations $y = \pm\lambda/16$. The projections are computed at uniform locations, $\lambda/8$ apart, along the line, OX′ at $\angle\varphi = \varphi_1 = 0°$ (starting from the first UC). The projected amplitude, Amp_i ($i = 1, 2, \ldots, N_R$), is given as:

$$\vdots$$

$$\vdots$$

$$\text{Amp}_1 \rightarrow p_1 + \sum_{n=2}^{N_R} p_n = A_{T1};\ \text{at } y_1 = \frac{\lambda}{16}$$

$$\text{Amp}_2 \rightarrow 2p_2 + \sum_{n=3}^{N_R} p_n = A_{T2};\ \text{at } y_2 = \frac{\lambda}{16} + \frac{\lambda}{8} = \frac{3\lambda}{16}$$

(The factor 2 indicates that there are 2 UCs in ring 2 with the same amplitude.)

$$\text{Amp}_3 \rightarrow 3p_3 + \sum_{n=4}^{N_R} p_n = A_{T3};\ \text{at } y_3 = \frac{3\lambda}{16} + \frac{\lambda}{8} = \frac{5\lambda}{16}$$

(The factor 3 indicates that there are 3 UCs in ring 2 with the same amplitude.)

$$\begin{aligned} &\text{Amp}_{N_R-1} \rightarrow N_{R-1}p_{N_R-1} + p_{N_R} = A_{TN_{R-1}};\ \text{at } y_{N_R-1} = \frac{2\left(N_{R-1} - 1\right)\lambda}{16} \\ &\text{Amp}_{N_R} \rightarrow N_R p_{N_R} = A_{TN_R};\ \text{at } y_{N_R} = \frac{2\left(N_R - 1\right)\lambda}{16} \end{aligned} \tag{7.26}$$

Or succinctly written as

$$A_n \rightarrow p_n + \sum_{k=n+1}^{N_R} p_k = A_{Tn} \quad \text{at} \quad y_n = \frac{2(n-1)\lambda}{16} \tag{7.27}$$

Thus, A_n is equivalent to p_n contributions from all of the rings equated to modal amplitude taper (Taylor's distribution here), A_{Tn} in the nth ring at a collapsed location, y_n. Similarly, A_n along $\varphi_c = \varphi_1 + 90 = \pi$ or 0° cut (refer Figure 7.17(c)) can be written as:

$$A_n \rightarrow p_n + \sum_{k=n+1}^{N_R} p_k = A_{Tn} \quad \text{at} \quad x_n = \frac{2(n-1)\lambda}{8} \tag{7.28}$$

The set of equations (7.27) or (7.28) can be rewritten in the form of a matrix formulation as:

$$\begin{bmatrix} 1 & 1 & \cdots & 1 \\ 0 & 2 & \cdots & 1 \\ \vdots & \vdots & \ddots & 1 \\ 0 & 0 & \cdots & N_R \end{bmatrix}_{N_R \times N_R} \begin{pmatrix} p_1 \\ p_2 \\ \vdots \\ p_{N_R-1} \\ p_{N_R} \end{pmatrix}_{1 \times N_R} = \begin{pmatrix} A_{T1} \\ A_{T2} \\ \vdots \\ A_{TN_R-1} \\ A_{TN_R} \end{pmatrix}_{1 \times N_R} \tag{7.29}$$

Or

$$\overline{\overline{\mathbb{C}}} \bullet \wp = \mathbb{A} \text{ or } \wp = \overline{\overline{\mathbb{C}}}^{-1}\mathbb{A} \quad (7.30)$$

In (7.30), $\mathbb{A}$ is the known model amplitude distribution matrix, $\overline{\overline{\mathbb{C}}}$ is the known coefficient matrix (highly sparse), and $\wp$ is the unknown localized or partial amplitude excitation matrix, which can be derived by solving (7.30). The term, p_n, is the average amplitude per unit cell. The relationship between p_n and the number of UCs in the nth ring, denoted as a_n, is given by:

$$\sum_{n=1}^{N_R} p_n \bullet a_n = A_S \quad (7.31)$$

In (7.31), A_S is the total sum of the amplitudes distributed across the concentric rings in the array aperture. Thus, the number of elements, Λ_n, distributed in the nth ring, is estimated as:

$$\Lambda_n = \frac{1}{A_S}\left(p_n \bullet a_n\right) \times N_a \quad (7.32)$$

In (7.32), N_a is the desired total number of active elements in the antenna array, such that the sparsity, η_s, of the sparse antenna array is defined as η_s (%) = 100 • (1 – N_a/K), where K is the total number of elements in the corresponding dense array. Thus, (7.31) and (7.32) provide the necessary mapping equation between amplitude tapering and the spatial tapering.

7.8.1 Synthesis of the Spatial Tapering Across the Array Aperture

Equation (7.32) establishes the fact that Λ_n is dictated by the model amplitude distribution (Taylor's) and the desired sparsity. Having chosen the number of elements per ring, Λ_n, it is required to place these elements at particular locations to minimize the SLL envelope. It implies choosing a spatial distribution emulating the model amplitude distribution closely. As an initial approach, the whole array aperture may be divided into angular sectors, as shown in Figure 7.18. Figure 7.18 shows 4 regions (I to IV) dividing the whole

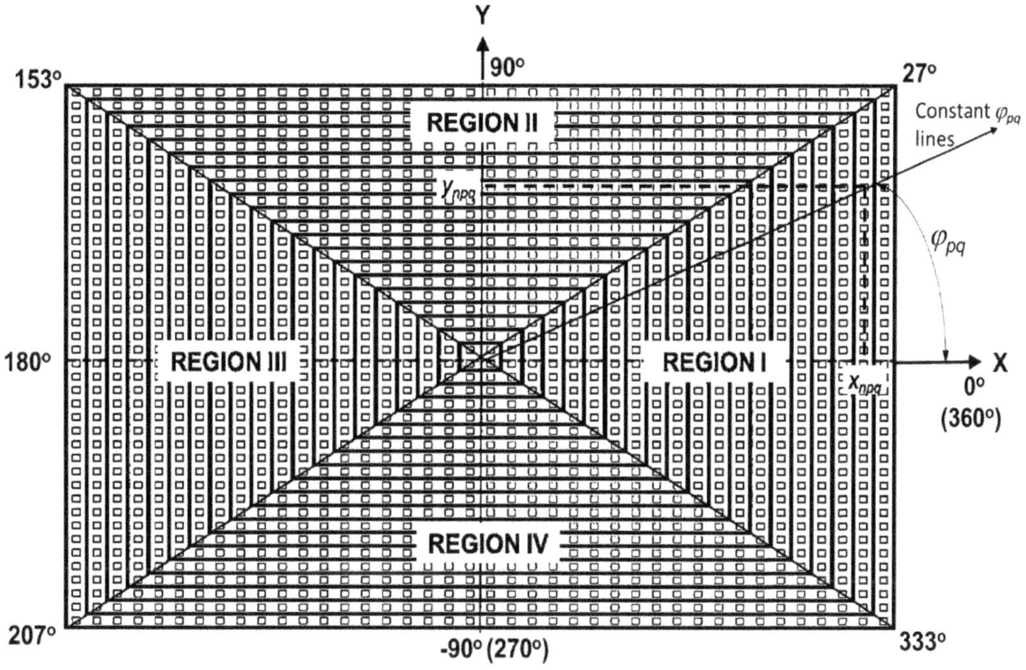

Figure 7.18
Angular grid region scheme for spatial distribution.

array aperture into the four regions: (333° to 27°), (27° to 153°), (153° to 207°), and (207° to 333°).

It is ensured that the distribution is symmetric (to facilitate difference beam formation for monopulse tracking in radars). Further, it is ensured that a maximal sparsity and a minimal SLL envelope is also achieved along the hemisphere. Henceforth, N_a number of φ-angles, denoted as $\varphi_{p,q}$, are selected ($p = 1, 2, \ldots, N_a$ and $q = 1, 2, \ldots, Q$, Q is the total number of different combinations possible for the element placement) in a random manner to destroy any sort of periodicity in the array. The choice of randomness or Q combinations of the angular locations may be based on different criteria that may be based on the deterministic approach (see Appendix 7A) or the stochastic approach. Care has to be taken while choosing the element locations such that their sum total should not exceed the number of elements fixed per ring (i.e., Λ_n and shall satisfy the average amplitude requirement, p_n). Thus, N_a and Λ_n are the two restrictions that are helpful in reducing the total number of possibilities in the solution domain to a feasible number (Q) that can be handled easily while optimizing. The choice of Q (see Appendix 7A) will be explained with the help of a numerical test example later. The choice of $\varphi_{p,q}$ can be made such that they are spread at N_a locations over the four designated regions in the aperture, ensuring a minimal distance constraint to reduce the mutual coupling and maintain symmetry across the four regions.

After choosing $\varphi_{p,q}$, the next task is the identification of the space coordinates of the elements in the four regions. The process of their estimation is given below (with reference to Figure 7.18).

For $\varphi = 0°$ (x-axis):

In Region I, $-\pi/2 < \varphi_{p,q} < \pi/2$, and $|\tan\varphi_{p,q}| < 0.5$ ($|\varphi_{p,q}| \sim 27°$)

$$A_{m_{npq}} = p_n; \text{ and } x_{npq} = \frac{(2n-1)\lambda}{8} \tag{7.33}$$

In Region II, $-\pi/2 < \varphi_{p,q} < \pi/2$, and $|\cot\varphi_{p,q}| < 2$

$$A_{m_{npq}} = p_n; \text{ and } x_{npq} = \frac{(2n-1)\cot\varphi_{p,q}\lambda}{16} \tag{7.34}$$

In Region III, $\pi/2 < \varphi_{p,q} < 3\pi/2$, and $|\tan\varphi_{p,q}| < 0.5$

$$A_{m_{npq}} = p_n; \text{ and } x_{npq} = \frac{-(2n-1)\lambda}{8} \tag{7.35}$$

In Region IV, $\pi/2 < \varphi_{p,q} < 3\pi/2$, and $|\cot\varphi_{p,q}| < 2$

$$A_{m_{npq}} = p_n; \text{ and } x_{npq} = \frac{-(2n-1)\cot\varphi_{p,q}\lambda}{16} \tag{7.36}$$

Here, $n = 1, 2, \ldots, N_R$, $p - 1, 2, \ldots, N_a$, and $q = 1, 2, \ldots, Q$ and $A_{m_{npq}}$ denotes the partial amplitude allotted to the pth element in the nth ring for the qth combination and (x_{npq}, y_{npq}) is the coordinate of the pq^{th} element in the nth ring. Similarly, for $\varphi = 90°$ (y-axis):

For Regions I and III,

$$A_{m_{npq}} = p_n;\ \text{and}\ y_{npq} = \frac{\pm(2n-1)\tan\varphi_{p,q}\lambda}{8} \tag{7.37}$$

and for Regions II and IV,

$$A_{m_{npq}} = p_n;\ \text{and}\ y_{npq} = \frac{\pm(2n-1)\lambda}{16} \tag{7.38}$$

Henceforth, (7.33) through (7.38) give the spatial locations as obtained from the angular locations. Thus, fixing either x_{npq} or y_{npq}, other coordinates can be estimated based on the available information. Referring to Figures 7.16 or 7.17, it can be seen that the x-coordinate is always at the middle of a UC, whereas the y-coordinate can be either in the lower or in the upper half of the UC. It is to be remembered that the element density per ring is constant. Thus, for example, if we consider ring 2 (see Figure 7.16), then in Region I, $x_{2pq} = 3\lambda/8$ and then $y_{2pq} = \tan(\varphi_{p,q}) \cdot 3\lambda/8$ gives various possibilities of choosing the y-locations corresponding to the fixed x-location. Like this, one has various possibilities of choosing y-locations and, in turn, the projected amplitude due to the contribution of different rings can be determined at the specific x-location (such as $3\lambda/8$, $5\lambda/8$, ...). Thus, at every $\lambda/4$ step, the contribution due to the projected amplitudes can be sampled from the concentric rings and summed up, increasing the overall projected amplitudes along the locations in the virtual linear array along a particular phi-cut.

7.8.1.1 Optimization Criteria

The optimization criteria for the element locations are as follows:

- Normalize the values of $A_{m_{npq}}$ and check the least mean square error (LMSE) between $A_{m_{npq}}$ and A_{Tn} ($n = 1, 2, \ldots, N_R$) for all of the (p, q) combinations and chose the best set of $\varphi_{p,q}$ with the least LMSE.
- Verify Λ_n with the p_n.

- Plot the array factor for all (p, q) combinations and check the worst PSLL. The combination with minimal PSLL may be chosen.

Thus, this scheme gives a geometrical way to optimize a maximally sparse antenna array design with a low SLL. The projected amplitudes along the φ_c-cuts are checked with respect to the modal amplitude taper (Taylor distribution may be used) distribution for LMSE for different probable combinations as considered and the one with low SLL may be chosen. This process should be followed with a minimal spacing constraint to curb the mutual coupling effects. Another method for optimization is given in the next section.

7.8.2 Synthesis of a Taylor-Like Distribution

The aim here is to select an apt amplitude taper (Taylor distribution) as a set target, discretized with respect to $\lambda/4$ (along the x-axis) or $\lambda/8$ steps (along the y-axis); see Figure 7.16. The total number of discrete steps is chosen equal to twice the number of concentric rings, N_R, as the distribution is symmetric about the origin. These discrete amplitude values are transformed to a spatial taper by assuming a minimum number of elements per $\lambda/4$ (along the x-axis) or $\lambda/8$ steps (along the y-axis) to ensure the desired sparsity. Thus, few design constraints are defined as:

- The projected or local amplitudes along the φ_c-cuts can be 2, 1, or 0, implying two, one, or nil numbers of elements in all of the rings at the respective projected locations.
- The sparsity should be more near the array edges to maintain a low PSLL; henceforth, the projected amplitudes can be either 1 or 0 near the edges of an array.
- Enforce a symmetricity in the taper (spatial), which also helps to reduce the design and optimization complexity and involved computation time.

As per the Nyquist sampling theorem and the array theory, the far-field pattern is completely determined by sampling the aperture current distribution at the locations spaced half a wavelength apart. However, the discrete steps chosen here are shorter ($\lambda/4$ or $\lambda/8$). The shorter sampling steps create a discontinuity in the spatial taper distribution and, henceforth, a smoothing operation needs to be applied over a wavelength step. A process called a box car filter or moving averages in statistics [69] can be used for smoothing the spatial distribution. A simple moving average is the unweighted mean of the previous n data points, given as:

$$\overline{x_n} = \frac{x_1 + x_2 + \cdots \bullet + x_n}{n} = \frac{1}{n}\sum_{i=1}^{n} x_i \tag{7.39}$$

Thus, in the present case, the amplitude at location n, x_n (corresponding to the nth ring) can be determined by summing x_{n-2}, x_{n-1}, x_n, x_{n+1}, and x_{n+2} values and dividing their sum by 5. Hence, in this manner, the desired amplitude distribution is transformed into a spatial taper yielding a low SLL.

If $W(x)$ and $W(y)$ are assumed as the projected amplitude distributions (in terms of 2, 1, or 0 values as decided earlier) along φ_c = 0° and 90° φ-cut planes, then, for a planar array, the amplitude distribution will be $W(x, y) = W(x) \bullet W(y)$, where

$$W(x,y) = \begin{cases} W(x+) \bullet W(y+) & \text{if } x,y > 0 \\ W(x-) \bullet W(y+) & \text{if } x < 0\ ;\ y > 0 \\ W(x+) \bullet W(y-) & \text{if } x > 0\ ;\ y < 0 \\ W(x-) \bullet W(y-) & \text{if } x,y < 0 \end{cases} \tag{7.40}$$

Thus, four sets of distributions are there in four regions that are totally uncorrelated and deterministic. For a given $\varphi_{p,q}$ (N_a in total), assuming region 1, we have $y_{npq} = x_{npq}\tan(\varphi_{p,q})$ and similar for negative x coordinates. Therefore, the distribution of the projections along the linear array at an angle $\varphi_{p,q}$ (with respect to a positive x-axis) in Regions I and II,

$$W_{pq}(x,y) = W\left(x_{npq_{\text{I,II}}}\right) \bullet W\left(y = y_{npq_{\text{I,II}}} = x_{npq_{\text{I,II}}}\tan\left(\varphi_{p,q}\right)\right) \tag{7.41}$$

Thus, in this manner, the projected linear arrays can be formed at different $\varphi_{p,q}$ angles and checked for a low SLL performance. This process may be repeated for all of the four regions to ensure a low SLL envelope across the hemisphere. For Regions III and IV, (7.35) becomes

$$W_{pq}(x,y) = W\left(x = x_{npq_{\text{III,IV}}} = y_{npq_{\text{III,IV}}} \cot\left(\varphi_{p,q}\right)\right) \bullet W\left(y_{npq_{\text{III,IV}}}\right) \tag{7.42}$$

Thus, we have in total N_a number of distributions and corresponding linear antenna array patterns. Thus, ensuring a low SLL output from all of these distributions, a good deterministic design is obtained with complete intuitiveness and knowledge of the design process.

There are N_a number of $\varphi_{p,q}$ values, which facilitates a total of N_a times N_r possible locations, that is, (x_s, y_s), where $y_s = x_s\tan(\varphi_s)$, and φ_s $(s = 1, 2, \ldots, N_R)$ is one of the possible angular locations out of N_a angles, φ_{pq}. The distance, z_s, of location, s, with respect to the origin of the coordinate system is given as:

$$z_s = \sqrt{x_s^2 + y_s^2} \tag{7.43}$$

(It is to be noted that z_s can be larger than the largest dimension of the aperture.)

If a projection is considered from the line $y = x \bullet \tan(\varphi_{pq})$, it leads to a distance to the origin given as:

$$Z_{pq}^{P} = z_s \cos\left(\varphi_s - \varphi_{pq}\right) = z_s\left[\cos\varphi_s \cos\varphi_{pq} + \cos\varphi_s \cos\varphi_{pq}\right] \tag{7.44}$$

$$Z_{pq}^{P} = x_s \cos\varphi_{pq} + y_s \sin\varphi_{pq} \tag{7.45}$$

$W\,(Z_{pq}^{P})$ can be computed to see the SLL distribution across various projections to assess the performance of the array.

These concepts can be utilized to design a highly sparse antenna array with a low sidelobe level performance. The concept can be well understood by considering a numerical test case highlighting the steps of design and optimization of a sparse antenna array in a deterministic manner.

7.8.3 Problem Statement

An antenna aperture array of size, 10λ by 5λ, is considered and the desired sparsity requires the total number of elements, N_a equal to 50. The desired amplitude distribution is a Taylor distribution along the x- and y-planes (with PSLL = –30 dB). The number of concentric rectangles chosen are N_R with size $2a \bullet a$ where $\lambda/4 < a < 5$ (see Figure 7.16). All rectangles make use of a unit cell having a size of $\lambda/4 \bullet \lambda/4$ to discretize the aperture current distribution. The total number of unit cells is 800, as shown in Figure 7.16. Table 7.3 shows the initial proposed design of concentric rectangular ring-based antenna array chosen for the further design and optimization process. Column 1 indicates the rectangular ring's index, n, column 2 indicates the size of the nth ring, column 4 indicates the number of UCs of the size $\lambda/4 \bullet \lambda/4$ in the nth ring, and column 5 indicates the fraction of the total aperture size shared by the UCs in the particular ring.

The projection method (for $y > 0$) is applied along the plane $\varphi = \pi/2$ (the area of the split cell in the center ring is $\lambda/2 \times \lambda/8 \sim \lambda^2/16$) (see Figure 7.17). The projected locations along the y-axis are given as (in terms of wavelength):

$$\begin{aligned} yn_{\text{proj}} = [&-2.4375 \;\; -2.3125 \;\; -2.1875 \;\; -2.0625 \;\; -1.9375 \;\; -1.8125 \\ &-1.6875 \;\; -1.5625 \;\; -1.4375 \;\; -1.3125 \;\; -1.1875 \;\; -1.0625 \;\; -0.9375 \\ &-0.8125 \;\; -0.6875 \;\; -0.5625 \;\; -0.4375 \;\; -0.3125 \;\; -0.1875 \;\; -0.0625 \\ &0.0625 \;\; 0.1875 \;\; 0.3125 \;\; 0.4375 \;\; 0.5625 \;\; 0.6875 \;\; 0.8125 \;\; 0.9375 \\ &1.0625 \;\; 1.1875 \;\; 1.3125 \;\; 1.4375 \;\; 1.5625 \;\; 1.6875 \;\; 1.8125 \;\; 1.9375 \\ &2.0625 \;\; 2.1875 \;\; 2.3125 \;\; 2.4375] \end{aligned} \tag{7.46}$$

The desired Taylor's amplitude distribution for 20 rings is:

$$\begin{aligned} A_{Tn} = [&0.2448 \;\; 0.2566 \;\; 0.2796 \;\; 0.3128 \;\; 0.3548 \;\; 0.4038 \;\; 0.4580 \\ &0.5155 \;\; 0.5745 \;\; 0.6336 \;\; 0.6913 \;\; 0.7465 \;\; 0.7981 \;\; 0.8453 \;\; 0.8874 \\ &0.9237 \;\; 0.9536 \;\; 0.9766 \;\; 0.9921 \;\; 1.0000 \;\; 1.0000 \;\; 0.9921 \;\; 0.9766 \\ &0.9536 \;\; 0.9237 \;\; 0.8874 \;\; 0.8453 \;\; 0.7981 \;\; 0.7465 \;\; 0.6913 \;\; 0.6336 \\ &0.5745 \;\; 0.5155 \;\; 0.4580 \;\; 0.4038 \;\; 0.3548 \;\; 0.3128 \;\; 0.2796 \;\; 0.2566 \\ &0.2448] \end{aligned} \tag{7.47}$$

The amplitude distribution is symmetric consisting of 40 values in total symmetric about the centre. Referring to (7.29) and (7.30), $\overline{\overline{\mathbb{C}}} \bullet \wp = \mathbb{A}$ or $\wp = \overline{\overline{\mathbb{C}}}^{-1} \mathbb{A}$, the values of the variables are given as:

$$\overline{\overline{\mathbb{C}}} = \begin{matrix} 1&1&1&1&1&1&1&1&1&1&1&1&1&1&1&1&1&1&1&1\\ 0&2&1&1&1&1&1&1&1&1&1&1&1&1&1&1&1&1&1&1\\ 0&0&3&1&1&1&1&1&1&1&1&1&1&1&1&1&1&1&1&1\\ 0&0&0&4&1&1&1&1&1&1&1&1&1&1&1&1&1&1&1&1\\ 0&0&0&0&5&1&1&1&1&1&1&1&1&1&1&1&1&1&1&1\\ 0&0&0&0&0&6&1&1&1&1&1&1&1&1&1&1&1&1&1&1\\ 0&0&0&0&0&0&7&1&1&1&1&1&1&1&1&1&1&1&1&1\\ 0&0&0&0&0&0&0&8&1&1&1&1&1&1&1&1&1&1&1&1\\ 0&0&0&0&0&0&0&0&9&1&1&1&1&1&1&1&1&1&1&1\\ 0&0&0&0&0&0&0&0&0&10&1&1&1&1&1&1&1&1&1&1\\ 0&0&0&0&0&0&0&0&0&0&11&1&1&1&1&1&1&1&1&1\\ 0&0&0&0&0&0&0&0&0&0&0&12&1&1&1&1&1&1&1&1\\ 0&0&0&0&0&0&0&0&0&0&0&0&13&1&1&1&1&1&1&1\\ 0&0&0&0&0&0&0&0&0&0&0&0&0&14&1&1&1&1&1&1\\ 0&0&0&0&0&0&0&0&0&0&0&0&0&0&15&1&1&1&1&1\\ 0&0&0&0&0&0&0&0&0&0&0&0&0&0&0&16&1&1&1&1\\ 0&0&0&0&0&0&0&0&0&0&0&0&0&0&0&0&17&1&1&1\\ 0&0&0&0&0&0&0&0&0&0&0&0&0&0&0&0&0&18&1&1\\ 0&0&0&0&0&0&0&0&0&0&0&0&0&0&0&0&0&0&19&1\\ 0&0&0&0&0&0&0&0&0&0&0&0&0&0&0&0&0&0&0&20 \end{matrix} \tag{7.48}$$

$$\mathbb{A} = \begin{bmatrix} 1 & 0.994 & 0.981 & 0.961 & 0.935 & 0.904 & 0.868 & 0.826 & 0.781 & 0.733 \\ 0.682 & 0.631 & 0.580 & 0.531 & 0.486 & 0.446 & 0.412 & 0.385 & 0.367 & 0.357 \end{bmatrix} \tag{7.49}$$

Therefore, solving the inner equation, $\wp$ is obtained as:

$$\wp = \begin{bmatrix} 1.0000 & 2.9843 & 4.9219 & 6.7832 & 8.5410 & 10.1707 & 11.6508 & 12.6935 & 14.0951 & 15.0366 \\ 15.7845 & 16.3429 & 16.7240 & 16.9498 & 17.0532 & 17.0768 & 17.0731 & 17.1009 & 17.2228 & 17.5002 \end{bmatrix} \tag{7.50}$$

According to (7.31), $A_s = 543.789$, which helps to compute the number of elements in each ring obtained using (7.32). These are obtained as:

$$\Lambda_n = \begin{bmatrix} 0 & 1 & 2 & 2 & 3 & 3 & 3 & 4 & 4 & 4 & 4 & 3 & 3 & 3 & 2 & 2 & 2 & 2 & 2 & 1 \end{bmatrix} \tag{7.51}$$

Table 7.3
Initial Design Proposal for Rectangular Sparse Array with 20 Rings

No.	Size ($2a \cdot a$)	(Number of UCs Up to Rectangle n – Number of UCs in Inner Rectangles)	Added Number of Cells per Rectangle a_n	Total Array Size per Rectangle
1	$\lambda/2 \cdot \lambda/4$	(2)	2	0.25%
2	$\lambda \cdot \lambda/2$	(8 – 2)	6	0.75%
3	$3/2\lambda \cdot 3\lambda/4$	(18 – 8)	10	1.25%
4	$2\lambda \cdot \lambda$	(32 – 18)	14	1.75%
5	$2.5\lambda \cdot 5/4\lambda$	(50 – 32)	18	2.25%
6	$3\lambda \cdot 3/2\lambda$	(72 – 50)	22	2.5%
7	$3.5\lambda \cdot 7/4\lambda$	(98 – 72)	26	3.25%
8	$4\lambda \cdot 2\lambda$	(128 – 98)	30	3.75%
9	$4.5\lambda \cdot 9/4\lambda$	(162 – 128)	34	4.25%
10	$5\lambda \cdot 5/2\lambda$	(200 – 162)	38	4.75%
11	$5.5\lambda \cdot 11/4\lambda$	(242 – 200)	42	5.25%
12	$6\lambda \cdot 3\lambda$	(288 – 242)	46	5.75%
13	$6.5\lambda \cdot 13/4\lambda$	(338 – 288)	50	6.25%
14	$7\lambda \cdot 14/4\lambda$	(392 – 338)	54	6.75%
15	$7.5\lambda \cdot 15/4\lambda$	(450 – 392)	58	7.25%
16	$8\lambda \cdot 4\lambda$	(512 – 450)	62	7.75%
17	$8.5\lambda \cdot 17/4\lambda$	(578 – 512)	66	8.25%
18	$9\lambda \cdot 18/4\lambda$	(648 – 578)	70	8.75%
19	$9.5\lambda \cdot 19/4\lambda$	(722 – 648)	74	9.25%
20	$10\lambda \cdot 5\lambda$	(800 – 722)	78	9.75%
			$\sum_{n=1}^{20} a_n = 800$	

The sum total of Λ_n is 50 (i.e., equal to N_a) as required. Thus, from the desired Taylor's amplitude distribution, we get the required number of elements in each of the concentric rings. It is seen that the central ring ($n = 1$) should not have any elements in it. Equation (7.31) gives p_n (average amplitude per UC) derived from the modal Taylor's distribution and the number of elements per ring, Λ_n is obtained from p_n, a_n, and N_A. Hence, it is observed that the desired sparsity and the modal amplitude distribution dictate the allocation of the number of elements in the rings and the average amplitude per unit cell in the rings.

Next, it is required to estimate the exact location of the elements, that is, (x_{npq}, y_{npq}) from φ_{pq} in the respective rings while maintaining Λ_n and p_n. Referring to Figure 7.18, it is observed that φ_{pq} represents the line of the constant angle passing from the origin through all of the constituent rings. There are fixed x-locations depending upon the discretization steps as chosen, which varies from $\lambda/8$ to $(N_R - 1)\lambda/8$ in the steps of $\lambda/4$. Fixing either of these locations, the corresponding y-coordinate can be chosen following (7.33) through (7.38). It is observed that there are multiple options available for y-locations. For instance, at $x = \lambda/8$, restricting to Region I only, two y-locations are possible, $y = \pm\lambda/16$; at $x = 3\lambda/8$, four y-locations, $y = \pm\lambda/16, \pm 3\lambda/16$ are possible, and so on. In the similar manner, all four regions have potential locations, x_{pq} and y_{pq}, proportional to φ_{pq}, from which the optimal ones may be chosen. The choice may be made using stochastic techniques as well. However, herein the proposed deterministic technique is chosen for the purpose.

The PT is used to map the projected amplitudes, $W(x)$ and $W(y)$, along the positive and negative y-axis ($W(x+) = W(x-)$ and $W(y+) = W(y-)$, that is, a symmetric amplitude distribution about the origin), as derived from the Taylor's distribution to the values in (7.43). The choice of (x_{npq}, y_{npq}) along the lines at a constant angle φ_{pq} passing through the rings shall be so chosen that the projected amplitudes at discrete x-locations, x_{npq}, match to the $W(x)$ and $W(y)$ distributions. Here, the numbering of (x, y) locations of the dense antenna array grid is defined as:

$$\begin{aligned} (x,y) &= \left(-(2N_R - 1)\cdot\frac{\lambda}{8} + r\cdot\frac{\lambda}{4}, -(2N_R - 1)\cdot\frac{\lambda}{16} + s\cdot\frac{\lambda}{8}\right) \\ r,s &= 0,1,\ldots,(2N_R - 1) \end{aligned} \tag{7.52}$$

Thus, choosing the locations as per (7.52) gives 1,600 possible locations over which antenna elements can be located. It is important to define a few design constraints, which will reduce the number of locations, thereby reducing the computational burden. These are:

- In the array, there are only 20 locations along a positive or negative x-axis ($N_R = 20$) and $N_a = 50$. Since there is no element

in the central ring, there are (N_R – 1) times N_a ~ 19 × 50 = 950 probable locations out of 1,600. This is the first elimination step where 750 locations are thus eliminated.
- A geometric elimination method may be chosen by truncating the rectangular aperture to an elliptical aperture and all of the element's locations lying outside the boundary may be eliminated. This is the second elimination step, which will further reduce the locations by 150 to 800 (i.e., nearly a 50% reduction in the number of possible locations out of 1,600 locations).
- Minimum distance constraint implies that no two elements shall be located closer than a half-wavelength to avoid issues such as mutual coupling. This is the third elimination step.
- Finally, the element density shall reduce as the distance from the array center increases towards the array's finite edges.

The discretized Taylor's amplitude distribution (PSLL = –30 dB; 52a = 4) sampled along the x-axis locations is:

- W_T = 0.2448 0.2566 0.2796 0.3128 0.3548 0.4038 0.4580 0.5155 0.5745 0.6336 0.6913 0.7465 0.7981 0.8453 0.8874 0.9237 0.9536 0.9766 0.9921 1.0000 1.0000 0.9921 0.9766 0.9536 0.9237 0.8874 0.8453 0.7981 0.7465 0.6913 0.6336 0.5745 0.5155 0.4580 0.4038 0.3548 0.3128 0.2796 0.2566 0.2448

  The Taylor's distribution being symmetric is separated along two halves of the x-axis given as:

- $W_T(x+)$ = 1.0000 0.9921 0.9766 0.9536 0.9237 0.8874 0.8453 0.7981 0.7465 0.6913 0.6336 0.5745 0.5155 0.4580 0.4038 0.3548 0.3128 0.2796 0.2566 0.2448
- $W_T(x-)$ = 0.2448 0.2566 0.2796 0.3128 0.3548 0.4038 0.4580 0.5155 0.5745 0.6336 0.6913 0.7465 0.7981 0.8453 0.8874 0.9237 0.9536 0.9766 0.9921 1.0000
- The sampled locations along the $+x$-axis (in terms of λ), that is, [1/8:1/4:39/8] = 0.1250 0.3750 0.6250 0.8750 1.1250 1.3750 1.6250 1.8750 2.1250 2.3750 2.6250 2. 8750 3.1250 3.3750 3.6250 3.8750 4.1250 4.3750 4.6 250 4.8750
- The sampled locations along the $-x$-axis (in terms of wavelength), that is, –0.1250 –0.3750 –0.6250 –0.8750

−1.1250 −1.3750 −1.6250 −1.8750 −2.1250 −2.3750
−2.6250 −2.8750 −3.1250 −3.3750 −3.6250 −3.8750
−4.1250 −4.3750 −4.6250 −4.8750

- Complete Taylor distribution for 40 elements along the x-axis (10λ length) is obtained as $W_{TX} = [W_T(x-)\ W_T(x+)]$ and for the complete y-axis as $W_{TY} = [W_T(y-)\ W_T(y+)]$. The respective array patterns as obtained using W_{TX} and W_{TY} are given as in Figure 7.19.

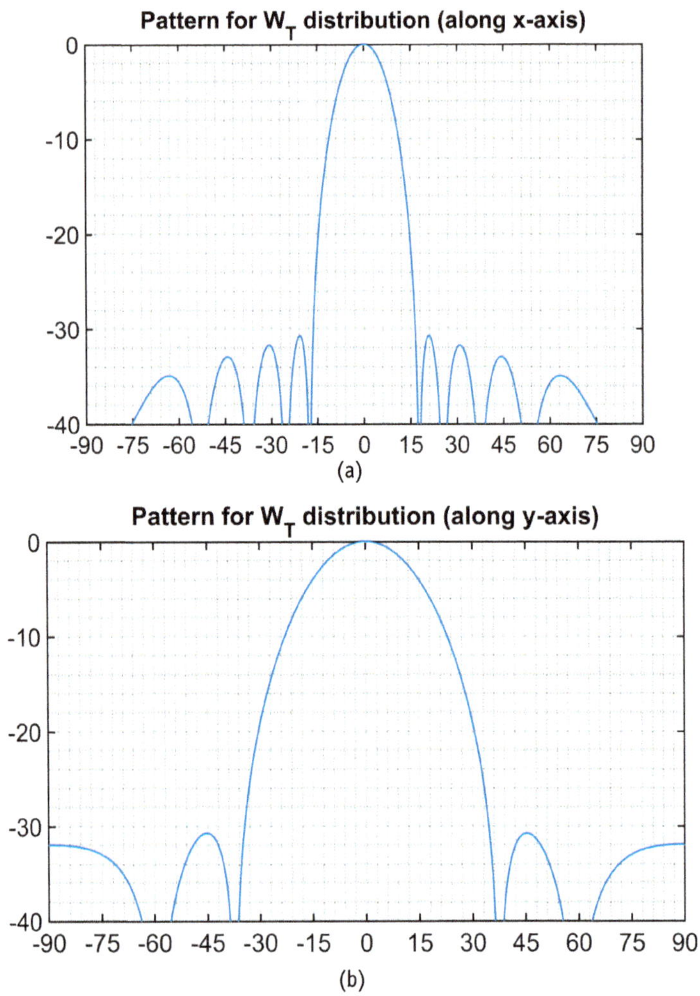

Figure 7.19
Array factor plots for Taylor's distribution along the x- and y-axis of array discretized at the sampled locations.

The amplitude distribution thus obtained is set as a target to achieve by synthesizing a spatial distribution of the uniformly fed elements.

7.8.3.1 Mapping of Model Amplitude Distribution to the Spatial (Density) Distribution

The discretized projected amplitude may be 2, 1, or 0 indicating the total number of overlapping elements along a particular projected location along x- or y-axes. Table 7.4 summarizes the various design steps leading to the final discrete spatial distribution. Columns 1 and 2 list the ring index and the discrete sampled locations along positive x ($+x$) direction. Column 3 lists the amplitude distribution, W_{Tn}, sampled at $+x$-locations based on the Taylor's distribution corresponding to n and x_n sampled at $\lambda/4$ locations and $\sum_{n=1}^{N_R} W_{Tn} = 12.8486$. Normalizing the W_{Tn} values as in (7.53), it is seen from (7.54) and (7.55) that W_{Tn} is normalized.

$$W_{T_{n,\text{norm}}} = \frac{2}{L}\frac{N_R}{2 \cdot \sum_{n=1}^{N_R} W_{T_n}} \cdot W_{T_n} \cdot \frac{\lambda}{4} \tag{7.53}$$

and

$$W_{T_N}(x+) = \frac{2}{L}\frac{N_R}{2 \cdot \sum_{n=1}^{N_R} W_{T_n}} \cdot \sum_{n=1}^{N_R} W_{T_n} \cdot \frac{\lambda}{4} = 0.5 = W_{T_N}(x-) \tag{7.54}$$

$$W_{T_N}(x+) + W_{T_N}(x-) = 1 \tag{7.55}$$

$W_{T_{n,\text{norm}}}$ are listed in column 5 as per (7.53) for the Taylor's distribution. From Table 7.4, $\sum_{n=1}^{N_R} W_{T_{n,\text{norm}}} = 24.9$ close to $N_a/2$, indicating that the total current distribution is proportional to the number of active elements, N_a.

Column 6 lists the spatial distribution as obtained from the amplitude distribution along the projected discrete location coordinates along the $+x$-axis, denoted as W_{Dn}. Initiating the design and choosing the first location (i.e., $n = 1$ ($\sim x = \lambda/8$)), it is assumed that the amplitude (projected) here is 2. While comparing with $W_{T_{n,\text{norm}}}$

values, whenever its value goes down below 0.9 with reference to 1 (maximum), the amplitude becomes 1, and again for next location the amplitude goes back to 2. In this way, the first five locations (corresponding to 5 rings) for the amplitudes are 2, 2, 1, 2, 2 and, using (7.39), taking the moving average, the amplitude becomes 1.8 or 0.9. Following the similar logical sequence, again whenever the values of $W_{T_{n,\text{norm}}}$ goes down below 0.8, it leads to the location for the next 1. Following this strategy, we chose locations with 2 or 1. Also, the total sum of amplitudes should not be more than $N_a/2$. Thus, at a few places where the variation in the Taylor's amplitude is very slow, a 0 can be inserted as well to match the total of $N_a/2$. Column 6 summarizes two options, type a and type b, obtained by varying the transition points chosen differently. Column 7 shows the corresponding values of discrete amplitudes, $W_{D_{N,\text{mov}}}$ obtained after taking moving averages of W_{Dn} and the sum of these values for both of the types is 12.5, which is very close in comparison to the value 12.8486, as obtained for ideal Taylor's distribution.

In the similar manner, distributions along the positive and negative x- and y-axes can be obtained. Henceforth, four uncorrelated deterministic discrete amplitudes are obtained and derived from the chosen ideal amplitude distribution. Next, the normalization factors for each of the distributions needs to be obtained. From the Taylor's distribution, $\sum_{n=1}^{N_R} W_T(x+) = 12.8486$ and from the discrete distribution, $\sum_{n=1}^{N_R} W_{D_{N,\text{mov}}}(x+) = 12.5$. Hence, types a and b distributions closely emulate the desired Taylor's distribution.

Thus, assuming a symmetric distribution across the aperture, the four sums along positive and negative coordinate axes may be written as:

$$\begin{aligned}
W_{D_{N,\text{mov}}}(x+) &= \sum_{n=1}^{N_R} W_{DM_n}\left(x = \frac{n\lambda}{4} - \frac{\lambda}{8}\right) = 12.5 \\
W_{D_{N,\text{mov}}}(x-) &= \sum_{n=1}^{N_R} W_{DM_n}\left(x = \frac{\lambda}{8} - \frac{n\lambda}{4}\right) = 12.5 \\
W_{D_{N,\text{mov}}}(y+) &= \sum_{n=1}^{N_R} W_{DM_n}\left(x = \frac{m\lambda}{8} - \frac{\lambda}{16}\right) = 12.5 \\
W_{D_{N,\text{mov}}}(y-) &= \sum_{n=1}^{N_R} W_{DM_n}\left(x = \frac{\lambda}{16} - \frac{m\lambda}{8}\right) = 12.5
\end{aligned} \tag{7.56}$$

Thus, the overall sum of these four discrete distributions (after taking the moving average) along the x- or y-axes is equal to $N_a \sim 50$, indicating the sparsity is achieved.

In the projections, we have $N_a = 50$ elements along the x-axis, so the amplitudes for element locations (1-D) are derived from the expressions given below and listed in column 8 in Table 7.4.

$$W_{D_{N,\text{norm}}}(x+) = \text{Int}\left[50\frac{1}{5\lambda}\frac{20}{25}W_n(x_n+)\frac{\lambda}{4}\right] \text{ for } x+ \tag{7.57}$$

$$W_{D_{N,\text{norm}}}(x-) = \text{Int}\left[50\frac{1}{5\lambda}\frac{20}{25}W_n(x_n-)\frac{\lambda}{4}\right] \text{ for } x- \tag{7.58}$$

The resulting sum of these values for types a and b is 25 (i.e., $N_a/2$ for x+). The same results is obtained for $-x$ distribution and the total number of elements is equal to $N_a = 50$. In a similar way, we may derive the amplitudes along the y-axis with y_m locations (with $\lambda/8$ discretization):

$$\text{Int}\left[50\frac{1}{2.5\lambda}\frac{20}{25}W_m(y_m+)\frac{\lambda}{8}\right] \text{ for } y+ \text{ and } \text{Int}\left[50\frac{1}{2.5\lambda}\frac{20}{25}W_m(y_m+)\frac{\lambda}{8}\right] \text{ for } y- \tag{7.59}$$

It is observed that the values of column 8 are not integers and it is required to convert them into integers, which will help to decide the number of elements x allocated to each of the rings. Hence, a thresholding technique is employed to convert the entries of columns 7 to whole numbers, $W_{D_{N,th}}$, as shown in column 9. The decision for the transition of amplitudes from 2 to 1 or from 1 to 0 is made based on (7.60) and (7.61).

$$\text{if } W_{D_{n,\text{norm}}} > \mathbb{Z}_{th,1};\ W_{D_{n,th}} \text{ is made 1 from 2} \tag{7.60}$$

$$\mathbb{Z}_{th,2} > W_{D_{n,\text{norm}}} > \mathbb{Z}_{th,1};\ W_{D_{n,th}} \text{ is made 0 from 1} \tag{7.61}$$

In (7.60) and (7.61), the choice of $\mathbb{Z}_{th,1}$ and $\mathbb{Z}_{th,2}$ may be taken as 1.5 and 0.4. These values may be experimented upon, but have been proposed based on my experience. However, in view of satisfying the criteria of p_n and N_a, a few of the values may be made 0 deterministically. It is seen that $\sum_{n=1}^{N_R} W_{D_{n,norm}} = 12.5$ for types a and b distributions. Thus, the discrete distributions obtained, W_{D_n} or $W_{D_{n,th}}$, may be used to decide the allocation of elements in rectangular rings.

This section explains the philosophy of how to correlate Λ_n with the optimized discrete distribution, W_{D_n}, while allocating elements to each of the rings in the array. The number of nonzero elements, Λ_n, per rectangle n ($n = 1, \ldots, 20$) is known from (7.51). In the case of $\Lambda_n = 2$ and $W_{D_n} = 1$, it implies that there are 2 elements in the opposite corner of the nth rectangle. In the case of $\Lambda_n = 3$ and $W_{Dn} = 1$, there is one element at one corner and the 2 remaining elements should be on the two sides opposite the element at the corner. In the case of $\Lambda_n = 4$ and $W_{Dn} = 1$, it means that all 4 elements should be on different sides and not at a corner. In the case of $\Lambda_n = 4$ and $W_{Dn} = 1$, 2 elements are at the different corners of the rectangle n. Thus, there are strict rules for placing the nonzero elements on the probable locations. An important design constraint is the adjustment of the concentric rings' dimensions such that $W_{D_n} \leq 2$ per ring. The selection of $\lambda/4$ and $\lambda/8$ step sizes for x- and y-directions satisfies this requirement. In the case of $\Lambda_n = 1$, it cannot be fit into a scheme with $W_{Dn} = 1$. Further, $\Lambda_n = 0$ implies nil elements in the particular ring.

Table 7.4
Amplitude Distribution Across the Rings Along Positive X-Axis

No.	$x_n = \frac{n\lambda}{4} - \frac{\lambda}{8}$	W_{Tn}	$W_{Tn,mov}$	$W_{Tn'norm}$	W_{Dn} Type a	W_{Dn} Type b	$W_{Dn,mov}$ Tywe a	$W_{Dn,mov}$ Type b	$W_{Dn,norm}$ Type a	$W_{Dn,norm}$ Type b	$W_{Dn,th}$ Type a	$W_{Dn,th}$ Type b
1	1/8	1	1.00	1.9	2	2	1	1	2	2	2	2
2	3/8	0.9921	0.99	1.9	2	2	1	1	2	2	2	2
3	5/8	0.9766	0.98	1.9	2	2	1	1	2	2	2	2
4	7/8	0.9536	0.95	1.9	1	1	1	1	2	2	2	2
5	9/8	0.9237	0.92	1.8	2	2	1	1	2	2	2	2
6	11/8	0.8874	0.89	1.7	2	2	1	1	2	2	2	2
7	13/8	0.8453	0.85	1.6	2	1	1	1	2	2	2	2
8	15/8	0.7981	0.80	1.6	2	2	0.9	0.9	1.8	1.8	2	2
9	17/8	0.7465	0.75	1.5	0	1	0.8	0.8	1.6	1.6	2	2
10	19/8	0.6913	0.69	1.3	2	1	0.7	0.7	1.4	1.4	1	1

Table 7.4
(continued)

No.	$x_n = \frac{n\lambda}{4} - \frac{\lambda}{8}$	W_{Tn}	$W_{Tn,\text{mov}}$	$W_{Tn'\text{norm}}$	W_{Dn} Type *a*	W_{Dn} Type *b*	$W_{Dn,\text{mov}}$ Type *a*	$W_{Dn,\text{mov}}$ Type *b*	$W_{Dn,\text{norm}}$ Type *a*	$W_{Dn,\text{norm}}$ Type *b*	$W_{Dn,\text{th}}$ Type *a*	$W_{Dn,\text{th}}$ Type *b*
12	23/8	0.5745	0.58	1.1	2	1	0.5	0.4	1	0.8	1	1
13	25/8	0.5155	0.52	1	1	1	0.4	0.4	0.8	0.8	1	1
14	27/8	0.458	0.47	0.9	0	1	0.4	0.4	0.8	0.8	1	0
15	29/8	0.4038	0.41	0.8	1	1	0.3	0.3	0.6	0.6	0	1
16	31/8	0.3548	0.36	0.7	0	0	0.3	0.3	0.6	0.6	1	1
17	33/8	0.3128	0.32	0.6	1	1	0.2	0.3	0.4	0.6	0	0
18	35/8	0.2796	0.29	0.5	1	0	0.2	0.2	0.4	0.4	1	1
19	37/8	0.2566	0.22	0.5	0	1	0.1	0.1	0.2	0.2	0	0
20	39/8	0.2448	0.16	0.5	1	1	0.1	0.1	0.2	0.2	0	0
	TOTAL	12.8486	12.80	24.9	25	25	12.5	12.5	25	25	25	25

Figure 7.20(a) shows the initial array configuration proposal showing the 50 locations as chosen based on the defined criterion overlapped with reference (Taylor) distribution regions. Figure 7.20(b) shows the array performance in different phi-planes showing a D_o = 20.54 dBi (assuming isotropic radiating element backed by a ground plane), HPBW ~ 6.9° and PSLL < –20.54 dB. However, the near SLL region is below –30 dB and the SLL is seen to rise near 60°. The mean spacing is found near 0.4λ indicating high level of coupling.

7.8.3.2 Further Optimization of the Rectangular Array

It is observed that PSLL in the diagonal planes (phi = 27° and phi = 153°) should be optimized, which is normally ignored leading to high PSLL along intercardinal planes. The observations are:

1. There is nonsymmetry in PSLL when comparing both planes.
2. In general, the phi = 27° and phi = 45° planes show better results than the phi = 135° and phi = 153° planes.
3. This observation probably caused by the corner elements in the concentric rectangular rings.

Therefore, the focus in optimization now is on the corner elements. It is proposed that:

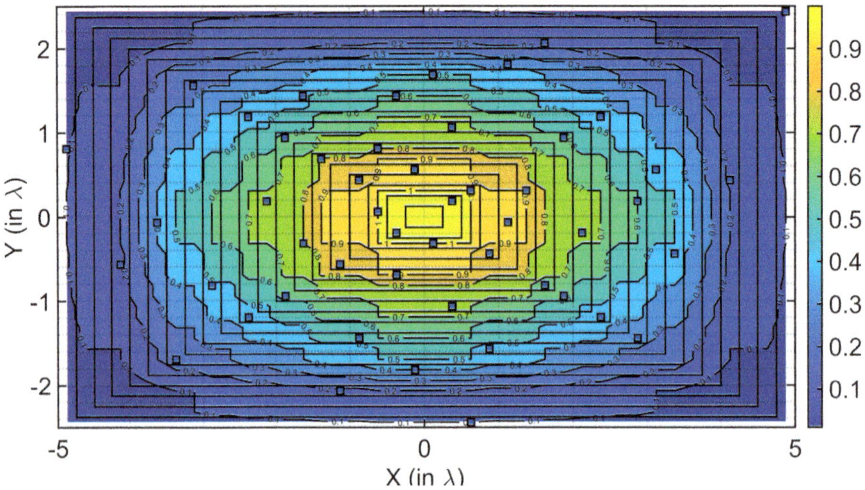

(a)

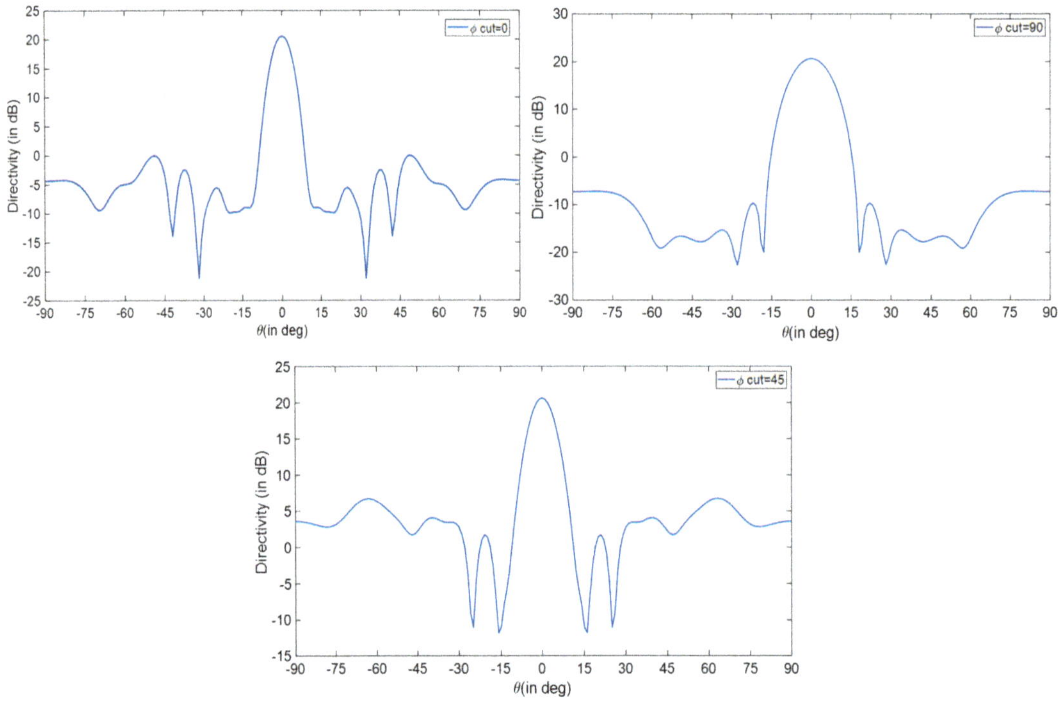

(b)

Figure 7.20

(a) Planar concentric ring sparse array antenna ($N_R = 20$; $N_a = 50$). (b) Array configuration directivity plots (in dB) across different phi-cuts.

1. Shift some of the corner elements that are near to $k/2$ (k is an integer) distance from the origin so that this distance becomes ideally an integer times $\lambda/2$.
2. Some corner elements with distances, which are not nearly $k/2$, are no longer placed at the corners; in other words, they are removed from the corners.
3. Lower the number of corner elements in the phi = 153° plane while keeping the same number of corner elements in the phi = 27° plane.
4. Apply a symmetric shift in the corner elements, which results into a shifted corner element that stays in the diagonal planes. For our rectangular array, it means that the shift in x (s_x) is 2 times the shift in y (s_y); in other words, $s_x = 2s_y$. For ring number r_n, n_s is the wanted spacing relative to λ (n_s = 0.5, 1.0, 1.5, ...). The corner element $c(r_n, c_n)$ where c_n is corner number in ring number r_n (e.g., $c(2, 1)$ is corner element 1 in ring 2) is derived as:

$$10s_y = -\frac{5 \cdot (2r_n - 1)}{8} + n_s\sqrt{20} \tag{7.62}$$

Figure 7.21(a) shows the optimized array configuration superimposed on the $Loc^D_{n,m}$ distribution, showing an optimal distribution of the elements across the aperture. Figure 7.21(b) shows the 2-D radiation pattern plots across three phi-cuts. Figure 7.21(c) shows the 3-D radiation pattern plot. It is seen that D_o = 20.7 dBi, HPBW = 6.9°, PSLL < –20 dB as a worst case across the principal planes and the diagonal planes. However, the PSLL rises along intercardinal planes up to –13 dB as expected as the optimization was done along the principal planes only. However, extending the optimization to intercardinal planes may increase the design cycle. Hence, it is suggested to combine stochastic techniques to further refine the design, if required.

Table 7.5 summarizes the isophoric sparse antenna array perfromance in principal and diagonal φ-cuts. The directivity as computed using (4.14) is 20.7 dBi at boresight and drops to 18.9 dBi at a 60° scan angle showing a scan loss of 1.8 dB assuming a half-isotropic element. A similar broadening effect following $\cos\theta$ behavior is shown in HPBW as well. PSLL better than –30.9 dB, –22.8 dB, and

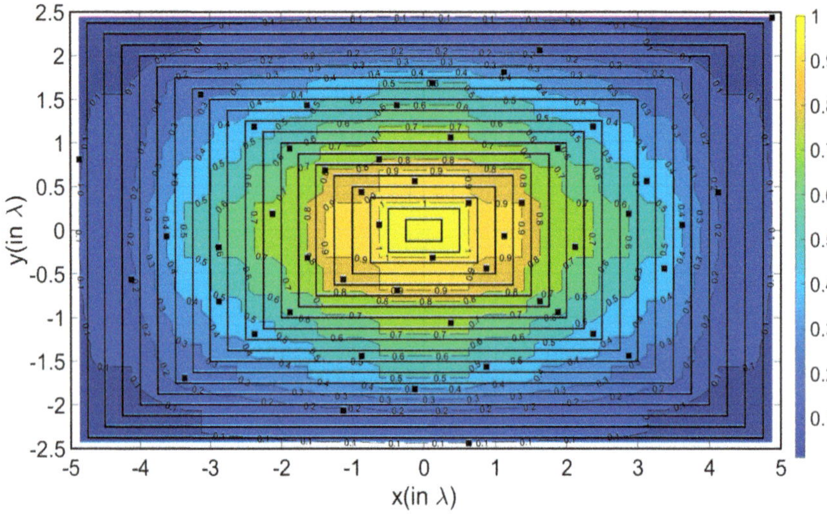

(a)

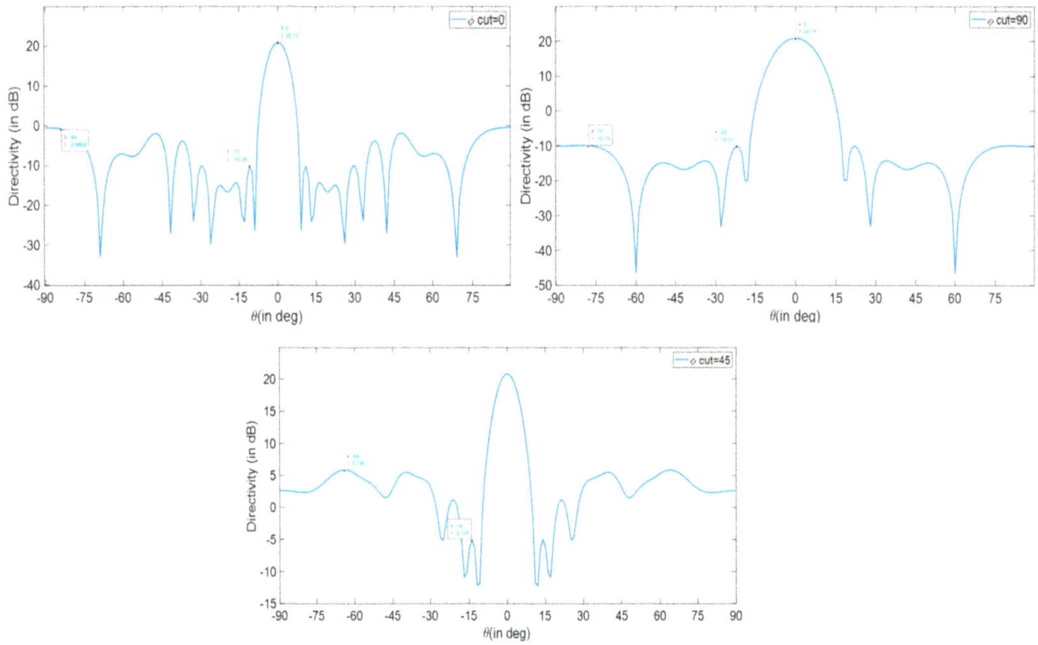

(b)

Figure 7.21

(a) Optimized planar concentric ring sparse array antenna (N_R = 20; N_a = 50). Array configuration: (b) directivity plots (in dB) across different phi-cuts, (c) 3-D radiation pattern plot (normalized) in the $\theta\varphi$ plane, and (d) directivity plot in decibels in the uv coordinate system.

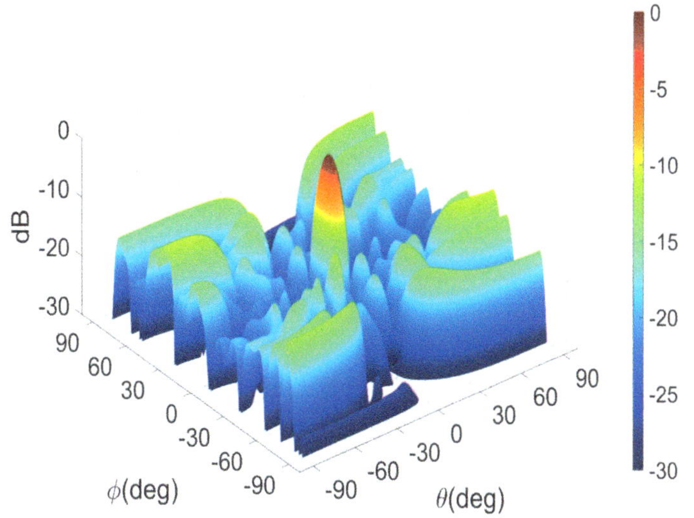

(c)

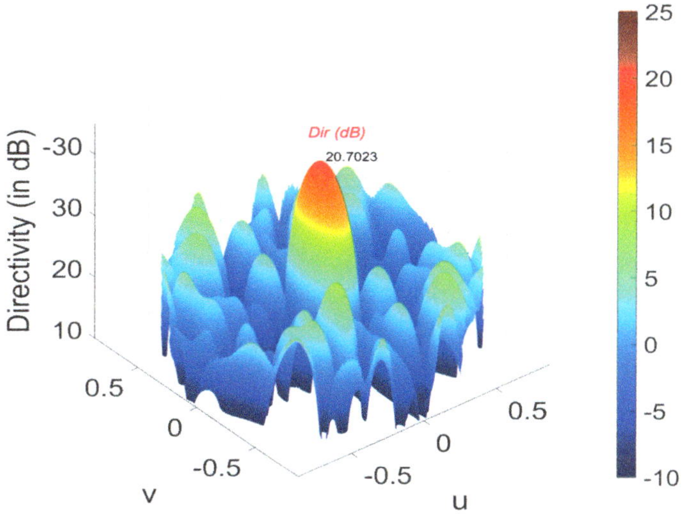

(d)

Figure 7.21
(Continued)

–12 dB is shown in three respective phi cuts. ASLL is observed less than –26.6 dB. The designed array shows a wide scan performance.

A computational electromagnetic-based experimentation is done using Ansys HFSS to validate the design realization using a practical antenna element. A sparse PAA (SPAA) is modeled by choosing

Table 7.5
Scan Performance of a 50-Element PSSA

	$\varphi_o = 0°$				$\varphi_o = 45°$				$\varphi_o = 90°$				ASLL
$\theta_o° \rightarrow$	0	30	45	60	0	30	45	60	0	30	45	60	(dB)
D_o (dBi)	20.7	20.3	19.3	18.9	20.7	20.3	19.3	18.9	20.7	20.3	19.3	18.9	
HPBW (°)	6.6	7.5	9.3	13.5	8.1	9.3	11.6	16.7	13.1	15.4	19.1	31.4	−26.6
PSLL (dB)	−31.1	−31.4	−30.9	−31.4	−12.4	−12.4	−12.4	−12.4	−30.9	−24.9	−24.7	−22.8	

a wideband U-slot-fed microstrip patch antenna (US-MPA) operating over 9 to 10 GHz, which restricts the operating bandwidth of the array. US-MPA has a HPBW of 90° and 80° in its E- and H-planes, which implies that a minimal scan loss (less than 3–4 dB) can only be achieved over an angular region of 90° (±45°) only. A series of experiments are carried out to assess the wide scan performance of the designed SPAA over the desired scan volume using HFSS. The simulation of one scenario consumed 53 GB of RAM and took approximately 38 minutes and 13 seconds on 32-core machines. Table 7.6 summarizes the simulated results from the various using computational experiments done using Ansys HFSS.

Figure 7.22 shows the array model overlapped with the 3-D radiation pattern plot (Figure 7.22(a)) and the various scanned array patterns for different scan arrangements. Figure 7.22(b) shows the 2-D radiation pattern plots in the principal (φ_o = 0° and 90°) and diagonal planes (φ_o = 45°). The overall sidelobe level demonstrated is better than −21 dB and −18.9 dB in the principal and diagonal planes, respectively. The simulation shows a D_o = 23.5 dBi, HPBW of 6.5°, 12.9°, and 8.2° in the principal and diagonal planes, respectively, at boresight. Figure 7.22(c) shows the scanned pattern at (30°,

Table 7.6
Scan Performance of a 50-Element PSAA Using Ansys HFSS

	$\varphi_o = 0°$				$\varphi_o = 30°$	$\varphi_o = 45°$		$\varphi_o = 90°$
$\theta_o° \rightarrow$	0	30	45	60	30	0	45	0
D_o (dBi)	23.5	22.0	18.7	14.8	18.9	18.7	17.2	23.5
HPBW (°)	6.5	7.4	9.8	13.8	11.1	9.8	15.6	12.9
PSLL (dB)	−21.1	−20.3	−15.6	−9.5	−9.6	−15.6	−9.2	−23.8

0°) and (45°, 0°) showing the beam broadening and scan loss with respect to the scan angle. Figure 7.22(d) shows the 3-D radiation pattern plot at an extreme scan angle of (45°, 45°). A clear observation demonstrates that there is no onset of any grating lobe on scanning to an extreme scan angle. The summarized results are in excellent agreement with the MATLAB simulations assuming a half isotropic radiator. The directivity is higher in comparison because the U-slot antenna has a higher gain around 4–6 dBi compared to a half-isotropic radiator with a 2–3-dBi gain. In general, the SLL behavior is in a very close approximation to the MATLAB simulations except for little bit degradation in near SLL (within 2–3 dB only) in the principal planes due to finite-edge considerations in

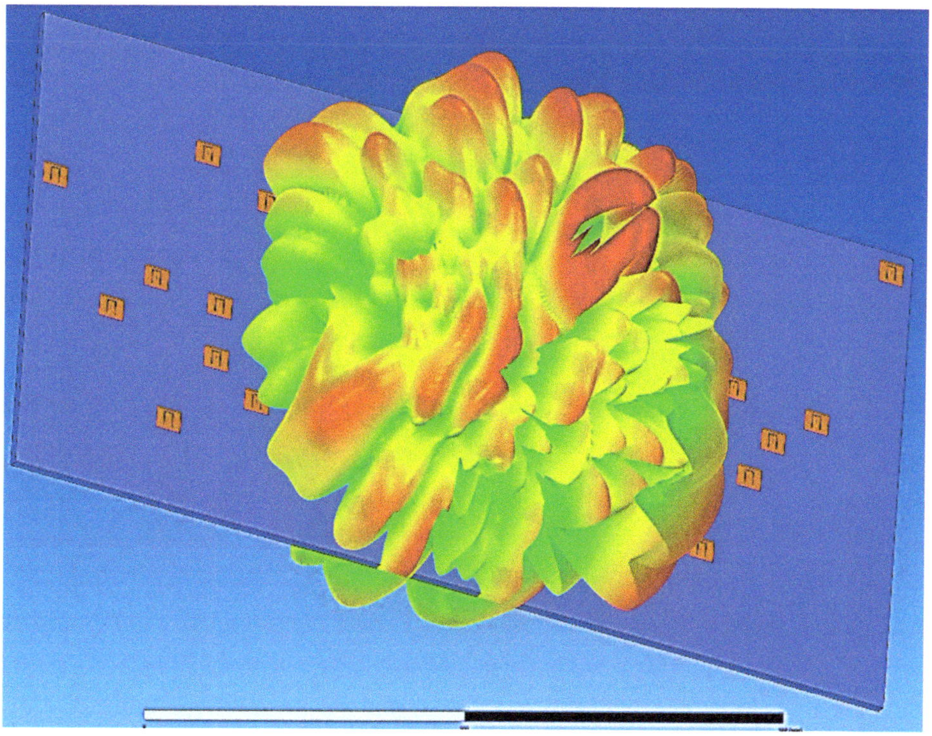

(a)

Figure 7.22

Radiation performance of a concentric ring sparse array antenna ($N_R = 20$; $N_a = 50$) using Ansys HFSS: (a) superimposed 3-D radiation pattern plot at (0°, 0°) overlapped with the model, (b) 2-D principal plane patterns at (0°, 0°), (c) scanned patterns at (30°, 0°) and (45°, 0°), and (d) 3-D radiation pattern plot at (45°, 45°).

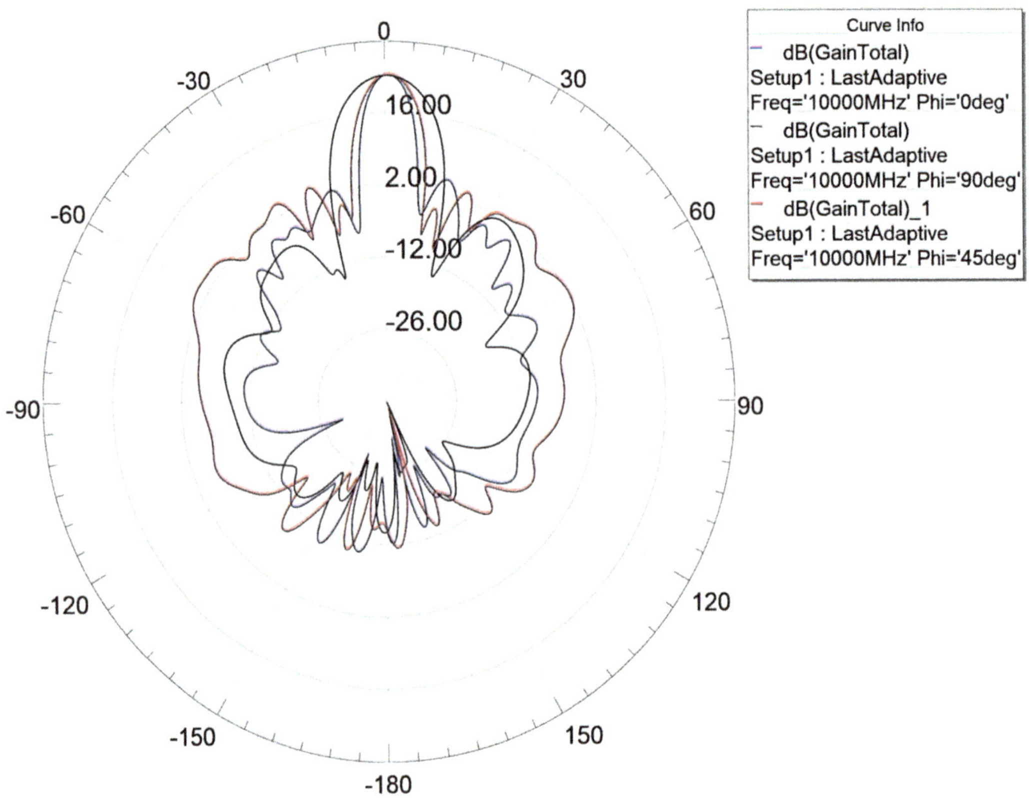

(b)

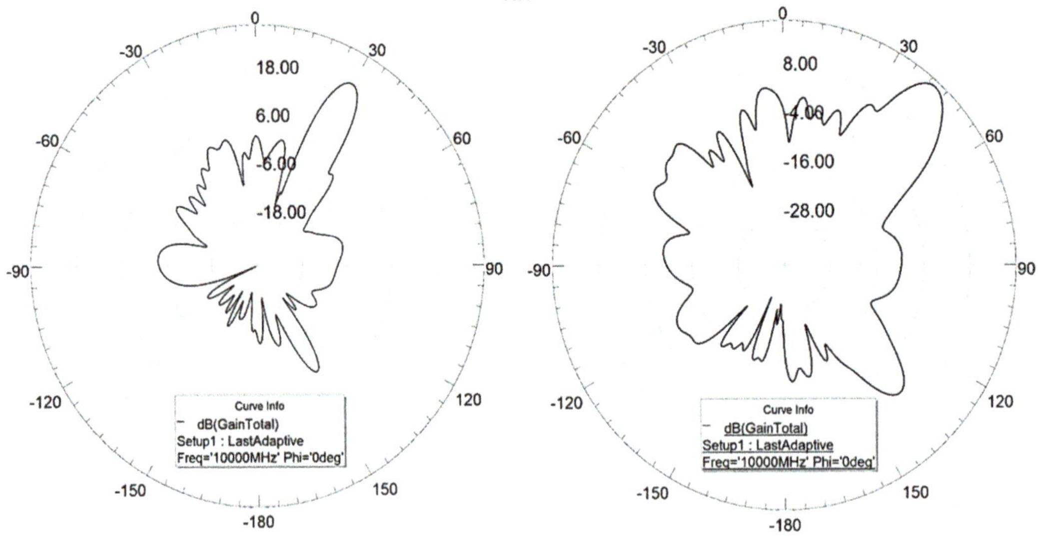

(c)

Figure 7.22
(Continued)

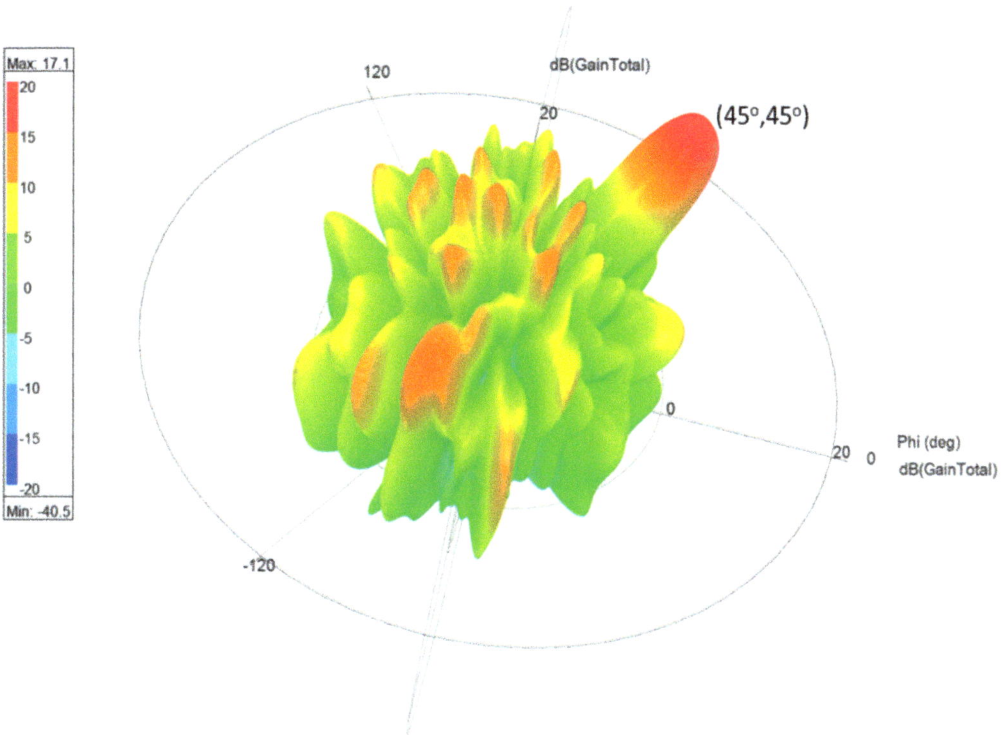

(d)

Figure 7.22
(Continued)

HFSS, which are ignored in the MATLAB simulations. Thus, the efficacy of the deterministic proposal in designing wideband and wide scan sparse antenna arrays is demonstrated using extensive computational experiments using HFSS. The proposed technique and the simulations using (4.14) are giving very accurate estimates and can be used for analyzing array parameters without resorting to HFSS simulations, which are time-consuming in further numerical experiments.

7.9 Conclusions

This chapter presented an overview of various techniques illustrated with examples to showcase the advantages and design of planar sparse array antennas. The techniques presented include stochastic, deterministic, and hybridization of these techniques. It is shown that sparse array antennas can result into Taylor-like patterns along with high sparsity levels of more than 50% with a performance quite close to the dense counterparts.

References

[1] Sandler, S., "Some Equivalences Between Equally and Unequally Spaced Arrays," *IRE Transactions on Antennas and Propagation*, Vol. 8, No. 5, 1960, pp. 496–500.

[2] Lindsay, J., "A Circular Loop Antenna with Nonuniform Current Distribution," *IRE Transactions on Antennas and Propagation*, Vol. 8, No. 4, 1960, pp. 439–441.

[3] Skolnik, M., "Resolution of Angular Ambiguities in Radar Array Antennas with Widely-Spaced Elements and Grating Lobes," *IRE Transactions on Antennas and Propagation*, Vol. 10, No. 3, 1962, pp. 351–352.

[4] Ishimaru, A., "Theory of Unequally-Spaced Arrays," *IRE Transactions on Antennas and Propagation*, Vol. 10, No. 6, 1962, pp. 691–702.

[5] Maffett, A., "Array Factors with Nonuniform Spacing Parameter," *IRE Transactions on Antennas and Propagation*, Vol. 10, No. 2, 1962, pp. 131–136.

[6] Lo, Y., "A Spacing Weighted Antenna Array," *1958 IRE International Convention Record*, New York, March 21–25, 1962, pp. 191–195.

[7] Lo, Y., "High Resolution Antenna Arrays with Randomly Spaced Elements," *1963 Antennas and Propagation Society International Symposium*, Boulder, CO, July 9–11, 1963, pp. 212–218.

[8] Maher, T., and D. Cheng, "Random Removal of Radiators from Large Linear Arrays," *IEEE Transactions on Antennas and Propagation*, Vol. 11, No. 2, 1963, pp. 106–112.

[9] Sherman, J., and M. Skolnik, "Thinning Planar Array Antennas with Ring Arrays," *1958 IRE International Convention Record*, New York, NY, March 21–25, 1963, pp. 77–86.

[10] Lo, Y., "A Mathematical Theory of Antenna Arrays with Randomly Spaced Elements," *IEEE Transactions on Antennas and Propagation*, Vol. 12, No. 3, 1964, pp. 257–268.

[11] Galejs, J., "Minimization of Sidelobes in Space Tapered Linear Arrays," *IEEE Transactions on Antennas and Propagation*, Vol. 12, No. 4, 1964, pp. 497–498.

[12] Snover, A., and A. Ferraro, "An Investigation of the Properties of Synthesized Nonuniformly Spaced Antenna Arrays," *IEEE Transactions on Antennas and Propagation*, Vol. 12, No. 1, 1964, pp. 132–133.

[13] Skolnik, M., G. Nemhauser, and J. Sherman, "Dynamic Programming Applied to Unequally Spaced Arrays," *IEEE Transactions on Antennas and Propagation*, Vol. 12, No. 1, 1964, pp. 35–43.

[14] Galindo, V., "Nonlinear Antenna Arrays," *IEEE Transactions on Antennas and Propagation*, Vol. 12, No. 6, 1964, pp. 782–783.

[15] Ma, M., "Application of Bernstein Polynomials and Interpolation Theory to Linear Array Synthesis," *IEEE Transactions on Antennas and Propagation*, Vol. 12, No. 6, 1964, pp. 668–677.

[16] Ma, M., "Another Method of Synthesizing Nonuniformly Spaced Arrays," *IEEE Transactions on Antennas and Propagation*, Vol. 13, No. 5, 1965, pp. 833–834.

[17] Larson, R., J. Ferris, and V. Powers, "Nonuniformly Spaced Linear Arrays of Unequal-Width Elements," *IEEE Transactions on Antennas and Propagation*, Vol. 13, No. 6, 1965, pp. 974–976.

[18] Ishimaru, A., and Y. Chen, "Thinning and Broadbanding Antenna Arrays by Unequal Spacings," *IEEE Transactions on Antennas and Propagation*, Vol. 13, No. 1, 1965, pp. 34–42.

[19] Tang, C., "An Approximate Method of Designing Nonuniformly Spaced Arrays," *IEEE Transactions on Antennas and Propagation*, Vol. 13, No. 1, 1965, pp. 177–179.

[20] Lo, Y., and S. Lee, "A Study of Space-Tapered Arrays," *IEEE Transactions on Antennas and Propagation*, Vol. 14, No. 1, 1966, pp. 22–30.

[21] Chow, Y., "On the Error Involved in Poisson's Sum Formulation of Nonuniformly Spaced Antenna Arrays," *IEEE Transactions on Antennas and Propagation*, Vol. 14, No. 1, 1966, pp. 101–102.

[22] Steinberg, B., "The Peak Sidelobe of the Phased Array Having Randomly Located Elements," *IEEE Transactions on Antennas and Propagation*, Vol. 20, No. 2, 1972, pp. 129–136.

[23] Lo, Y., and W. Richards, "Pattern Synthesis Based on Optimization in a Probabilistic Sense," *1973 Antennas and Propagation Society International Symposium*, Boulder, CO, April 22–24, 1973, pp. 267–270.

[24] Lin, Z., et al., "Synthesis of Sparse Linear Arrays Using Vector Mapping and Simultaneous Perturbation Stochastic Approximation," *IEEE Antennas and Wireless Propagation Letters*, Vol. 11, 2012, pp. 220–223.

[25] Saudi, E., et al., "A Hybrid Technique Linear Sparse Array Antenna Design Approach," *2015 International Symposium on Antennas and Propagation (ISAP)*, Hobart, TAS, Australia, November 9–12, 2015, pp. 1–3.

[26] Bencivenni, C., et al., "Design of Maximally Sparse Antenna Arrays in the Presence of Mutual Coupling," *IEEE Antennas and Wireless Propagation Letters*, Vol. 14, 2015, pp. 159–162.

[27] Caratelli, D., M. C. Viganó, and A. Yarovoy, "A Novel Deterministic Synthesis Technique for Non-Uniformly Spaced Linear Antenna Arrays," *2011 41st European Microwave Conference*, Manchester, U.K., October 10–13, 2011, pp. 1324–1327.

[28] Poli, L., et al., "Compressive Sensing Technique for Multi-Frequency Sparse Linear Array Design," *2015 9th European Conference on Antennas and Propagation (EuCAP)*, Lisbon, Portugal, April 13–17, 2015, pp. 1–3.

[29] Caratelli, D., M. C. Viganó, and A. Yarovoy, "A Novel Deterministic Synthesis Technique for Non-Uniformly Spaced Linear Antenna Arrays," *2011 8th European Radar Conference*, Manchester, U.K., October 12–14, 2011, pp. 436–439.

[30] Deparateanu, D., et al., "Sparse Array Antenna Optimization Using Genetic Algorithms," *2016 8th International Conference on Electronics, Computers and Artificial Intelligence (ECAI)*, Ploiesti, Romania, June 30–July 2, 2016, pp. 1–4.

[31] Caratelli, D., and M. C. Vigano, "A Novel Deterministic Synthesis Technique for Constrained Sparse Array Design Problems," *IEEE Transactions on Antennas and Propagation*, Vol. 59, No. 11, 2011, pp. 4085–4093.

[32] Viganö, M. C., et al., "Sparse Antenna Array for Earth-Coverage Satellite Applications," *Proceedings of the Fourth European Conference on Antennas and Propagation*, Barcelona, Spain, April 12–16, 2010, pp. 1–4.

[33] Coman, C. I., I. E. Lager, and L. P. Ligthart, "A Deterministic Solution to the Problem of Interleaving Multiple Sparse Array Antennas," *2005 European Microwave Conference*, Paris, France, October 4–6, 2005.

[34] Dai, D., "An Asymmetric Mapping Method for the Synthesis of Sparse Planar Arrays," *IEEE Antennas and Wireless Propagation Letters*, Vol. 17, No. 1, 2018, pp. 70–73.

[35] Dickhof, J. H., et al., "Pattern Synthesis of Linear, Sparse Array Antennas," *2006 European Radar Conference*, Manchester, U.K., September 13–15, 2006, pp. 84–87.

[36] Kedar, A., "Deterministic Synthesis Approach for Linear Sparse Array Antennas," *IEEE Transactions on Antennas and Propagation*, Vol. 68, No. 9, 2020, pp. 6667–6674.

[37] Zhao, X., Q. Yang, and Y. Zhang, "A Hybrid Method for the Optimal Synthesis of 3-D Patterns of Sparse Concentric Ring Arrays," *IEEE Transactions on Antennas and Propagation*, Vol. 64, No. 2, 2016, pp. 515–524.

[38] Siakavara, K., "Hybrid-Fractal Direct Radiating Antenna Arrays with Small Number of Elements for Satellite Communications," *IEEE Transactions on Antennas and Propagation*, Vol. 58, No. 6, 2010, pp. 2102–2106.

[39] Bekers, D. J., et al., "A Ka-Band Spaceborne Synthetic Aperture Radar Instrument: A Modular Sparse Array Antenna Design," *IEEE Antennas and Propagation Magazine*, Vol. 61, No. 5, 2019, pp. 97–104.

[40] Angeletti, P., and G. Toso, "Array Antennas with Jointly Optimized Elements Positions and Dimensions Part I: Linear Arrays," *IEEE Transactions on Antennas and Propagation*, Vol. 62, No. 4, 2014, pp. 1619–1626.

[41] Angeletti, P., G. Toso, and G. Ruggerini, "Array Antennas with Jointly Optimized Elements Positions and Dimensions Part II: Planar Circular Arrays," *IEEE Transactions on Antennas and Propagation*, Vol. 62, No. 4, 2014, pp. 1627–1639.

[42] Kedar, A., "Deterministic Synthesis of Wide Scanning Sparse Concentric Ring Antenna Arrays," *IEEE Transactions on Antennas and Propagation*, Vol. 67, No. 12, 2019, pp. 7387–7395.

[43] Chen, K., et al., "Synthesis of Sparse Planar Arrays Using Modified Real Genetic Algorithm," *IEEE Transactions on Antennas and Propagation*, Vol. 55, No. 4, 2007, pp. 1067–1073.

[44] Jiang, Y., S. Zhang, and Q. Guo, "An Effective Two-Step Approach to the Synthesis of Uniform Amplitude Linear Arrays," *IEEE Antennas and Wireless Propagation Letters*, Vol. 16, 2017, pp. 437–440.

[45] Blockley, R., and W. Shyy, *Encyclopedia of Aerospace Engineering*, New York: John Wiley & Sons, 2010.

[46] Rahmat-Samii, Y., and E. Michielssen, *Electromagnetic Optimization by Genetic Algorithms*, New York: Wiley, 1999.

[47] Rahmat-Samii, Y., J. M. Kovitz, and H. Rajagopalan, "Nature-Inspired Optimization Techniques in Communication Antenna Designs," *Proceedings of the IEEE*, Vol. 100, No. 7, 2012, pp. 2132–2144.

[48] Aslan, Y., M. Candotti, and A. Yarovoy, "Synthesis of Multi-Beam Space-Tapered Linear Arrays with Side Lobe Level Minimization in the Presence of Mutual Coupling," *2019 13th European Conference on Antennas and Propagation (EuCAP)*, Krakow, Poland, March 31–April 5, 2019, pp. 1–5.

[49] Abdelhay, M. A., N. O. Korany, and S. E. El-Khamy, "Synthesis of Uniformly Weighted Sparse Concentric Ring Arrays Based on Off-Grid Compressive Sensing Framework," *IEEE Antennas and Wireless Propagation Letters*, Vol. 20, No. 4, 2021, pp. 448–452.

[50] Sun, B., et al., "A Novel Sparse Array Synthesis Method Based on Two-Step Convex Optimization," *2016 10th European Conference on Antennas and Propagation (EuCAP)*, Davos, Switzerland, April 10–15, 2016, pp. 1–3.

[51] Prisco, G., and M. D'Urso, "Maximally Sparse Arrays Via Sequential Convex Optimizations," *IEEE Antennas and Wireless Propagation Letters*, Vol. 11, 2012, pp. 192–195.

[52] Yan, C., et al., "Synthesis of Planar Sparse Arrays with Minimum Spacing Constraint," *IEEE Antennas and Wireless Propagation Letters*, Vol. 17, No. 6, 2018, pp. 1095–1098.

[53] Zhao, X. W., Q. S. Yang, and Y. H. Zhang, "Synthesis of Sparse Concentric Ring Antenna Arrays by Convex Optimization," *IET International Radar Conference 2015*, Hangzhou, China, October 14–16, 2015, pp. 1–5.

[54] D'Urso, M., G. Prisco, and R. M. Tumolo, "Maximally Sparse, Steerable, and Nonsuperdirective Array Antennas Via Convex Optimizations," *IEEE Transactions on Antennas and Propagation*, Vol. 64, No. 9, 2016, pp. 3840–3849.

[55] Keizer, W. P., "Linear Array Thinning Using Iterative FFT Techniques," *IEEE Transactions on Antennas and Propagation*, Vol. 56, No. 8, 2008, pp. 2757–2760.

[56] Keizer, W. P. M. N., "Large Planar Array Thinning Using Iterative FFT Techniques," *IEEE Transactions on Antennas and Propagation*, Vol. 57, No. 10, 2009, pp. 3359–3362.

[57] Liu, Y., et al., "Synthesizing Beam-Scannable Thinned Massive Antenna Array Utilizing Modified Iterative FFT for Millimeter-Wave Communication," *IEEE Antennas and Wireless Propagation Letters*, Vol. 19, No. 11, 2020, pp. 1983–1987.

[58] du Plessis, W. P., "Weighted Thinned Linear Array Design with the Iterative FFT Technique," *IEEE Transactions on Antennas and Propagation*, Vol. 59, No. 9, 2011, pp. 3473–3477.

[59] Keizer, W. P. M. N., "Low-Sidelobe Pattern Synthesis Using Iterative Fourier

Techniques Coded in MATLAB [EM Programmer's Notebook]," *IEEE Antennas and Propagation Magazine,* Vol. 51, No. 2, 2009, pp. 137–150.

[60] Cui, C., and Y. Huang, "Effective Nonuniform Fast Fourier Transform for Calculating Sparse Array Patterns," *2019 International Symposium on Antennas and Propagation (ISAP)*, Xi'an, China, October 27–30, 2019, pp. 1–3.

[61] Visser, H. J., *Array and Phased Array Antenna Basics,* New York: John Wiley & Sons, 2005.

[62] Vigano, M. C., et al., "Projection Based Methodology for Designing Non-Periodic, Planar Arrays," *2008 38th European Microwave Conference,* Amsterdam, Netherlands, October 27–31, 2008, pp. 1624–1627.

[63] Elliott, R. S., *Antenna Theory and Design,* New York: Wiley-IEEE Press, 2003.

[64] Abramowitz, M., and I. A. Stegen, *Handbook of Mathematical Functions: with Formulas, Graphs, and Mathematical Tables,* Mineola, NY: Dover Publications, 1972.

[65] Watson, G. N., *A Treatise on the Theory of Bessel Functions/Theory of Bessel Functions,* Second Edition, Cambridge, U.K.: Cambridge University Press, 1944.

[66] Skolnik, M. I., J. W. Sherman, and F. C. Ogg, "Statistically Designed Density-Tapered Arrays," *IEEE Transactions on Antennas and Propagation,* 1964, pp. 408–417.

[67] Peitgen, H. O., H. Jurgens, and D. Saupe, *Chaos and Fractals: New Frontiers of Science,* Berlin, Germany: Springer, 1992.

[68] Johnson, J. M., and V. Rahmat-Samii, "Genetic Algorithms in Engineering Electromagnetics," *IEEE Antennas and Propagation Magazine,* Vol. 39, No. 4, 1997, pp. 7–21.

[69] https://robjhyndman.com/papers/movingaverage.pdf.

APPENDIX 7A

Randomization Approach: Geolocating Method

Referring to Figure 7.14, there are four regions (angular) in which probable angular location need to be determined deterministically. The procedure can be described point-wise referring to the specific example of the antenna aperture array of size, 10λ by 5λ, having the total number of active elements, N_a equal to 50, as:

1. In order to set up statistics, split the number of elements in the different angular sectors as a function of φ. For Region I, $-\pi/2 < \varphi_{p,q} < \pi/2$, there are 25 elements into 8.33 elements for $|\tan\varphi_{p,q}| < 0.5$ (i.e., $|\varphi_{p,q}| \sim 27°$) and 16.66 elements for $|\tan\varphi_{p,q}| > 0.5$ (the angles are divided in the ratio of 2:1, that is, 16.66:8.33 ~ 126°:54° ~ 2:1).
2. For Region II, $\pi/2 < \varphi_{p,q} < 3\pi/2$, again divide the rest of the 25 elements in the ratio of 2:1. This ratio may be changed depending upon the considered case and has been chosen here just to illustrate the randomization process. The regions may be divided in the steps of 1/20 times $d(\tan\varphi)$.
3. In Region III, 8.33 angles may be chosen by selecting randomly 3 times (i.e., 2 times set of 8 angles, 1 time set of 9 angles because with this selection we get (28 + 19)/(2 + 1) = (16 + 9)/3 = 25/3 = 8.33 angles).

4. In a similar fashion, we chose 16.33 angles in Region IV by selecting randomly 3 times (i.e., 2 times set of 17 angles, 1 time set of 16 angles because with this selection we get (217 + 116)/(2 + 1) = (34 + 16)/3 = 50/3 = 8.33 angles).
5. Thus, from these randomly chosen angle sets consisting of 8, 9, 16, and 17 angles, we may choose n total, a set consisting of 50 angles corresponding to the desired value of N_a. For example, 8 + 8 + 17 + 17 = 50 or 8 + 9 + 16 + 17 = 50 or 9 + 9 + 16 + 16 = 50. In this manner, we have Q = 81 combinations (~3!) from randomization.
6. Let's assume that the random start value (SV) near 0° is actually 0. This may be chosen as the first angle or the start angle. Now $|\varphi_{p,q}|$ ~ 27° implies a total of 54° angular extent, which may be divided into 8.33 angles with a step size of 54/8.33 ~ 6.5°. Thus, other angles may be chosen as SV+4 times 6.5 or SV-4 times 6.5 if SV < 1° giving total 9 angles or SV+3 times 6.5 or SV-4 times 6.5 if SV > 1° giving total 8 angles. Similarly, if SV < –1°, it gives SV+3 times 6.5 or SV-4 times 6.5 yielding 8 angles. If –1° < SV < 1°, it gives 9 angles as SV+4 times 6.5 or SV-4 times 6.5.
7. Similarly, while choosing 16.33 angles in other sectors, the angular step here is 126/16.66 ~ 7.56°. Thus, for 16 angles, 16 × 7.56 = 121° or 60.5° relative to SV. For 17 angles, it becomes 17 × 7.56 = 128.5° or 64.25° relative to SV. When SV > 92.5°, it gives 16 angles in the sector (7 angles between SV and 153°, and 8 angles between 27° and SV). It yields 16 angles when SV < 87.5° and 17 angles when 87.5° < SV < 92.5°. A similar consideration can be given for Region II, where SA > 181° or SA < 179° lead to 8 angles and 179° < SV < 181° gives 9 angles.

The choice of angles should be such that a greater number of elements near the center of the zone spread out (lead to the increasing sparsity) towards the edge of the sector. This is fulfilled by the above randomization approach.

An alternate approach in the form of empirical expressions is summarized below. Considering Region I (–27° to 27°) and Region II (153° to 207°) referring to Figure 7.14, the procedure may be described as:

1. Choice of start values: Take 4 random values (rnd_n; n = 1, 2, 3, and 4) between –3.25° and 3.25° corresponding to the four

chosen sets of 8 angles, 8a, 8b, 8c, and 8d, that is, rnd_1 is chosen as the start angle (SA_{8a}) for set 8a; rnd_2 is chosen as the start angle (SA_{8b}) for set 8b; rnd_3 is chosen as the start angle (SA_{8c}) for set 8c; and rnd_4 is chosen as the start angle (SA_{8d}) for set 8d.

Similarly, for a set of 9 angles, choose random SAs, that is, rnd_5 and rnd_6 between –1° and 1°. Thus, rnd_5 is SA_{9a} for set 9a and rnd_6 is SA_{9b} for set 9b.

2. Determine the deterministic center values (CV) of the angles for different sets. For set 8a and $SA_{8a} > 0$, CV becomes

$$CV_{8at} = SA_{8a} + 6.5t$$

where $t = -4, -3, -2, -1, 0, 1, 2, 3$.

Putting $n = 5 + t$, CV_{8at} becomes

$$CV_{8an} = SA_{8a} + 6.5(n - 5)$$

where $n = 1, 2, \ldots, 8$ for set 8a, and

$$CV_{8bn} = SA_{8b} + 6.5(n - 4)$$

when $SA_{8b} < 0$.

Similarly, set 8c will give

$$CV_{8cn} = SA_{8c} + 6.5(n - 5) \text{ for } SA_{8c} > 180 \text{ and}$$
$$CV_{8cn} = SA_{8c} + 6.5(n - 4) \text{ for } SA_{8c} < 180$$

Further, set 8d will give

$$CV_{8dn} = SA_{8d} + 6.5(n - 5) \text{ for } SA_{8d} > 180 \text{ and}$$
$$CV_{8dn} = SA_{8d} + 6.5(n - 4) \text{ for } SA_{8c} < 180$$

For set 9a,

$$CV_{9an} = SA_{9a} + 6.5(n - 5) \text{ for } -1 < SA_{9a} < 1 \text{ and}$$
$$CV_{9bn} = SA_{9b} + 6.5(n - 5) \text{ for } -179 < SA_{9b} < 181$$

Thus, all CVs are calculated analytically except for the random calculation of the start angles.

3. The randomization around deterministically obtained CVs: The range in randomization is between –3.25 and 3.25. Thus, the number of randomization cycles becomes 7 for set 8a plus

7 for set 8b plus 7 for set 8c and 7 for set 8d plus 8 for set 9c and 8 for set 9b. Therefore, in total, there are 44 values that are added to 44 CV values (i.e., all CVs except 6 number of SA values). Therefore, in total, we get 50 values of random angles.

An example, if we chose start angles as

$$
\begin{aligned}
\text{rnd}_1 &= -1.8\left(\text{SA}_{8a} < 0\right)\\
\text{rnd}_2 &= +2.0\left(\text{SA}_{8b} > 0\right)\\
\text{rnd}_3 &= +0.9\left(\text{SA}_{8c} > 180\right)\\
\text{rnd}_4 &= -2.5\left(\text{SA}_{8d} < 180\right)\\
\text{rnd}_5 &= -0.6\left(-1 < \text{SA}_{9a} < 0\right)\\
\text{rnd}_6 &= -0.6\left(180 < \text{SA}_{9b} < 181\right)
\end{aligned}
$$

Then CV becomes

$$
\begin{aligned}
\text{CV}_{8an} &= \text{SA}_{8b} + 6.5(n-4)\\
\text{CV}_{8bn} &= \text{SA}_{8b} + 6.5(n-5)\\
\text{CV}_{8cn} &= \text{SA}_{8c} + 6.5(n-5)\\
\text{CV}_{8dn} &= \text{SA}_{8d} + 6.5(n-4)\\
\text{CV}_{9an} &= \text{SA}_{8a} + 6.5(n-5)\\
\text{CV}_{9bn} &= \text{SA}_{9b} + 6.5(n-5)
\end{aligned}
$$

Thus, set 8b becomes –24°, –17.5°, –11°, –4.5°, 2° (SA), 8.5°, 15°, 21.5°. After randomization, it becomes –25.2°, –16°, –12.3°, –4.9°, 2°, 8.1°, 16.8°, 20.7°. Similarly, the procedure can be adopted for the other regions, leading to a total of 50 random angles.

8

Use Cases for Sparse Array Antennas

8.1 Introduction

The previous chapters in this book discussed the fundamentals of sparse array antennas, the various design strategies to design these antenna arrays, their advantages, and the limitations illustrated with numerous case studies. This chapter will briefly discuss a few of my indigenous case studies that will illustrate the realization aspects of these arrays.

8.2 Global Scenario on Sparse Array Antennas

Although the work on the sparse array antennas started many decades back, the transition from the conceptualization towards realization started only two decades ago. In the late 1950s, the first instance of a nonuniformly spaced linear array was presented by Unz [1]. Later, during the 1960s to the 1980s, more methods for design of nonuniformly spaced linear arrays were presented, which mainly included the orthogonal methods [2–4], dynamic programming [5], finite impulse response (FIR) filter concept-based thinning technique [6], and probability-based thinning approaches [7–9]. In the last 30 years, due to the huge progress made in multiple aspects such as numerical computation, optimization theory, signal processing, and meta-heuristic intelligence led to advent of much more advanced design concepts for the design of the nonuniformly spaced antenna arrays or sparse array antennas. However, the irregularity in the sparse array designs due to the aperiodicity or randomness in the array grid leads to the complication in the realization of front-end electronics [10]. There are very few case studies available discussing the realized sparse array antenna architectures. In this chapter, a few references will be discussed that will showcase the difficulties in the realization of the sparse array antennas and how they can be overcome by undertaking efficient, realization-oriented design procedure.

The realization of sparseness in the array architecture is multifold and can be achieved in various ways such as:

1. *At the individual element level:* Aperiodicity at the element level will be difficult for realization of systems such as active phased array antennas because of the active radio frequency (RF) front-end electronics required in the arrays for dynamic beamforming applications. It requires assembling of lots of RF cables, transmit receive modules (TRMs), and beamformer integrated circuits (ICs) in the back end of array, which are difficult to assemble in the view of aperiodicity. In regular arrays, we can make groups or tiles of these electronics components that help in achieving modularity. Modularity helps to productionize the tiles in numbers and can be connected to antenna array in a uniform manner. However, in a sparse array antenna, due to the aperiodicity, the modularity cannot be achieved. Digital beamforming (DBF) provides a solution to this problem and digitizing the RF circuitry except for the power amplifier (transmit chain) and low noise amplifier (receive chain). This will reduce the hardware complexity to some extent.
2. *At the subarray/cluster of elements level:* The arrangement of elements in a subarray introduces a modularity in the antenna array and it allows packaging of RF circuitry in the subarray housings. These subarrays can be of different forms (in terms of antennas arrangement within a subarray) allowing uniformity in their structure, which helps in their production and assembly. These subarrays can be arranged in an overlapped manner, introducing aperiodicity in terms of their phase centers. Thus, the aperiodicity, so introduced in the array, leads to sparse array formation at the subarray level. A careful design

mechanism may lead to a modular sparse array antenna as shown in [10].

3. *By orienting the elements in a different sense from each other:* Aperiodicity can be achieved in a uniform antenna array by changing orientation of each of the elements. Thus, in this type of antenna array, the arrangement of RF circuitry is uniform and the randomness is achieved in the terms of the spatial orientation of the element.
4. *By ensuring the interelement spacing more than 0.5λ:* This type of sparse array antenna has a periodic array grid but with large interelement spacing ($\gg 0.5\lambda$). This will reduce mutual coupling but will limit the scan volume due to the early onset of grating lobes. Thus, this type can be used for the applications requiring limited scanning.

The few choices are mentioned above and either of these can be used based on the trade-off between system requirements, cost, and complexity of the hardware.

8.3 Use Case Studies

This section presents a few case studies based on my work, illustrating the usefulness of sparse array antennas.

8.3.1 A Stochastic Technique-Based Low SLL Sparse Array Antenna

This case study considers a dense antenna array (DAA) with 2,048 (32×64) elements arranged over a regular (periodic) grid with period, $0.53\lambda_0$ by $0.53\lambda_0$, along the x- and y-axes direction as shown in Figure 8.1. A uniformly fed DAA will exhibit a peak SLL < –13 dB as seen in a 3-D plot of array factor in Figure 8.1. Array efficiency (~100%) is achievable from DAA with uniformly fed elements yielding PSLL < –13 dB, which is high and susceptible to clutter and jammers. Further, in form phased array considerations, a fully active phased array implies a number of transmit/receive modules (TRMs) being equal to the number of antenna elements. This has impact on the size, weight, aperture size, and power-cost (SWAP-C) parameters. Moreover, a tight array grid (interelement spacing chosen to avoid grating lobes) leads to a tight mutual coupling, which

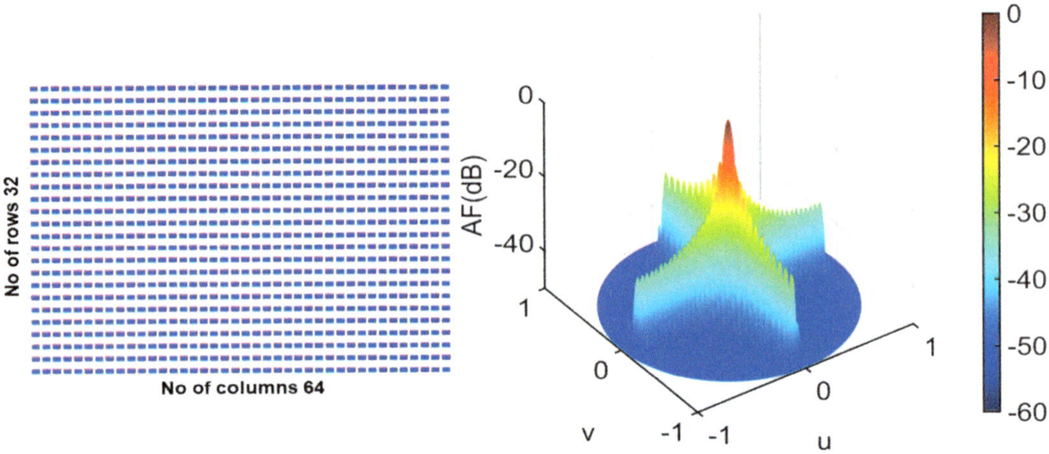

Figure 8.1
DAA geometry and its array factor at boresight.

may lead to the scan blindness. Thus, the solution is sparse array antenna.

A sparse array solution [11] is provided here by choosing the genetic algorithm (GA) technique. Figure 8.2 shows the GA setup for the GA parameters [12] and the fitness function chosen for the algorithm convergence. Here, the worst SLL along two principal j-planes is chosen to check the convergence criterion as these two cuts will have the worst SLL scenario.

Further, a zoning technique [11] is applied to segment the array aperture into different zones based on the magnitude of the current

❑ Genetic algorithm parameter

- Population size :100
- Mutation : 0.05
- Cross over single point

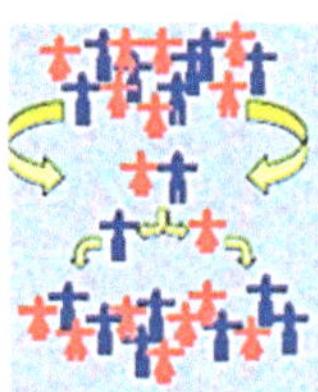

- Fitness function :

$$\text{»}F_1 = w_1 \times \max\{|SLL|_{\varphi=0^o}\} + w_2 \times \max\{|SLL|_{\varphi=90^o}\}$$

Figure 8.2
GA setup and fitness equation.

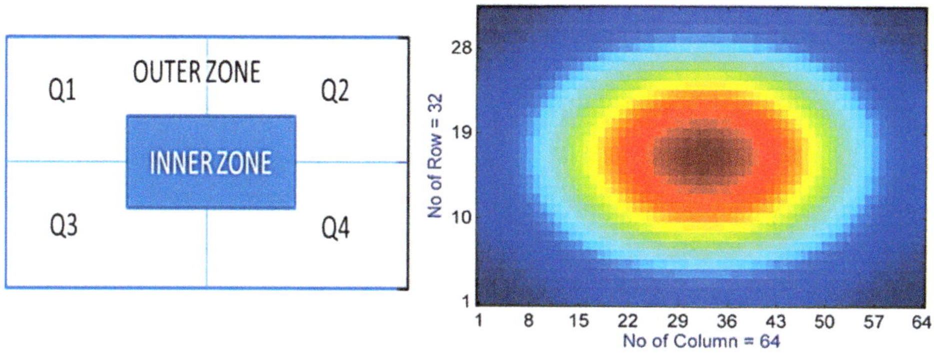

Figure 8.3
Zoning technique. (© 2018 IEEE. Reprinted with permission from [11].)

distribution (see Figure 8.3). One can refer to [11] for more details; henceforth, here the focus will be upon the results to show the advantages of the sparse array antennas. In the present case, two zones are considered, that is, zone 1 contains elements with amplitude < 3 dB and the inner zone and zone 2 contain elements with amplitude > 3 dB elements (i.e., outer zone). This technique helps in accelerating the convergence criterion of GA. Figure 8.4 shows the sparse array layout and the 3-D array factor plot as simulated. The array consist of 1,122 elements (i.e., a fill factor of almost 46% is achieved with PSLL < –20 dB by uniformly feeding all of the array elements).

Figure 8.5 shows the realized antenna array as an inset of the figure showing the measured amplitude distribution on the array aperture. The radiating elements chosen for the realization are cavity-backed microstrip patch antennas with a bandwidth of 1 GHz in X-band. The elements in array are connected to TRMs and the ones with zero excitations (thinned) are fed with a high attenuation. Thus, a uniformity in RF electronics is achieved and sparsity is achieved by the virtue of the array thinning. A clear, dense region in the middle is visible with the higher sparsity in the edge regions. Table 8.1 summarizes the measured array performance compared with the performance of the dense counterpart. Clearly, except the loss in directivity, the sparse array antenna excels and shows the low SLL with similar angular resolution while reducing the elements by nearly half.

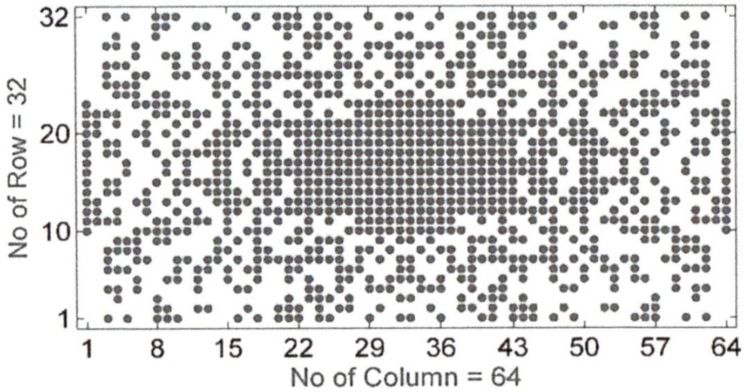

(a)

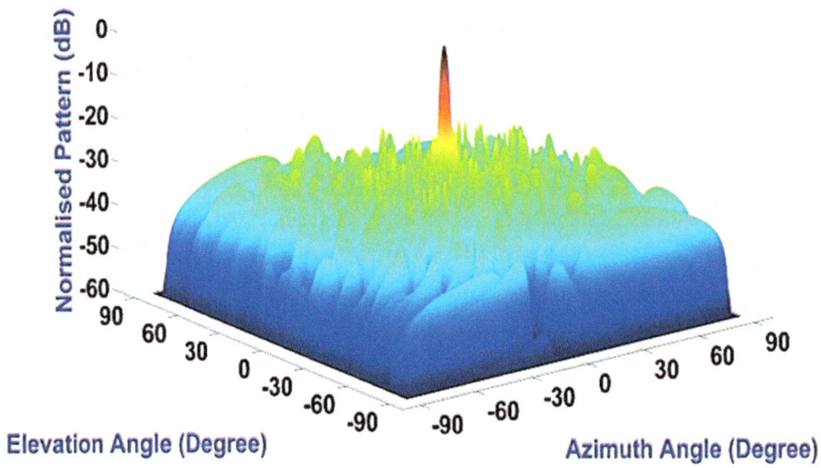

(b)

Figure 8.4
Sparse array antenna layout and 3-D AF plot. (© 2018 IEEE. Reprinted with permission from [11].)

8.3.2 Multifunctional Dual-Band Sparse Array Antenna

This case study discusses the design procedure of a dual-band sparse array antenna collocated on the same aperture, in a shared aperture configuration designed using theory of collapsed distributions (TCD) as explained in Chapter 7. The shared aperture configuration presented here is required to operate in two bands, namely, S-band and X-band with the elements arranged in an interleaved manner. The arrays shall have high sparsity with desired directivity

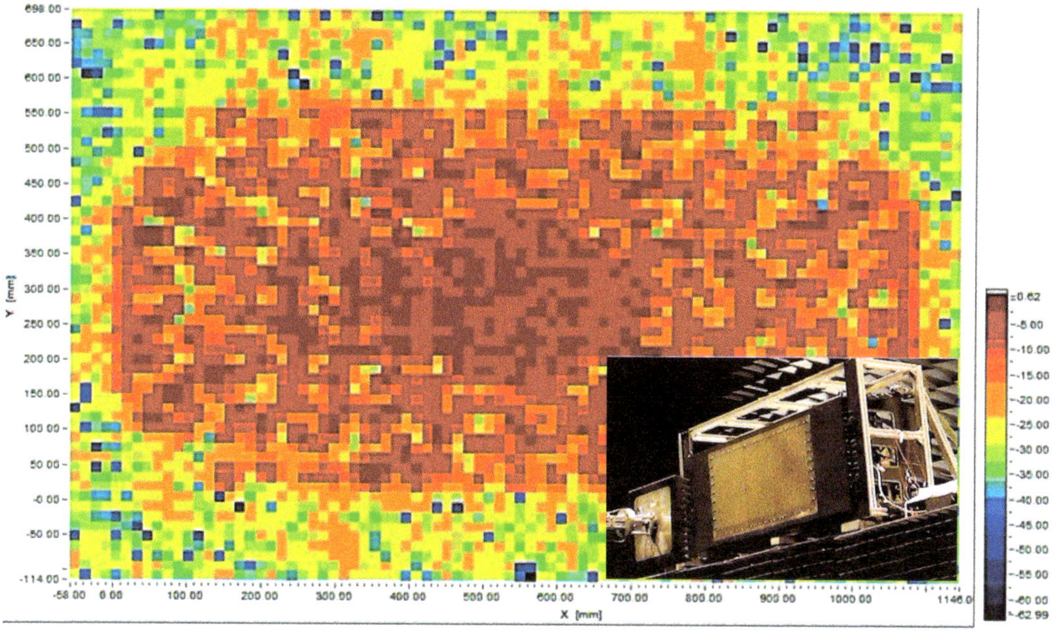

Figure 8.5
Realized sparse array antenna layout and 3-D AF plot. (Partially reproduced from [11] with permission from IEEE.)

> 20 dBi and low cross-polarization. The desired angular beam width is 4.6° and 14° in X-band and Sw-band, respectively. Both of the arrays need to have wide angular scan volume better than 90°.

8.3.2.1 Design

The array element chosen for the purpose is a U-slot microstrip patch antenna (USMPA) as shown as an inset in Figure 8.7. Figure 8.7 shows the unit cell model of the USMPA as modeled in Ansys HFSS

Table 8.1
Summarized Measured Performance of the Sparse Array Antenna

Array Characteristic Parameters	Uniform	Sparse
Efficiency	100%	56%
SLL	−13 dB	−22 dB
Directivity	38 dBi	34.4 dBi
HPBW	1.4°/2.8°	1.6°/3.4°
ASLL	−45 dB	−35 dB

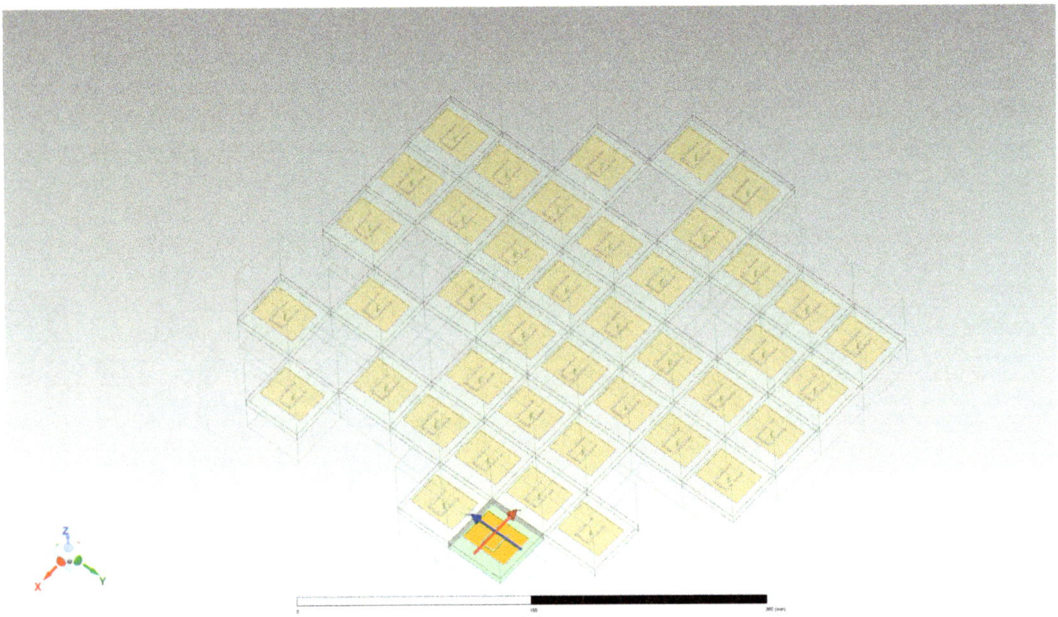

(a)

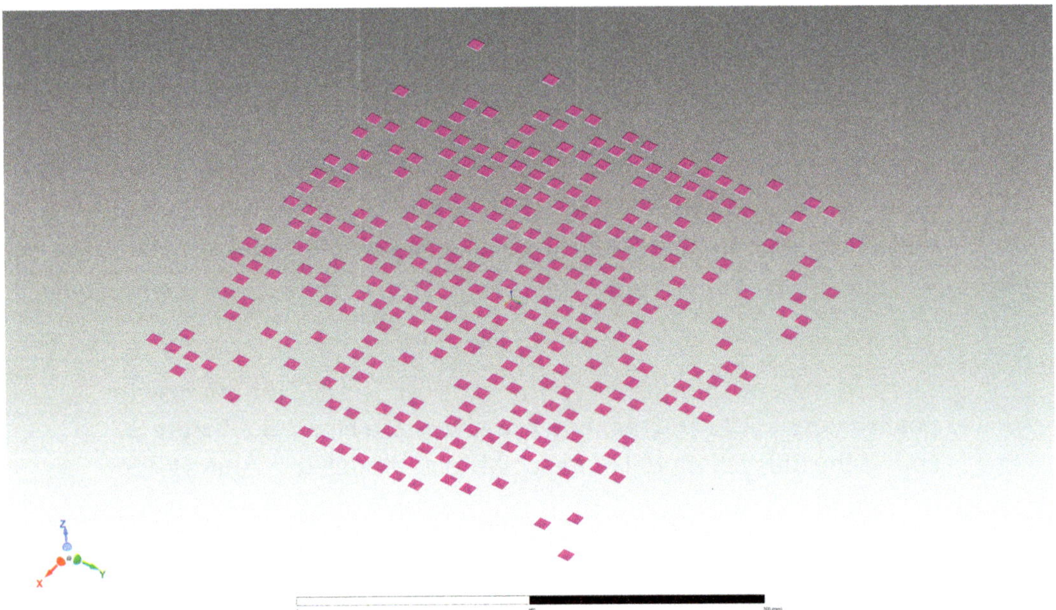

(b)

Figure 8.6
Realized sparse array antenna layout. (a) S-band. (b) X-band.

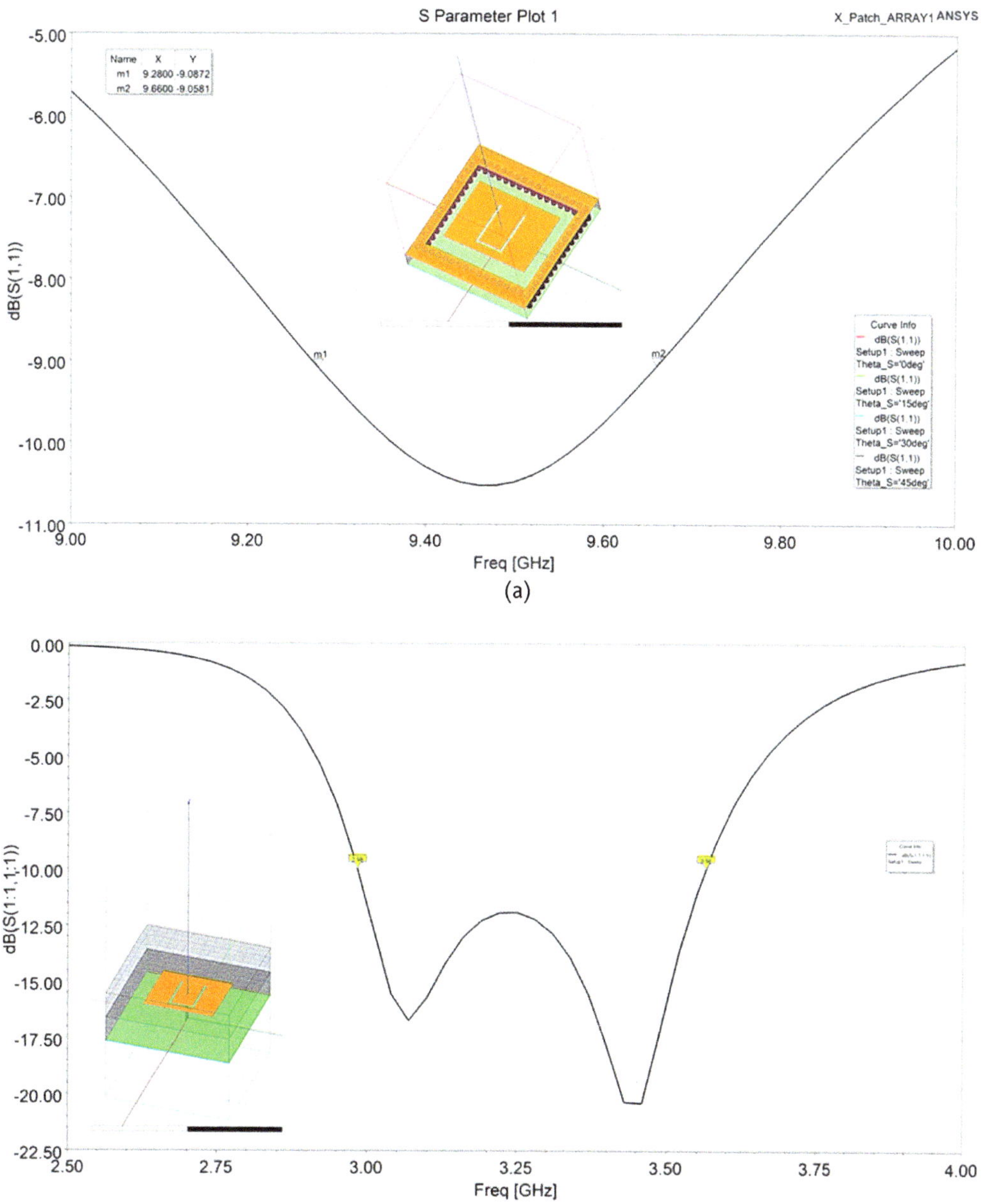

Figure 8.7
Unit cell reflection coefficient performance. (a) X-band. (b) S-band.

software. TCD is utilized and a Taylor's distribution with PSLL < −20 dB is chosen as a reference taper. Initially, TCD is used to model sparse array in X-band with high sparsity of 50% and then, choosing the same reference taper the S-band array, elements are located in the vacant slots in the X-band aperture. It shall be noted here that the basic array grid is chosen such that the interelement separations in the two grids, $d_{\text{S-band}}$ and $d_{\text{X-band}}$ have a ratio of 3:1. This allows the accommodation of a maximum three elements between two S-band elements without overlapping each other. This puts a constraint on the element design as well that the size of the element in the respective bands shall be lesser than $0.5d_{\text{S-band}}$ and $0.5d_{\text{X-band}}$.

Figures 8.6(a, b) show the S-band and X-band sparse array antennas' locations as simulated using TCD and modeled in HFSS [13] using USMPA. An average interelement spacing is much larger than $0.7\lambda_0$, where λ_0 is the wavelength corresponding to the highest frequency in the respective band of operations. Figure 8.7 shows the simulated USMPA unit cell's reflection coefficient in decibels as obtained from HFSS simulations in an infinite array environment using Floquet mode analysis [14]. Thus, the effect of the mutual coupling is accounted for in the analysis.

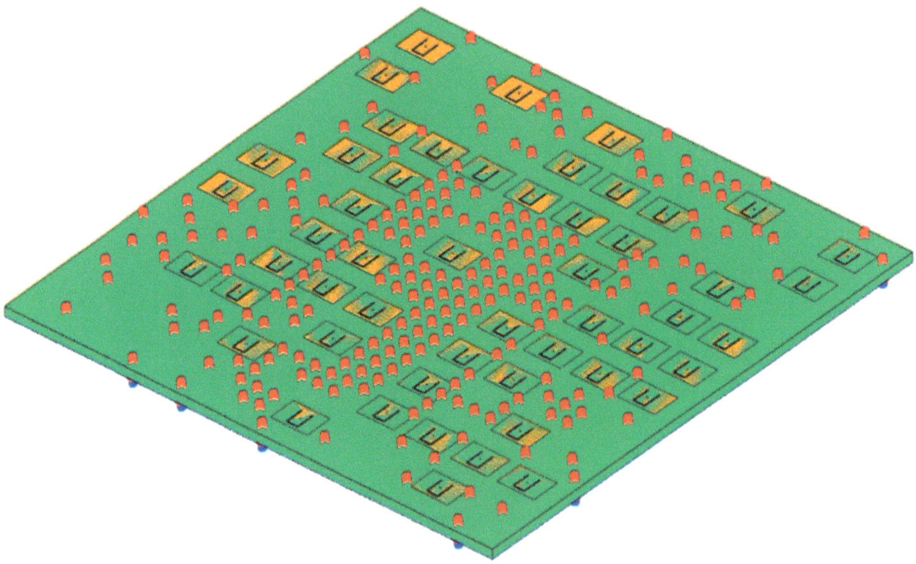

Figure 8.8
Dual-band sparse array antenna layout.

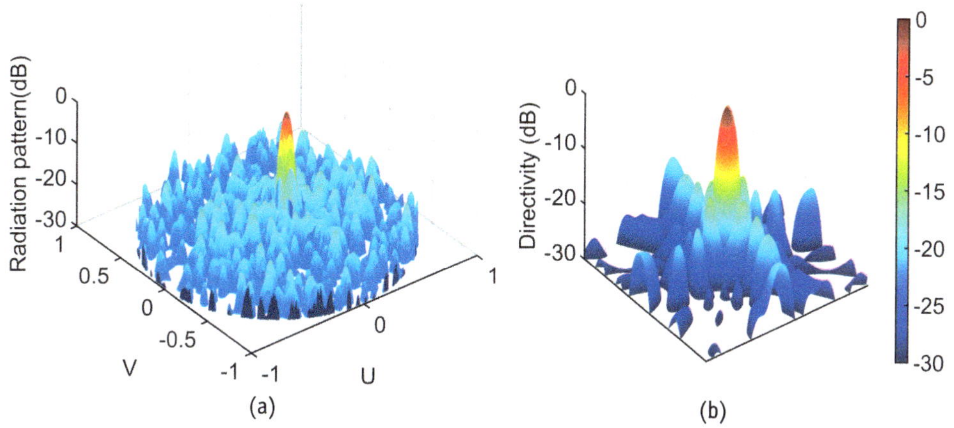

Figure 8.9
Simulated performance of the sparse array antenna.

Figure 8.8 shows the interleaved arrangement of S-band and X-band arrays on a common aperture. Figure 8.9 shows the simulated array factor plot obtained from MATLAB simulations assuming an isotropic antenna element. The plots clearly show lowered SLL as a result of sparse arrangement of element using TCD. In X-band, the worst PSLL < –20 dB and, in S-band, the worst PSLL < –13.5 dB are observed. However, along the principal planes, PSLL is much lower than –20 dB in both the cases.

Figure 8.10 shows the modeled simulations using HFSS-based on USMPA element utilizing the finite array domain decomposition

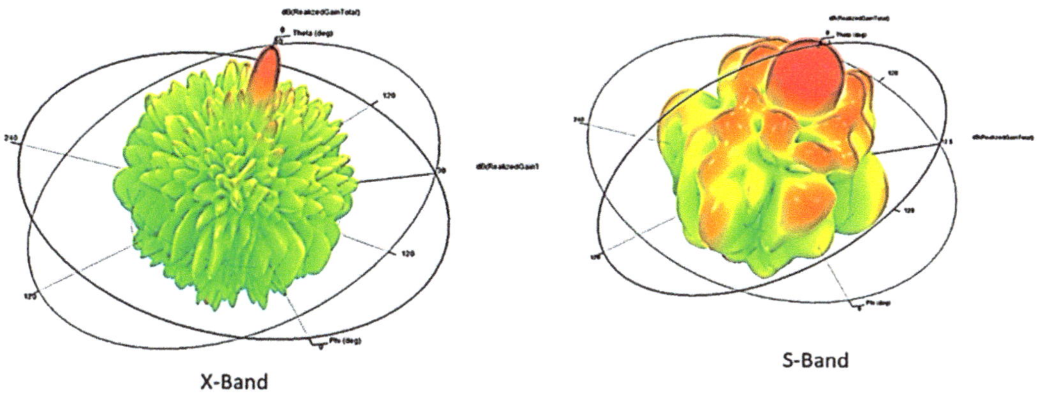

Figure 8.10
Antenna array's radiation pattern using HFSS.

technique (FA-DDM) [13], which helps in simulating large antenna arrays with the least computational resources. The results obtained using HFSS are close to the MATLAB simulations. The realization of this array can be done in a modular fashion as the basic underlying grid in S-band and X-band arrays is uniform. Thus, each element can be connected to a TRM for beamforming application. This case study shows a novel application of sparse array antennas in the realization of interleaved antenna arrays, which may facilitate dual-band and dual-polarization operation and is quite useful for phased array antenna applications.

8.4 Conclusions

A few case studies describing the design of sparse array antennas, their realization, and evaluation were presented. There is an abundance of literature available as brought out in this book related to the design methodologies of sparse array antenna; however, there is not much literature available on their realization while considering the practical considerations such as complexity, cost, and hardware fabrication. I noticed that there are a few instances available, mainly for space-borne applications [10, 15, 16], square kilometer array (SKA) [17], and FMCW radar [18, 19].

References

[1] Unz, H., "Linear Arrays with Arbitrarily Distributed Elements," *IRE Transactions on Antennas and Propagation*, Vol. 8, 1960, pp. 222–223.

[2] Unz, H., "Nonuniformly Spaced Arrays: The Orthogonal Method," *Proceedings of the IEEE*, Vol. 54, No. 1, 1966, pp. 53–54.

[3] Sahalos, J. N., "The Orthogonal Method of Nonuniformly Spaced Arrays," *Proceedings of the IEEE*, Vol. 62, No. 2, 1974, pp. 281–282.

[4] Sahalos, J., "A Solution of Nonuniformly Linear Antenna Array with the Help of Chebyshev Polynomials," *IEEE Transactions on Antennas and Propagation*, Vol. 24, No. 2, 1976, pp. 239–242.

[5] Skolnik, M., G. Nernhauser, and J. Sherman, "Dynamic Programming Applied to Unequally Spaced Arrays," *IEEE Transactions on Antennas and Propagation*, Vol. 12, No. 1, 1964, pp. 35C43.

[6] Jarske, P., et al., "On Properties and Design of Nonuniformly Spaced Linear Arrays (Antennas)," *IEEE Transactions on Antennas and Propagation*, Vol. 36, No. 3, 1988, pp. 372–380.

[7] Skolnik, M. I., J. W. Sherman, III, and F. C. Ogg, Jr., "Statistically Designed Density-Tapered Arrays," *IEEE Transactions on Antennas and Propagation*, Vol. 12, No. 4, 1964, pp. 408–417.

[8] Lo, Y. T., "A Probabilistic Approach to the Problem of Large Antenna Arrays," *Radio Sci.*, Vol. 68, No. 9, 1964, pp. 1011–1019.

[9] Steinberg, B., "The Peak Sidelobe of the Phased Array Having Randomly Located Elements," *IEEE Transactions on Antennas and Propagation*, Vol. 20, No. 2, 1972, pp. 129–136.

[10] Bekers, D. J., et al., "A Ka-Band Spaceborne Synthetic Aperture Radar Instrument: A Modular Sparse Array Antenna Design," *IEEE Antennas and Propagation Magazine*, Vol. 61, No. 5, 2009, pp. 97–104.

[11] Kumar, P., A. Kedar, and N. K. Vishwakarma, "Genetic Algorithm Aided Synthesis of Low SLL Antenna Array Utilizing Position Optimization Based on Zoning Technique," *2018 IEEE Indian conference on Antennas and Propagation (INCAP)*, Hyderabad, India, December 16–19, 2018, pp. 1–4.

[12] Haupt, R. L., and S. E. Haupt, *Practical Genetic Algorithms*, New York: John Wiley & Sons, 2003.

[13] Ansys HFSS Manual, www. Ansys.com/hfss/manual.

[14] Bhattacharya, A. K., *Phased Array Antennas: Floquet Analysis, Synthesis, BFNs and Active Array Systems*, New York: Wiley-Interscience, 2006.

[15] Vigano, M. C., et al., "Sparse Array Antenna for Ku-Band Mobile Terminals Using 1 Bit Phase Control," *IEEE Transactions on Antennas and Propagation*, Vol. 62, No. 4, 2014, pp. 1723–1730.

[16] Siakavara, K., "Hybrid-Fractal Direct Radiating Antenna Arrays with Small Number of Elements for Satellite Communications," *IEEE Transactions on Antennas and Propagation*, Vol. 58, No. 6, 2010, pp. 2102–2106.

[17] Hall, P. J., *The Square Kilometre Array: An Engineering Perspective*, Berlin, Germany: Springer, 2010.

[18] Lager, I. E., et al., "Interleaved Array Antennas for FMCW Radar Applications," *IEEE Transactions on Antennas and Propagation*, Vol. 57, No. 8, 2009, pp. 2486–2490.

[19] Trampuz, C., et al., "Low Sidelobe Interleaved Transmit-Receive Antennas for FMCW Radar Applications," *2009 IEEE Antennas and Propagation Society International Symposium*, North Charleston, SC, June 1–5, 2009, pp. 1–4.

About the Author

Ashutosh Kedar received a PhD in electronics science, an MTech in microwave electronics, an MSc in physics, and a BSc in physics in 2003, 1998, 1995, and 1993, respectively, from University of Delhi, India. He has been a lecturer for over 3.5 years at the Department of Electronic Science (DOES), University of Delhi, teaching microwave courses, and as a research associate at Council of Scientific and Industrial Research (CSIR) for 6 months at the University of Delhi. He was a scientist in the Electronics and Radar Development Establishment (LRDE), Defense Research and Development Organization (DRDO), Bangalore, in 2003. Currently, he is leading the research and development activities related to active phased array antennas for various radar applications at LRDE.

He has contributed toward the design and development of antenna/antenna arrays and RF and microwave systems for various AESA radars. He has also contributed to the electromagnetic modeling and analysis of normal and high-temperature superconducting microwave passive components using the method-of-moments technique during his PhD tenure. His interests include computational electromagnetics, evolutionary optimization techniques, and wide scanning active phased array antennas. He has authored more than 80 research papers in different peer-reviewed national and international journals and conferences and has a patent, *Radiating Element for Active Phased Array Radars*. He has also written a few book chapters. He received DRDO technology group awards in 2010 and 2018 for indigenous development of single- and four-walled active phased array radars, respectively. He received the Best Paper Award at ICMARS-08 (Indian Conference on Microwaves, Antennas, and Remote Sensing) and ATMS18 (Antenna Test & Measurement Society). He is an active member of the IEEE conferences local chapter and contributes to various initiatives of the chapter. He is a reviewer for journals and conferences of the IEEE, IET, PIERS, Wiley, IETE, AEUE, ACES, and Elsevier, and also has chaired several sessions in various national and international forums. He is serving as Secretary of IEEE AP-MTT, Joint Chapter, Bangalore. He is a Senior Member of the IEEE and IEEE AP-S, a Life Member of ACES, and a Fellow of the IETE.

Index

Recent Titles in the Artech House Antennas and Propagation Library

Christos Christodoulou, Series Editor

Adaptive Array Measurements in Communications, M. A. Halim

Antenna-on-Chip: Design, Challenges, and Opportunities, Hammad M. Cheema, Fatima Khalid, and Atif Shamim

Antenna Design for Cognitive Radio, Youssef Tawk, Joseph Costantine, and Christos Christodoulou

Antenna Design for CubeSats, Reyhan Baktur

Antenna Design with Fiber Optics, A. Kumar

Antenna Engineering Using Physical Optics: Practical CAD Techniques and Software, Leo Diaz and Thomas Milligan

Antenna Measurement Techniques, Gary E. Evans.

Antennas and Propagation for Body-Centric Wireless Communications, Second Edition, Peter S. Hall and Yang Hao, editors

Antennas and Site Engineering for Mobile Radio Networks, Bruno Delorme

Antennas for Small Mobile Terminals, Kyohei Fujimoto and Koichi Ito, editors

Analysis, Design and Measurement of Small and Low-profile Antennas, Kazuhiro Hirasawa

Analysis of Radome-Enclosed Antennas, Second Edition, Dennis J. Kozakoff

Applications of Geographic Information Systems for Wireless Network Planning, Francisco Saez de Adana, Josefa Gómez Pérez, Abdelhamid Tayebi Tayebi, and Juan Casado Ballesteros

The Art and Science of Ultrawideband Antennas, Second Edition, Hans G. Schantz

AWAS for Windows Version 2.0: Analysis of Wire Antennas and Scatterers, Antonije R. Djordjević, et al.

Broadband Microstrip Antennas, Girsh Kumar and K. P. Ray

Broadband Patch Antennas, Jean-François Zürcher and Fred E. Gardiol

Circularly Polarized Dielectric Resonator Antennas, Raghvendra Kumar Chaudhary, Rajkishor Kumar, and Rakesh Chowdhury

Design and Applications of Active Integrated Antennas, Mohammad S. Sharawi and Oualid Hammi

Designer Notes for Microwave Antennas, Richard C. Johnson

Dielectric Resonator Antenna Handbook, Aldo Petosa

Dielectric Resonator Antennas, Biswajeet Mukherjee and Monika Chauhan

Electromagnetics and Antenna Technology, Alan J. Fenn

Electromagnetics, Microwave Circuit and Antenna Design for Communications Engineering, Peter Russer

Engineering Applications of the Modulated Scatterer Technique, Jean-Charles Bolomey and Fred E. Gardiol

Fixed & Mobile Terminal Antennas, Akhileswar Kumar

Foundations of Antenna Engineering: A Unified Approach for Line-of -Sight and Multipath, Per-Simon Kildal

Four-Arm Spiral Antennas, Joseph A. Mosko

Frequency-Agile Antennas for Wireless Communications, Aldo Petosa

Fresnel Zones in Wireless Links, Zone Plate Lenses and Antennas, Hristo D. Hristov

Handbook of Antennas for EMC, Second Edition, Thereza Macnamara with John McAuley

Handbook of Reflector Antennas and Feed Systems, Volume I: Theory and Design of Reflectors, Satish Sharma, Sudhakar Rao, and Lotfollah Shafai, editors

Handbook of Reflector Antennas and Feed Systems, Volume II: Feed Systems, Lotfollah Shafai, Satish Sharma, and Sudhakar Rao, editors

Handbook of Reflector Antennas and Feed Systems, Volume III: Applications of Reflectors, Sudhakar Rao, Lotfollah Shafai, and Satish Sharma, editors

Interference Suppression Techniques for Microwave Antennas and Transmitters, Ernest Freeman

Introduction to Antenna Analysis Using EM Simulators, Hiroaki Kogure, Yoshie Kogure, and James C. Rautio

Introduction to Antennas and RF Propagation Analysis, Dean James Friesen

LONRS: Low-Noise Receiving Systems Performance and Analysis Toolkit, Charles T. Stelzried, Macgregor S. Reid, and Arthur J. Freiley

Measurement of Mobile Antenna Systems, Second Edition, Hiroyuki Arai

Microstrip Lines and Slotlines, Third Edition, Ramesh Garg Inder J. Bahl, Maurizio Bozzi

Microstrip Antenna Design, K.C Gupta and Abdelaziz Benalla

Microwave and Millimeter-Wave Remote Sensing for Security Applications, Jeffrey A. Nanzer

Microwave Cavity Antennas, Akhileswar Kumar and Hristo D. Hristov

Millimeter Wave Microstrip and Printed Circuit Antennas, Prakash Bhartia

Mobile Antenna Systems Handbook, Third Edition, Kyohei Fujimoto, editor

Moment Methods in Antennas and Scattering, Robert C. Hansen

Multiband Integrated Antennas for 4G Terminals, David A. Sánchez-Hernández, editor

Near-Field Antenna Measurements, Dan Slater

Noise Temperature Theory and Applications for Deep Space Communications Antenna Systems, Tom Y. Otoshi

Phased Array Antenna Handbook, Third Edition, Robert J. Mailloux

Phased Array Antennas, Arthur A. Oliner

Phased Array Antennas with Optimized Element Patterns, Sergei P. Skobelev

Plasma Antennas, Second Edition, Theodore Anderson

Polarization in Electromagnetic Systems, Second Edition, Warren Stutzman

Practical Antenna Design for Wireless Products. Henry Lau

Practical Microstrip and Printed Antenna Design, Anil Pandey

Practical Phased Array Antenna Systems, Eli Brookner

Practical Simulation of Radar Antennas and Radomes, Herbert L. Hirsch

Printed MIMO Antenna Engineering, Mohammad S. Sharawi

Radiowave Propagation and Antennas for Personal Communications, Third Edition, Kazimierz Siwiak

Reconfigurable Antenna Design and Analysis, Mohammod Ali

Reflectarray Antennas: Analysis, Design, Fabrication and Measurement, Jafar Shaker, Mohammad Reza Chaharmir, and Jonathan Ethier

Reflector Lens Antennas: Analysis and Design Using Personal Computers, Carlyle J. Sletten, editor

RF Coaxial Slot Radiators: Modeling, Measurements, and Applications, Kok Yeow You

Shipboard Antennas, Second Edition, Preston Law

Significant Phased Array Papers, Robert C. Hansen

Smart Antenna Engineering, Ahmed El Zooghby

Solid Dielectric Horn Antennas, Carlos Salema, Carlos Fernandes, and Rama Kant Jha

Sparse Phased Array Antennas: Theory and Applications, Ashutosh Kedar

Ultrawideband Antennas for Microwave Imaging Systems, Tayeb A. Denidni and Gijo Augustin

Ultrawideband Short-Pulse Radio Systems, V. I. Koshelev, Yu. I. Buyanov, and V. P. Belichenko

Waveguide Components for Antenna Feed Systems: Theory and CAD, Jaroslaw Uher, Jens Bornemann, and Uwe Rosenberg

Wearable Antennas and Electronics, Asimina Kiourti and John L. Volakis, editors

For further information on these and other Artech House titles, including previously considered out-of-print books now available through our In-Print-Forever® (IPF®) program, contact:

Artech House
685 Canton Street
Norwood, MA 02062
Phone: 781-769-9750
Fax: 781-769-6334
e-mail: artech@artechhouse.com

Artech House
16 Sussex Street
London SW1V HRW UK
Phone: +44 (0)20 7596-8750
Fax: +44 (0)20 7630 0166
e-mail: artech-uk@artechhouse.com

Find us on the World Wide Web at: www.artechhouse.com